FLUID MECHANICS OF MIXING:
MODELLING, OPERATIONS AND EXPERIMENTAL TECHNIQUES

FLUID MECHANICS AND ITS APPLICATIONS
Volume 10

Series Editor: R. MOREAU
MADYLAM
Ecole Nationale Supérieure d'Hydraulique de Grenoble
Boîte Postale 95
38402 Saint Martin d'Hères Cedex, France

Aims and Scope of the Series

The purpose of this series is to focus on subjects in which fluid mechanics plays a fundamental role.

As well as the more traditional applications of aeronautics, hydraulics, heat and mass transfer etc., books will be published dealing with topics which are currently in a state of rapid development, such as turbulence, suspensions and multiphase fluids, super and hypersonic flows and numerical modelling techniques.

It is a widely held view that it is the interdisciplinary subjects that will receive intense scientific attention, bringing them to the forefront of technological advancement. Fluids have the ability to transport matter and its properties as well as transmit force, therefore fluid mechanics is a subject that is particulary open to cross fertilisation with other sciences and disciplines of engineering. The subject of fluid mechanics will be highly relevant in domains such as chemical, metallurgical, biological and ecological engineering. This series is particularly open to such new multidisciplinary domains.

The median level of presentation is the first year graduate student. Some texts are monographs defining the current state of a field; others are accessible to final year undergraduates; but essentially the emphasis is on readability and clarity.

For a list of related mechanics titles, see final pages.

Fluid Mechanics of Mixing:

Modelling, Operations and Experimental Techniques

edited by

R. King

KLUWER ACADEMIC PUBLISHERS
DORDRECHT / BOSTON / LONDON

ISBN 0-7923-1720-3

Published by Kluwer Academic Publishers,
P.O. Box 17, 3300 AA Dordrecht, The Netherlands.

Kluwer Academic Publishers incorporates
the publishing programmes of
D. Reidel, Martinus Nijhoff, Dr W. Junk and MTP Press.

Sold and distributed in the U.S.A. and Canada
by Kluwer Academic Publishers,
101 Philip Drive, Norwell, MA 02061, U.S.A.

In all other countries, sold and distributed
by Kluwer Academic Publishers Group,
P.O. Box 322, 3300 AH Dordrecht, The Netherlands.

Printed on acid-free paper

ACKNOWLEDGEMENTS

This volume contains selected papers presented at the 7th European Congress on Mixing, September 1991, Brugge, Belgium.

Organized on behalf of the Working Party on Mixing of the European Federation of Chemical Engineering (EFCE) by the Royal Flemish Society of Engineers (KVIV) and the Belgian Branch of the Société de Chimie Industrielle.

CONTENTS

PREFACE

This volume is a selection of the material presented at the 7th European Mixing Congress. It is concerned exclusively with mixing in circular section vessels, using centrally mounted paddles or similar impellers.

The contents are arranged under three classifications: **Modelling of Mixing Processes, Mixing Operations** and **Experimental Techniques**. The classifications result in the original material appearing in a different order to that of the Congress. This arrangement is intended to assist the reader in identifying the topic area by function or application, rather than by technology.

In this book the section on **Modelling** contains papers which focus on the representation of the mixing process, whether by equation, scale-up criteria, or fluid dynamic simulation.

Similarly, **Mixing Operations** are concerned with the application or function of the mixing process, such as mass transfer, heat transfer or mixing time.

Experimental Techniques addresses the tools the researcher needs to use at the data gathering experimental stage. It collects together advances made in the various methods used by some of the foremost researchers, and indicates those areas still in need of additional instrumentation or methods of data reduction.

The book is intended for researchers, designers and users of mixing equipment, and for those planning research and development programmes and who wish to keep up to date with advances in the basic technology and its applications.

It is published with the full permission of the Royal Flemish Society of Engineers.

SECTION 1: MODELLING OF MIXING PROCESSES

EFFECT OF MACROMIXING ON A COMPETITIVE/CONSECUTIVE REACTION IN A SEMI-BATCH STIRRED REACTOR : PAUL'S IODINATION EXPERIMENTS INTERPRETED BY NETWORKS-OF-ZONES

R. MANN, A. M. EL-HAMOUZ, UMIST, Manchester, GB

ABSTRACT

The classic experiments of Paul and Treybal (1971) on the competitive/ consecutive iodination of L-tyrosine exhibit mixing affected yields of the mono-iodo intermediate product. The network of backmixed zones model has been applied to provide a comprehensive simulation of Paul's semi-batch experiments which showed yields in the 50% to 70% range. It is shown that the reactions are fast relative to internal impeller generated mixing rates, so that iodine concentrations show strong partial segregation. A localised excess of iodine leads to a loss of yield of the intermediate relative to perfect mixing. The intensity of this partial segregation changes markedly with stirrer speed and so therefore does the yield. Details of the internal concentration fields in space and time are presented as sectional image reconstructions.

INTRODUCTION

Problems of imperfect mixing in chemical reactors have been traditionally addressed through the classical theory of residence time distributions allied to the stimulus response technique. Unfortunately this is of no use in assessing batch and semi-batch reactors for the simple reason that, by definition, there is no through flow.

Imperfect mixing in batch operations, most usually carried out in stirred vessels, is manifested through interactions between rates of mixing caused by the impeller generated internal flow and rates of chemical reactions. If the assumption of perfect backmixing is avoided, unless it can be proven to be valid, the internal convective flows can be referred to as macromixing and turbulence can be related to micromixing phenomena. The relative contributions of macro- and micromixing will rely on the relative magnitude of reaction speeds. In general, macromixing creates the environment within which the ultimate micromixing takes place. Thus macromixing determines the boundary conditions for any subsequent micromixing. Whenever one reagent is added to another, if the mixing is less than perfect, the result is a so-called partial segregation as shown in Fig. 1 (Levenspiel, 1972). This kind of reactant-segregation is sometimes visible as a reagent plume and the relative disposition of such plumes may differ qualitatively on scaling-up (Bourne and Dell'ava, 1987) as in Fig.2.

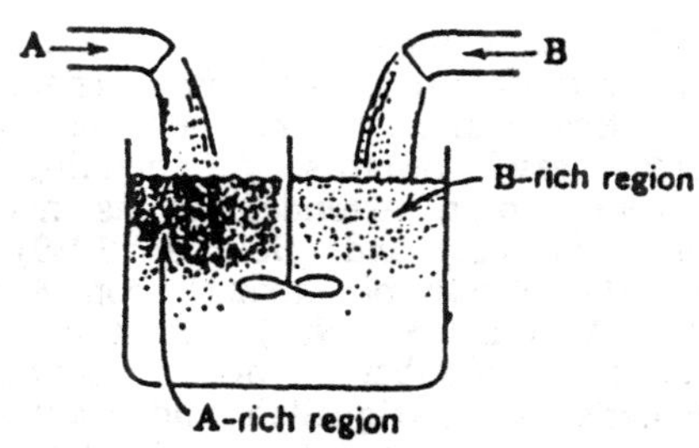

FIG. 1 Partial segregation in a semi-batch reactor

Computational fluid dynamics (CFD) seeks to solve the complex fluid motions inside a stirred vessel by direct solutions of the Navier-Stokes equations (Pericleous and Patel, 1986, Harvey and Greaves, 1982). Often the massive computing effort required to give a reasonably complete solution of the fluid mechanics leaves little room for incorporating sequences of reactions of practical interest. An early pioneering study (Middleton et al, 1985), whilst recognising the potential for yield losses for a competitive/consecutive semi-batch reactor, gave insufficient information to judge how properly the rates of reaction had been combined with the complex 3-D flow field computed.

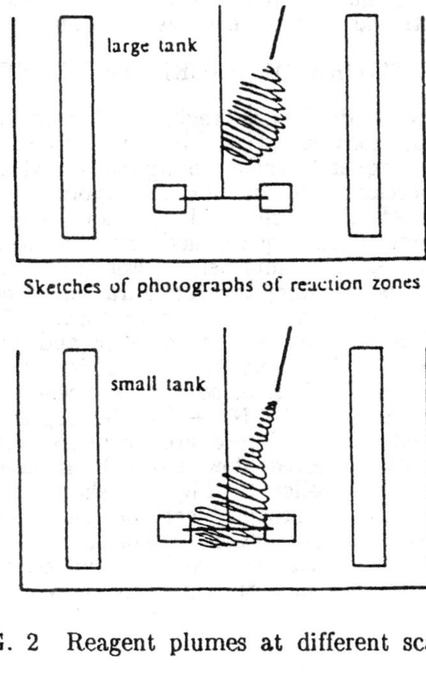

FIG. 2 Reagent plumes at different scales

Approaches based upon networks of backmixed zones sacrifice the need to solve the internal flows with great precision in favour of reasonable approximations to the stirrer generated flow. Unsteady mixing problems, even with complex reactions, are reduced to simple sets of first order ordinary differential equations. Networks–of–zones have been applied to flow follower simulation (Mann and Mavros, 1981), inert tracer mixing (Mann and Knysh, 1983, Mann et al, 1987), inert tracer visualisation (Mann et al, 1990a), reactive tracer visualisation (Mann et al, 1990b), and yield problems for a pair of parallel reactions (Mann and Wang, 1990, Mann and Wang, 1991). Networks–of–zones have also been applied to gas–liquid mixing (Mann, 1986, Mann and Hackett, 1988) as well as to solids suspension in stirred vessels (Brucato, et al, 1988).

Applications to complex sets of reactions are of vital importance to most fine chemical and pharmaceutical manufacturers, who make great use of stirred vessel reactors. Micromixing analyses have recently been extended to as many as 4 reactions (Baldyga and Bourne, 1989). Also recently, the effects of mixing upon 8 reactions (Fox and Villermaux, 1990) and the potentially semi–infinite number of reactions for co–polymerisation (Atiquallah and Nauman, 1990) have received attention, but both these last two cases refer to a CSTR and plug flow reactor respectively and are thus not directly related to batch processing.

In this work we propose to examine the case of the pair of reactions A+B→R and R+B→S. For this competitive/consecutive reaction, using L–tyrosine as A and iodine as B, the yield of the mon–iodo intermediate R has long been known to be sensitive to stirring speed and was the subject of extensive experimental investigation some twenty years ago (Paul and Treybal, 1971). It will be shown that the network–of–zones model can predict the internal concentration yields of A, B, R and S inside a stirred vessel and thereby provide a quantitative framework for explaining this particular mixing sensitive reaction.

NETWORK OF BACKMIXED ZONES

In line with Paul and Treybal (1971) the impeller is to be located at a clearance c=H/3 for a standard 'square' liquid configuration with H=T. It is necessary to modify the geometry of the network of zones since all previous applications used symmetrical square networks on either side of a mid–placed impeller. For this one–third clearance positioning, the configuration of networks applied uses an NxN arrangement for the convection loop below the impeller and an Nx2N set above the impeller. This is shown schematically in Fig. 3, so that the whole fluid is represented by an (NxN) + (Nx2N) aggregation of backmixed zones. There are therefore exactly N concentrically nested flow loops both above and below the impeller. Fig. 3 shows the two innermost loops around each of the upper and lower focii of circulation. The outermost flow loop is shown partially complete, indicative of the general case for an arbitrary value of N.

The impeller is located between j=N and j=N+1. The focus of the lower circulation loop is between i=N/2 and N/2+1 and j=N/2 and N/2+1. The upper loop focus is then between i=N/2 and N/2+1 and j=2N and 2N+1.

It is possible to introduce any form of internal velocity profile by allocating the overall impeller generated flow equally among the flow loops and adjusting zone volumes appropriately. However,

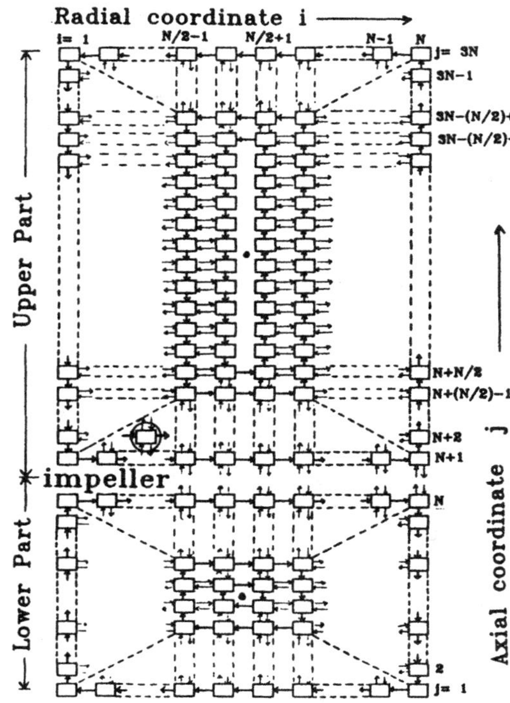

FIG. 3 Schematic of network–of–zones model

earlier work (Mann and Knysh, 1983) showed that volume adjustments to manipulate the velocity profiles made negligible differences to mixing behaviour in comparison with the simple equal volume case. For this present case with N=20, the set of zones appear as in Fig. 4. This (20x20) + (20x40) assembly contains 1200 zones. Each zone undergoes an equal and opposite exchange flow with adjacent zones in the lateral inner and outer flow loops, which mimics the effect of turbulence.

SEMI–BATCH OPERATION

The vessel is initially charged with reagent A and a small volume of concentrated reagent B (giving negligible additional volume) will be added semi–batchwise. Each zone in the vessel undergoes changes in concentration as mixing and reaction proceed.

For an arbitrary i, j zone, with main flow outward along a row of constant j, so that the flow in the loop is in the direction of increasing i, such as the cell (3, N+2) shown ringed in Fig. 3, an unsteady state material balance for any component I is written

$$\text{accumulation} = \begin{matrix}\text{input by flow}\\ \text{and exchange}\end{matrix} - \begin{matrix}\text{output by flow}\\ \text{and exchange}\end{matrix} \pm \begin{matrix}\text{rate of}\\ \text{reaction}\end{matrix}$$

so that

$$\frac{dC_{I\,i,j}}{dt} = \frac{q}{V_{i,j}}\left[C_{i-1,j} - (1+2\beta)C_{i,j} + \beta(C_{i,j-1} + C_{i,j+1})\right] \pm r_{I\,i,j} \quad (1)$$

For this competitive/consecutive reaction the reaction rate terms are

$$\begin{aligned} r_{A\,i,j} &= -\,k_1\, C_{A\,i,j} C_{B\,i,j} \\ r_{B\,i,j} &= -\,(k_1 C_{A\,i,j} C_{B\,i,j} + k_2\, C_{R\,i,j} C_{B\,i,j}) \\ r_{R\,i,j} &= k_1 C_{A\,i,j} C_{B\,i,j} - k_2\, C_{R\,i,j} C_{B\,i,j} \\ r_{S\,i,j} &= k_2 C_{R\,i,j} C_{B\,i,j} \end{aligned} \quad (2)$$

For the semi-batch addition of B to a zone i*, j*, there is an additional input term for the component B. For a linear addition rate of a volume V_B at a constant flow rate q_B^f, the input by semi-batch addition is given by the product $C_B^f q_B^f$ over an addition time t_a.

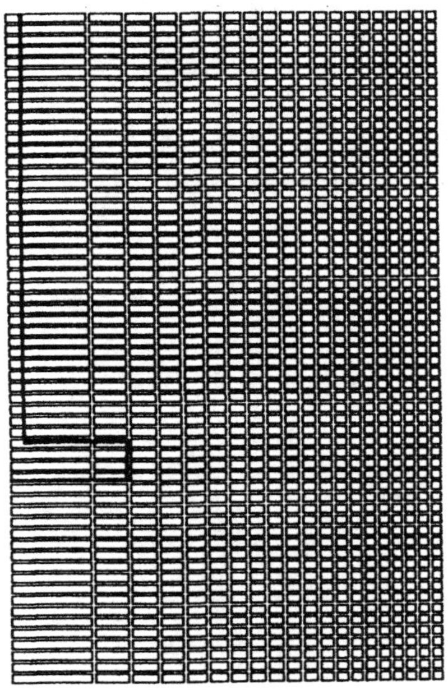

FIG. 4 **Equal volume zones for (20x20)+(20x40) configuration**

The initial conditions at commencement of operation are

$$C_{A\,i,j} = C_A^{\circ}$$

$$C_{B\,i,j},\ C_{R\,i,j},\ C_{S\,i,j} = O \qquad \begin{matrix}\text{for } O < i < N \\ \text{and } 0 < j < 3N\end{matrix} \quad (3)$$

The set of equations (1) for all i,j can be integrated from these initial conditions using any one of several integration sub-routines.

EXPERIMENTS ON IODINATION

Paul and Treybal's (1971) experiments on L–tyrosine involve the double iodination:–

L–tyrosine+I_2 $\rightarrow$ mono–iodo–L–tyrosine

mono–iodo– L–tyrosine+I_2 $\rightarrow$ di–iodio–L–tyrosine

Experimental details and rate constants are given in Table 1 for a semi–tech scale (36 litres) example at a stirrer speed of 95 rpm. A concentrated solution of I_2 was added over 60 seconds to a relatively weak solution of L–tyrosine in water. The molar ratio was 1:1.

CONCENTRATION FIELDS OF A, B, R and S

A set of image reconstructions of the axisymmetrical sectional concentration fields are presented in Figs 5(a), (b), (c) and (d), as calculated from a (20x20) + (20x40) network representation. These axisymmetric fields of concentration apply for the addition of iodine (B) reasonably close to the central impeller shaft. As the graphics in Fig 5(a) show, iodine is added linearly from the feed reservoir.

The overall internal flow has been estimated from Q=2.75 nD^3 (Oldshue, 1987) for n=95 rpm. This has been apportioned proportional to the upper and lower volumes (above and below the impeller) so as to give equal turnover times for the upper and lower flows. Also, as before, the flow has been equally allocated to each of the N/2 (in this case 10) nested flow loops in both the upper and lower flows. A value of the exchange flow parameter β=0.2 has been used (Mann et al, 1987), except for the impeller region where β=10 reflects the intensive turbulence immediately adjacent to the impeller. A uniform value of β=0.2 throughout the vessel (except for the impeller) does not imply that the energy dissipation is uniform everywhere, since β is the ratio of the local exchange flow rates relative to the local main convective flow.

The addition of iodine in Fig 5(a) is into zone (1, 58) representative of an addition just below the fluid surface. The shading convention in the whole of Fig. 5 uses between 0 and 5 horizontal lines spaced within each zone. With no line $O < C_I < 0.1\ C_A^{\circ}$ and with the maximum five lines $0.9C_A^{\circ} < C_I \leq C_A^{\circ}$. For the intermediate values, the convention is one line for $0.1C_A^{\circ} < C_I < 0.3C_A^{\circ}$, two lines for $0.3C_A^{\circ} < C_I < 0.5C_A^{\circ}$, three lines for $0.5\ C_A^{\circ} < C_I < 0.7C_A^{\circ}$ and four lines for $0.7\ C_A^{\circ} < C_I < 0.9C_A^{\circ}$.

Fig 5(a) then shows the sequence of the developing concentration field of iodine in solution in the vicinity of the addition point. The extent of visibility is quite limited throughout the addition period. It is significant that $C_{B\,i,j} \leq 0.7C_A^{\circ}$ since there are never more than three lines per zone. As addition proceeds, visibility (here taken to be $\geq$0.1 C_A° ie. one line per zone) is

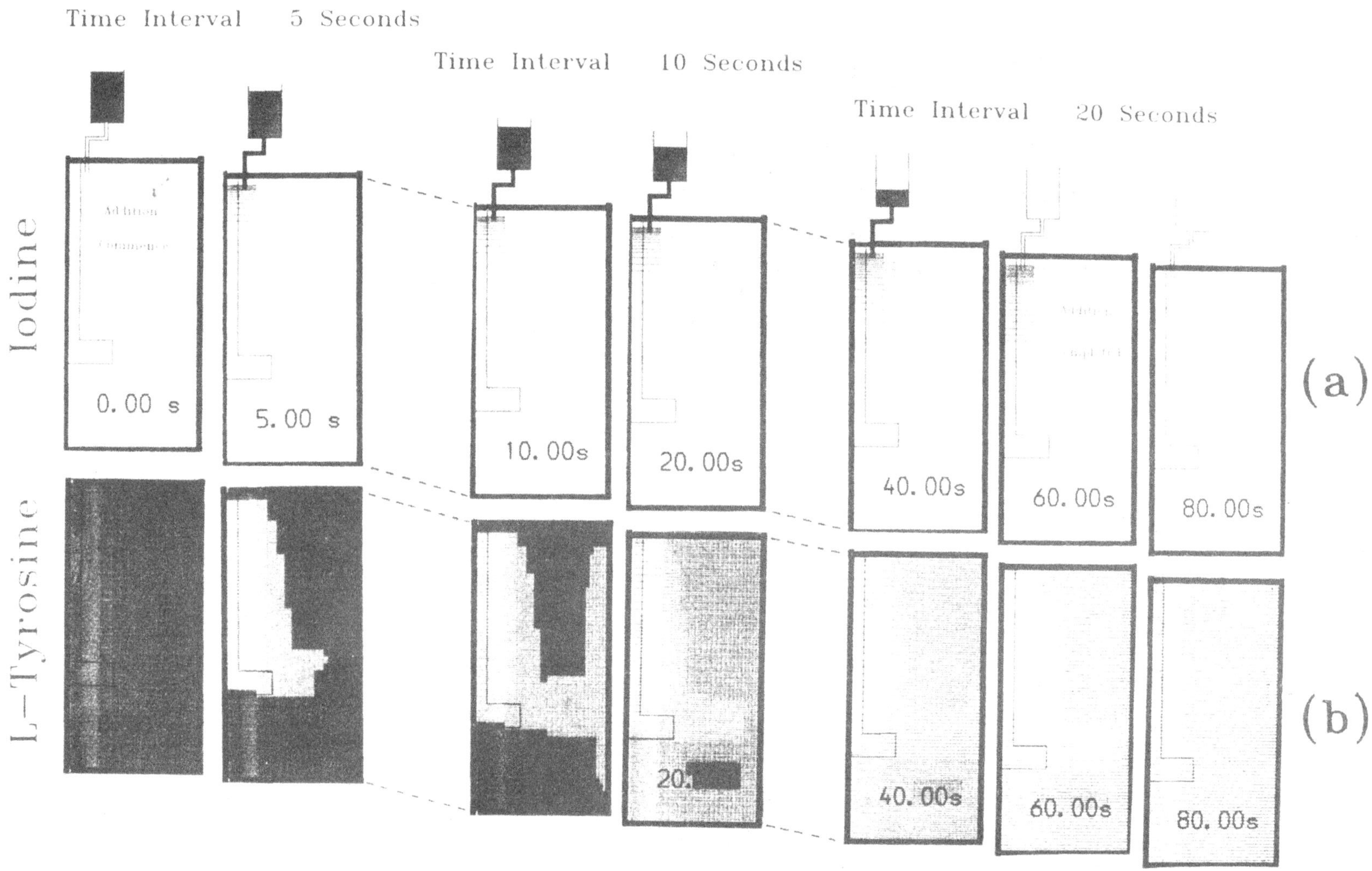

FIG. 5(a) and (b) Concentration fields for reagents

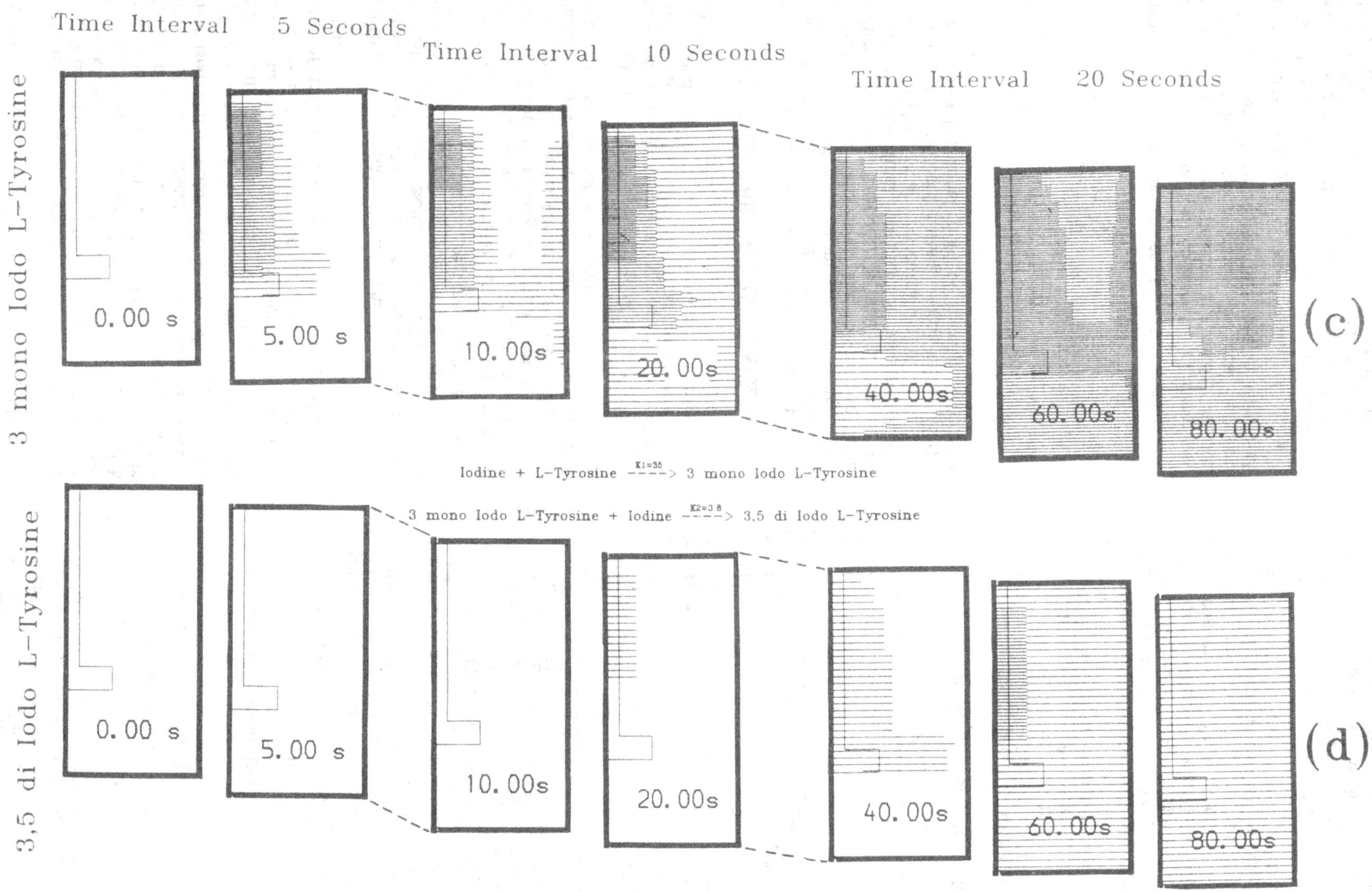

FIG. 5(c) and (d) Concentration fields for products.

confined to 5 zones after 5s enlarging to 23 zones at 60s on completion of the semi–batch addition.

Although B is not particularly evident by this visualised shading, the impact upon the reagent A (L–tyrosine) is relatively much more pronounced. As the impeller drives fluid through the B rich zones, reaction with A takes place and results in depletion of A. Thus even in the initial period after 5s, the extent of depletion is clearly visualised and a reduction below 0.9 C_{A_0} (a local 10% conversion of A) extends downwards towards, into and well beyond the impeller region (outwards up to i=10 and between j = 15 and j = 58) – this is approximately 100 zones out of a total of 1200 amounting to some 8% of the fluid volume.

Continuing circulation throughout the addition period of 60s shows up a developing depletion of A throughout the vessel. After 20s, just a small portion of the fluid is still within 10% of the initial C_{A_0}. The partial segregation of A and B thus becomes more marked as the addition of B proceeds. However, when addition ceases, the rates of internal mixing are sufficient to almost homogenise the vessel contents (at least within the shading convention used to indicate relative concentrations and/or visibility).

The evident partial segregation of A and B during semi–batch addition also shows up in non–uniform concentration fields for the first product R and the second product S as displayed in Figs 5(c) and 5(d).

YIELDS OF THE INTERMEDIATE R

The net result is that the primary reagent A is materially depleted in those zones where the flow is downwards and away from the immediate point of addition of B. Moreover, the R which is being produced by local reaction is more likely to be reacted to the final product S because of the relatively higher local concentration of B. This effect can be deduced through local values of the instantaneous fractional yield of R with respect to A given by

$$\varphi\left[\frac{R}{A}\right]_{i,j}=\frac{(dC_R/dt)_{i,j}}{(-dC_A/dt)_{i,j}}=\frac{k_1C_{A_{i,j}}C_{B_{i,j}}-k_2C_{R_{i,j}}C_{B_{i,j}}}{k_1C_{A_{i,j}}C_{B_{i,j}}}$$

$$= 1 - \frac{k_2}{k_1}\frac{C_{R_{i,j}}}{C_{A_{i,j}}} \qquad (4)$$

which shows that whenever C_A is lowered, the local yield of R is also lowered. Furthermore, because of the imperfection of the mixing caused by the finite local mixing rates, the local build–up of R also contributes to a reduction in the local instantaneous yield $\varphi\left[\frac{R}{A}\right]_{i,j}$ as Equation (4) shows. It is thus evident that the local build–up of reagent B reduces the potential yield of R by simultaneously causing a local depletion of A and a local excess of R. This means that the nature of the internal distribution of B – caused either by rate of addition of B or by rate of stirring – affects the product distribution . Under perfect mixing it is clearly the case that the addition of B has no effect. However,although C_B cancels from Equation (4), it still influences $\varphi\left[\frac{R}{A}\right]_{i,j}$ locally by indirectly causing changes in C_A and C_R which arise from the local imperfect mixing of B.

The overall deleterious effect of this partial segregation caused by imperfect macro–mixing can be seen in Fig. 6 which shows the overall yield of R with respect to A $\Phi\left[\frac{R}{A}\right]$ versus the overall conversion of A. Under hypothetical perfect mixing, a high overall yield of R is maintained until around 70% conversion (at which it is 95%). This should be contrasted with the result predicted at 190 rpm. The interaction of flow and mixing at this speed reduces the overall yield to 86%. This may seem a small yield penalty, but in fact the yield of by–product S (the di–iodo product) has in effect been almost doubled from a theoretical minimum of 8% to an imperfect mixing value of 14%.

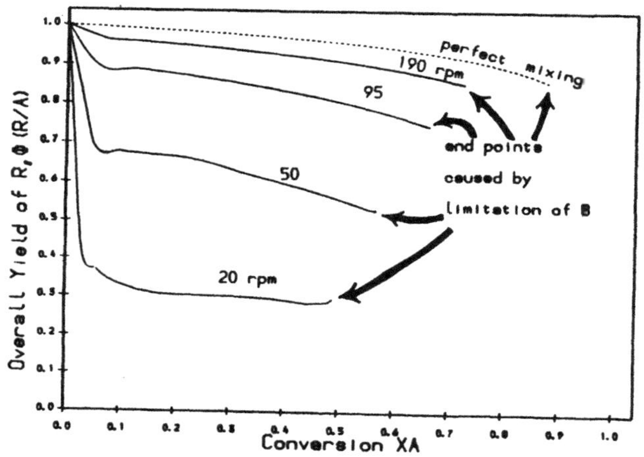

FIG. 6 Overall yield of intermediate product

This effect rapidly becomes worse as stirring speed is further reduced. Indeed at 95 rpm (a halving of the stirring speed) it is impossible to reach 70% conversion of A because so much of the reagent B has been consumed to make the extra undesirable S. In fact at 95 rpm at 60% conversion of A, the overall yield of R has reduced to about 77%. As stirrer speed is even further reduced, the yield of R suffers huge reductions. This is especially the case at the lower conversions. Thus at only 5% conversion, at 50 rpm the yield of R is just below 70%. By contrast, if perfect mixing were assumed, there would be a yield loss of only a fraction of 1%.

Finally, Fig. 7 shows the predictions of this (20x20)+(20x40) networks in a form which is directly comparable with Paul's experimental results on the amount of R produced. In Fig. 7, the ordinate is the final value of C_R/C_{A_0} (at various stirrer speeds) when all the B added semi-batchwise has reacted. With perfect mixing this would be just less than 0.76. The network predictions are similar, though the experimental amount of R is lower. This discrepancy may be attributable to the several simplifying assumptions that have been used. Further refinements of the model need to be undertaken. Nevertheless, it is concluded that the ability to predict and understand the full details of the internal concentration fields as semi-batch operation proceeds, provides new insights into the control of yield loss under partial segregation caused by imperfect mixing.

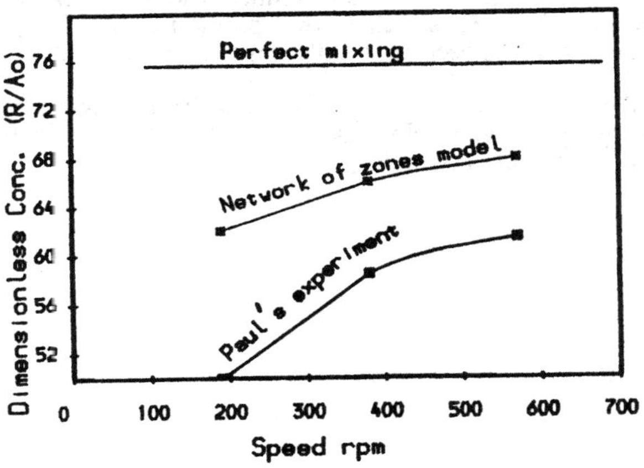

FIG. 7 Comparison with experiments

CONCLUSIONS

1. Modifications to the previous network-of-zones formulation allow for the calculation of mixing behaviour with the vessel impeller placed at one-third clearance (c=H/3)

2. The modified network has been successfully applied to a consecutive/competitive reaction. This reaction system has two reactions and four components, three of which act as reagents, which represents the most complex set of reactions so far analysed using networks-of- zones.

3. The kinetic data of Paul and Treybal (1971) for iodination of L-tyrosine show that for their semi-batch experiments, the rate of internal mixing was so slow as to give strong partial segregation of the reagent iodine. The computed concentration fields in space and time can be conveniently presented as coloured sectional image reconstructions.

4. The predicted levels of imperfect mixing result in a severe loss of yield of the intermediate mono-iodo product, caused by localised excess of the iodine. This yield loss becomes progressively worse as stirrer speed is reduced.

Table 1: Experimental parameters

Initial concentration of A	C_A^o	0.2 mol m^{-3}
Feed concentration of B	C_B^f	2.0 mol m^{-3}
Semi-batch addition rate	q^f	5.5×10^{-5} m^3 s^{-1}
Semi-batch addition time	t_a	60s
Volume of component B	V_B	3.3×10^{-3} m^3
Vessel volume	V	36 dm^3
Standard Rushton turbine	D	0.12 m
Rate constant (A+B)	k_1	35 m^3 mol^{-1} s^{-1}
Rate constant (R+B)	k_2	3.8 m^3 mol^{-1} s^{-1}
Addition feed position	(i*,j*)	(1,58)

NOMENCLATURE

C_A^o	initial concentration of A
C_B^f	feed concentration of B
C_I	concentration of component I
c	impeller clearance
D	impeller diameter
H	liquid depth
i	radial zone position
j	axial zone position
$k_{1,2}$	reaction rate constants
N	network size
n	impeller rotation speed
Q	total internal flow rate
q	zone loop flow rate
r_I	reaction rate
T	tank diameter
t	time
$V_{i,j}$	zone volume
β	turbulent exchange flow rate
$\psi(R/A)$	instantaneous yield
$\Phi(R/A)$	overall yield

REFERENCES

Atiquallah, M and Nauman, E. B., 1990 "A Model and Measurement Technique for Micromixing in Polymerisation Reactions. "Chem. Eng. Sci., 45, 1267

Baldyga, J. and Bourne, J. R., 1989, "Simplification of Micromixing Calculations 1: Derivation and Application of a New Model chem. Eng. Jnl., 42, 83

Bourne, J. R., and Dell'ava, P., 1987, "Micro and Macro-Mixing in Stirred Tank Reactors of Different Sizes". Chem. Eng. Res. Des. 65, 180

Brucato, A. and Rizzuti, L., 1988 "The Application of the Networks-of-Zones Model to Solid-Liquid Suspension" Proc. 6th European Conference on Mixing, 273

Fox, R. O. and Villermaux, J., 1990, "Micromixing Effects on the CLO_2 + I Reaction: Perturbation Analysis and Numerical Simulation of the Unsteady State IEM Model", Chem. Eng. Sci., 45, 2857

Harvey, P. S., and Greaves, M. 1982, "Turbulent Flow in an Agitated Vessel, part I, A predictive Model", Trans. IChemE., 60 195

Levenspiel, O., 1972 "Chemical Reaction Enginering", 2nd Ed., Wiley

Mann, R., Mavros, P. and Middleton, J. C., 1981. "A Structured Stochastic Flow Model for Interpreting Flow Follower Data from a Stirred Vessel, "Trans. IChemE., 59, 271

Mann, R. and Knysh, P. 1984, "Utility of Interconnected Networks of Backmixed Zones to represent Mixing in a Closed Stirred Vessel, "IChemE. Symp. Series, 89, 127

Mann, R., "Gas-Liquid Stirred Vessel Mixers: Towards a Unified Theory based on Networks of Zones", Ch.E.R.D., 64, 23

Mann, R., Knysh, P., Rasekoala, E. A. and Didari, M., 1987, "Mixing of a Tracer in a Closed Stirred Vessel: A Network of Zones Analysis of Mixing Curves Acquired by Fibre Optic Photometry", IChemE Symp. Series, 108, 55

Mann, R. and Hackett, L. A., 1988, "Fundamentals of Gas-Liquid Mixing in a Stirred Vessel: An Analysis Based on Networks of Zones", Proc. 6th Europ. Conf. Mixing, 321

Mann, R., Knysh, P., Pillai, S. K. and Motlagh, M. D. K., 1990a, "Mixing of a Tracer in a Batch Stirred Vessel: Visualisation Interpreted by Networks of Zones", submitted to Can. Jnl. of Chem. Eng.

Mann, R., Pillai, S. K. and Wang. Y. D., 1990b, "Characterisation of Mixing of Inert and Reactive Tracers in a Batch Stirred Vessel : An Image Reconstruction Approach", submitted to Chem. Eng. Commun.

Mann, R. and Wang, Y. D., 1990, "Mixing in a Stirred Semi-Batch Reactor: Partial Segregation for a Pair Competing Reactions Analysed via Networks of Zones", IChemE Symposium Series, 121, 241.

Mann, R, and Wang. Y. D., 1991, "Mixing and Partial Segregation in a Semi-Batch Stirred Vessel Reactor: Effect of Scale-Up on the Yield of a Pair of Reactions", submitted to Ch.E.R.D.

Middleton, J. C., Pierce, F. and Lynch, P. M., 1986, "Computations of Flow Fields and Complex Reaction Yield in Turbulent Stirred Reactions, and Comparison with Experimental Data", Chem. Eng. Res. Des., 64, 18

Oldshue J. Y., 1983, "Fluid Mixing Technology" McGraw-Hill Pub. Co., p174

Paul, E. L. and Treybal, R. E., 1971, "Mixing and Product Distribution for a Liquid Phase, Second-order, Competitive-Consecutive Reaction." AIChE J. 17, 719.

Pericleous, K. A. and Patel, A., 1986, "The Modelling of Tangential and Axial Agitators in Chemical Reactors", Int. Jnl. of P.C.H., 2, 105

Note added in response to referee's comment

It has been pointed out that our assumption that the upper and lower circulation flows are proportional to the volumes involved is erroneous – Voncken (1966)* has shown that for a one third positioning of the impeller, the two flow rates are approximately equal. We are presently checking this using a flow visualization method. If the flow rates are roughly equal, this would intensify the non-uniform concentration of B in the upper part of the vessel. This would reduce the yield of the intermediate R so that the prediction in fig. 7 would come closer to Paul's experimental results!

*Voncken, M., 1966, "Circulatiestromung en merging in geroerde vaten," Ph.D. Thesis, Delft University of Technology.

THE VARIATION OF FLOW PATTERN AND MIXING TIME WITH IMPELLER SPACING IN STIRRED VESSELS WITH TWO RUSHTON IMPELLERS

S.M. MAHMOUDI, M. YIANNESKIS, King's College London, Centre for Heat Transfer and Fluid Flow Measurement, School of Physical Sciences and Eng., GB

ABSTRACT

The flow patterns and mixing times in a stirred vessel of diameter T=294 mm with two Rushton impellers were investigated using flow visualization and conductivity measurement techniques. Experiments were carried out for a range of impeller spacings, rotational speeds, liquid levels and tracer insertion times.

Three different flow patterns were observed with different impeller spacings. The impeller streams from the two Rushton turbines were nearly horizontal and parallel when the spacing between the two impellers (C2) was large. For small values of C2 the two impeller streams were directed towards each other, merging together half-way between the impellers. For small clearances of the lower impeller from the bottom of the vessel (C1) the impeller stream from the lower turbine was directed towards the base of the vessel. For intermediate spacings the flow was unstable, changing from one pattern to the other with time.

The variation of the flow pattern was observed to affect the mixing times measured in the vessel. On average, merging flow was found to result in smaller mixing times.

INTRODUCTION

A number of mixing processes make use of two or more impellers mounted on the same shaft to stir the fluid in the tank, especially in fermenters. Most applications employ baffled tanks, although unbaffled tanks are sometimes used for the mixing of sensitive polymers. The flow patterns resulting from a combination of impellers can be quite complex and may affect the power numbers obtained significantly. The present investigation is concerned with the mixing and flow pattern characteristics in vessels stirred by two Rushton turbines which are frequently used in such combinations. Previous investigations have dealt primarily with aerated vessels (see, for example, 1-3) and/or non-Newtonian fluids (4). Flow patterns in unaerated tanks with two Rushton turbines have been studied by, among others, (5,6) and power numbers have been measured by (6,7).

The total power input is known to depend on the impeller spacing (2). Cronin (8) found that the power number in unaerated vessels with a liquid level H = 2 T and two Rushton impellers of diameter D = T/2 increased sharply as the clearance between the two impellers increased from about 0.25 T to 0.375 T. Similar increases in power consumption have been measured by Kuboi and Nienow (1), Hudcova et al (6) and Smith et al (9) in tanks with various diameters, impeller sizes and liquid levels. Bader (7) found a similar increase in power number as the level of liquid above the upper impeller increased. There is some evidence that the mixing efficiency of double impellers is inferior to that of single impellers (10). However, although the flows in stirred vessels with a single Rushton turbine have been extensively documented (see, for example, 11-13), there is little information on the effect of impeller spacing or tracer insertion time on the flow patterns and mixing time in tanks with two turbines. The present investigation was undertaken to quantify the latter effect and to examine in detail the variation of the flow patterns with liquid level and impeller spacing.

An experimental investigation of the mixing times and flow patterns in a vessel stirred by two Rushton turbines was carried out using conductivity and laser-sheet flow visualization techniques respectively. The visualization results were used to determine the overall flow pattern in the vessel for a range of conditions, and to aid the selection of the locations where the conductivity measurements were made. The latter were carried out primarily in order to identify the effect of the flow pattern on the mixing times in the vessel.

EXPERIMENTAL ARRANGEMENT AND TECHNIQUES

The mixing vessel used for all the experiments described in this paper was cylindrical with a diameter T = 294 mm, four flat baffles of width w = 0.1 T, and a flat bottom. The tank was made from a transparent cast acrylic cylinder in order to permit complete optical access to the flowfield for the visualization tests. Two standard Rushton impellers (D = T/3) were used to mix the fluid. The working fluid was water, and the liquid depth in the vessel was varied from 294 mm (H = T) to 441 mm (H = 1.5 T). Experiments were carried out over a range of impeller rotational speeds, from 60-260 rev/min. The outline of the vessel is shown in Figure 1. The cylindrical vessel was located inside a trough of square cross-section filled with water, in order to minimise light refraction and therefore optical distortion at the curved acrylic surface. In the coordinate system used below r is the radial distance from the axis, z the vertical distance from the bottom of the vessel and θ denotes the radial plane considered (with θ = 0° and 90° being the planes located half-way between two baffles, and θ = 45° and -45° being the planes of the baffles).

Flow visualization recordings were obtained by inserting small acrylic particles (clear Diakon of density ρ_p = 1180 kg/m^3 and mean diameter d_p = 270 μm) as tracers in the vessel and illuminating the flow in a plane in the vessel by means of a sheet of light; the light-sheet was produced either by a 2.4 W Argon-Ion laser operating on all lines, or by a 1 kW white light source. The characteristic response-time parameter for the particles $t = d_p^2 \rho_p/18\mu$ was approximately 4 ms and their sink velocity was 10 mm/s; the particles can therefore be expected to follow faithfully most fluid motions in the vessel.

Mixing times in the vessel were determined by inserting a passive scalar as tracer and measuring the conductivity of the liquid. The conductivity technique made use of potassium chloride solutions as the tracer, and their dispersion in the liquid was measured with a probe of the Khang and Fitzerald design (14). The tracer was inserted at a point located at z = T, r = 80 mm and θ = - 45°. The probe output was processed by a signal acquisition unit interfaced to a microcomputer. The conductivity-time recordings were transferred to a mainframe VAX computer for subsequent processing and analysis. Measurements were obtained with a wide range of tracer insertion times. The mixing times for the parameters investigated were determined from conductivity-time traces such as those presented below. The response of the probe was entirely linear throughout the range of concentration measured and a linear conversion of the voltage output of the probe to conductivity and therefore concentration was possible. In view of the different homogeneity degrees used to define mixing time in the literature, each mixing time value was calculated as an average: it is the arithmetic mean of the mixing times determined from a single trace corresponding to 13 different homogeneity degrees (92.0%, 92.5%, 93.0%, ..., 97.5%, 98%). The conductivity-time traces recorded could be processed to obtain the fluctuations of the instantaneous concentration value from the mean, the frequency content of the trace and other quantities of interest. For economy of presentation, only mixing time results are presented and discussed in this paper.

RESULTS AND DISCUSSION

EFFECT OF IMPELLER SPACING, ROTATIONAL SPEED AND LIQUID DEPTH ON FLOW PATTERN

A large number of photographic and video recordings of the flow patterns were made for impeller rotational speeds (N) of 60-260 rev/min, for liquid depths varying from 1.0-1.5 T, and for various impeller spacings. The three impeller spacings are defined in Figure 1: C1 is the clearance between the lower impeller and the vessel bottom, C2 the spacing between the two impellers and C3 the liquid depth above the upper impeller.

As these four parameters (N, C1, C2 and C3) were varied, three different flow patterns were observed. For relatively large values of C2 the two impeller streams produced by the turbines were almost horizontal and parallel to each other and the flow pattern from each turbine was very similar to that produced by a single Rushton turbine. This flow pattern comprised four main ring vortices as shown in Figure 1 (A) and is termed ‘parallel flow’ in the following text.

For relatively small values of C2, the impeller streams from the two turbines were inclined at a large angle from the horizontal, directed towards each other and merged at some point half-way between the two impellers. This pattern comprised two main ring vortices and is termed 'merging flow' below. The fluid trapped between the two impeller streams near the shaft was observed to simply rotate around the shaft with little evidence of any lateral motion. The merging flow pattern is shown in Fig. 1 (B).

For both of the above flow patterns, C1 was greater than about 0.20 T. With smaller C1 spacings the impeller stream from the lower turbine was directed towards the bottom of the vessel. A single ring vortex was thus formed by the lower impeller, whereas the upper impeller produced a flow similar to that of a single Rushton turbine. Therefore, the flow comprised three main ring vortices, as indicated in Figure 1 (C). This pattern is referred to as 'diverging flow' in the following text.

The instantaneous recordings varied from one instant to another, as expected in a turbulent flow. However, it was observed that for some impeller spacings and liquid levels there were longer variations which were not caused only by turbulence, but also by a kind of unsteadiness in the flow. In fact, it was found that for intermediate impeller spacings, the flow patterns were unstable, changing from one type of flow to another over periods of seconds or minutes. The flow changed between two or three flow patterns and all possible combinations of patterns were observed: parallel/merging (abbreviated to P.M. below), parallel/merging/diverging (P.M.D.), parallel/diverging (P.D.) and merging/ diverging (M.D.) flows. Experiments were repeated a few times to confirm the presence of instabilities.

The results are summarised in Figures 2 (a) - (f). These figures show liquid level H plotted against C1 impeller spacing for a range of C2 spacings. The type of flow pattern observed for different values of the parameters is indicated in the figures by the different symbols. Solid lines separate the regions where different flow regimes were observed. Figures 2 (a) - (e) show results at N = 260 rev/min. It was found that rotational speed did have an effect on the flow patterns and a sample of the results at a lower speed (N = 100 rev/min) is shown in Fig. 2 (f) for comparison.

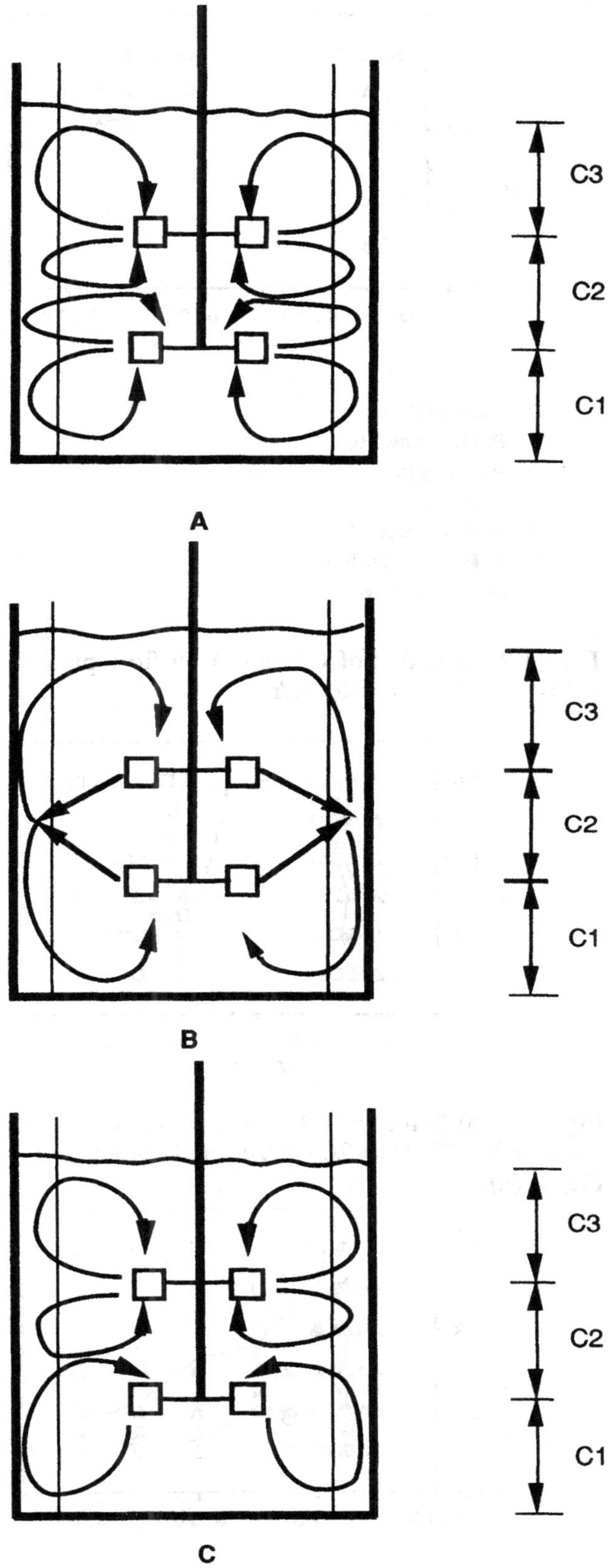

Figure 1 Mixing vessel configuration and diagram of flow patterns: (A) parallel; (B) merging; and (C) diverging flow.

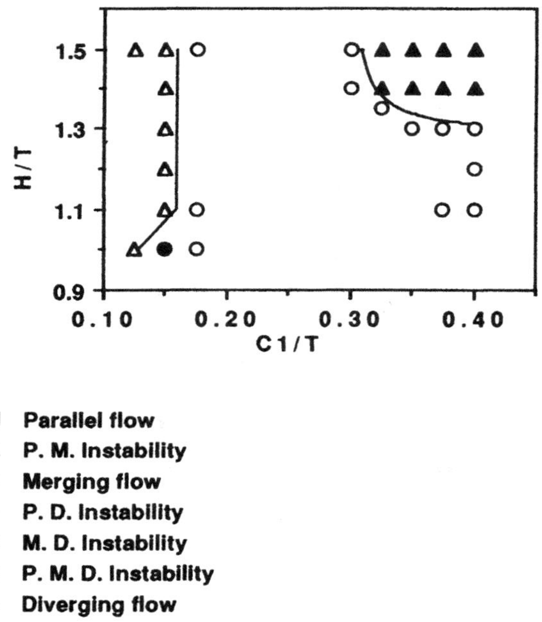

Figure 2 (a) Effect of C1 and H on flow pattern: C2 = 0.35 T, N = 260 rev/min.

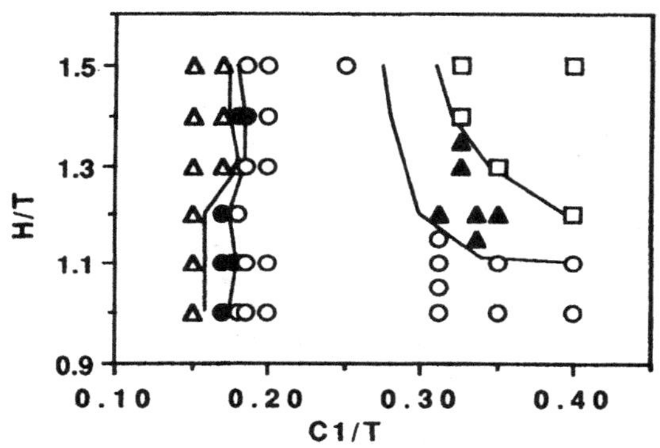

Figure 2 (b) Effect of C1 and H on flow pattern: C2 = 0.365 T, N = 260 rev/min. Legend as for Fig. 2 (a).

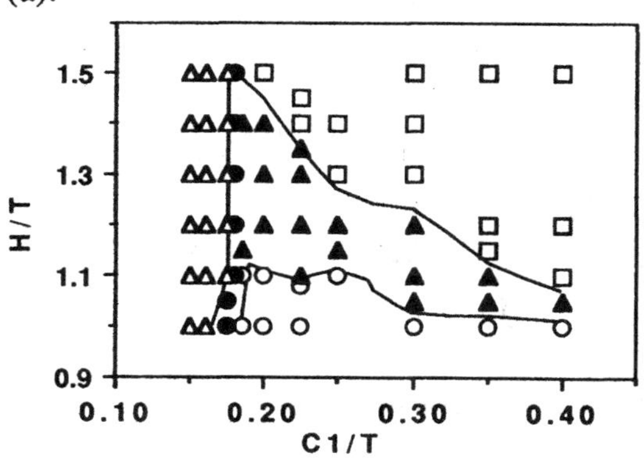

Figure 2 (c) Effect of C1 and H on flow pattern: C2 = 0.375 T, N = 260 rev/min. Legend as for Fig. 2 (a).

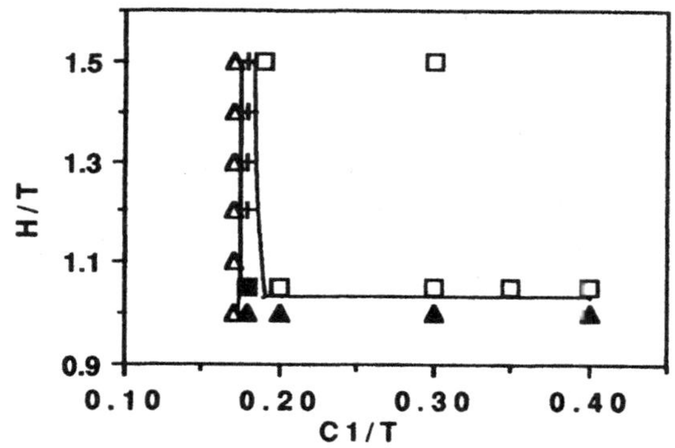

Figure 2 (d) Effect of C1 and H on flow pattern: C2 = 0.385 T, N = 260 rev/min. Legend as for Fig. 2 (a).

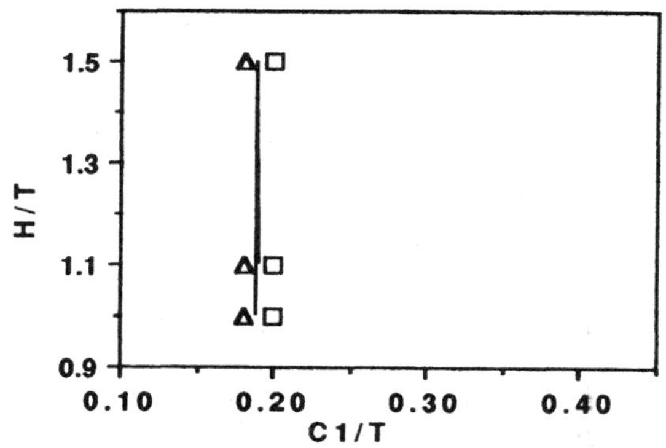

Figure 2 (e) Effect of C1 and H on flow pattern: C2 = 0.40 T, N = 260 rev/min. Legend as for Fig. 2 (a).

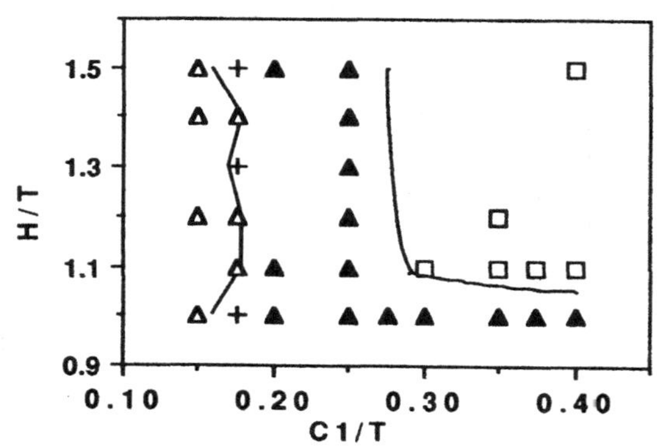

Figure 2 (f) Effect of C1 and H on flow pattern: C2 = 0.375 T, N = 100 rev/min. Legend as for Fig. 2 (a).

With the smallest C2 clearance, 0.35 T (Fig. 2 (a)), there was no stable parallel flow regime observed. Stable merging flow occurred for most C1 and H values, with diverging flow at the lower C1 values and unstable parallel/merging flow at high C1 and H values. Tests with C2 = 0.34 T (not shown here) revealed only stable merging flow present for C1 values above 0.17 T. As C2 is increased to 0.365 T, stable parallel flow occurs for the higher C1 and H values (Fig. 2 (b)) and diverging flow at the lower C1 values as before. Merging flow occurs over most of the range examined; in this case there are significant regions of unstable flow separating the stable flow regions.

In Fig. 2 (c) the results with C2 = 0.375 T are presented. There are large regions of unstable flow in the centre of the plot, separating the three stable flow regimes. With C2 = 0.385 T, Fig. 2 (d), parallel flow was observed over most of the parameter range examined, but the flow for H = T was unstable for C1 greater than about 0.175. Stable flow patterns were observed with C2 = 0.4 (Fig. 2 (e)); parallel flow occured for C1 spacings above 0.185 T and diverging flow for smaller C1 values.

The effect of impeller rotational speed on the flow patterns can be observed by comparison of Figs. 2 (c) and (f), which show results for C2 = 0.375 T at N = 260 and 100 rev/min respectively. The unstable flow region extends to higher values of H and no stable merging flow region was observed for the lower speed.

The above results help to quantify the complex variation of flow pattern with C1, C2, H and N. Although the flow variations are complicated, stable merging flow was observed with C2 less than 0.34 T and C1 greater than about 0.17 T; correspondingly, stable parallel flow was observed with C2 greater than 0.4 T and C1 greater than about 0.19 T. There are similarities in the results presented, e.g. stable diverging flow occurs almost always for C1 less than around 0.17 - 0.19 T. Further experiments are required though in order to allow the prediction of the flow pattern for different combinations of spacings etc.

The findings presented above reveal a complexity of the flow patterns which was little known: only two of the three stable flow patterns (parallel and merging) have been previously identified (6) and the authors could find no mention of unstable flows in the published literature. More importantly, the flow instabilities reported here could affect mixing times and process design significantly.

EFFECT OF IMPELLER SPACING ON MIXING TIME

In order to quantify the effect that different flow patterns in the vessel have on the mixing times, a series of concentration measurements was performed, using the conductivity technique outlined earlier.

Recordings of the concentration variation at one location in the vessel with time are shown in Figs. 3 (a)-(b). Both recordings were made with the probe positioned exactly half-way between the two impeller streams at a radius of 80 mm, and are presented in non-dimensional terms, with the concentration C normalised by the final concentration C_f and the time T normalised by the mixing time, i.e. the time from the start of the tracer insertion to when the fluid is fully-mixed, T_f.

The impeller spacing and liquid level for the first recording in Fig. 3 (a), C2 = 0.3 T and H = T, were such that a stable merging flow pattern was obtained. The probe was located exactly at the position where the two impeller streams merged, at z = T/2 and r = 80 mm. The corresponding impeller spacing and liquid level (C2 = 0.4 T and H = T) for the second recording (Fig. 3 (b)) resulted in a stable parallel flow pattern. The probe was at the same location, which now was at the boundary between the two ring vortices. The fluctuations are much larger with the merging flow as the flow can be expected to be considerably more turbulent when the two streams impinge upon each other.

A more appropriate representation of the level of concentration fluctuations can be found if the mean concentration throughout the vessel at any instant is subtracted from the instantaneous value shown in Figs. 3 (a) and (b). The fluctuations (Cinst - Cmean) corresponding to Figs. 3 (a) and (b) are shown in Figs. 3 (c) and (d) respectively. It is clear that the fluctuations with the merging flow pattern are almost 2.5 times larger than with the parallel flow pattern. In both cases the fluctuations decay with time, as expected from similar investigations with single turbines (15).

The variation of mixing time with C2 impeller spacing is shown in Figures 4 and 5. Mixing times determined from single recordings at three locations in the vessel are shown in Figure 4. In all cases the mixing time increases significantly

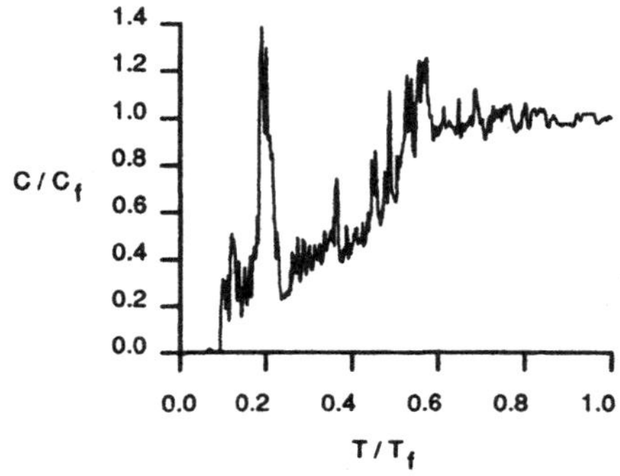

(a)

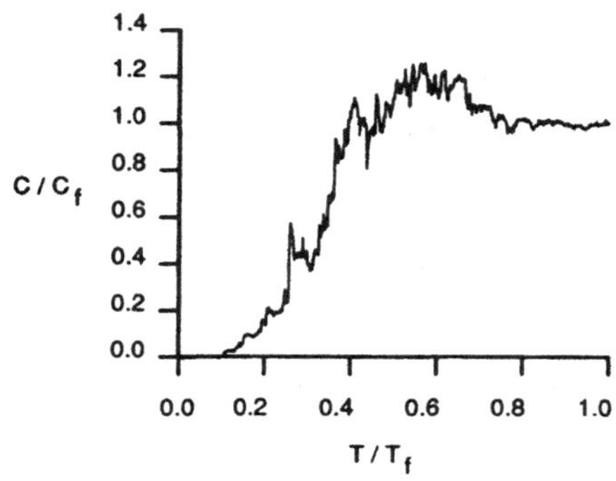

(b)

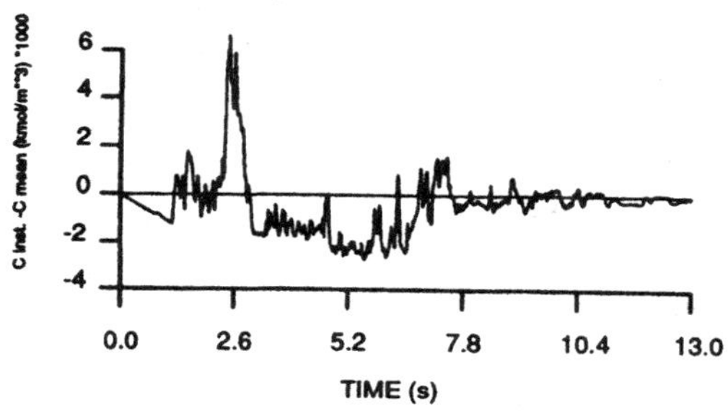

(c)

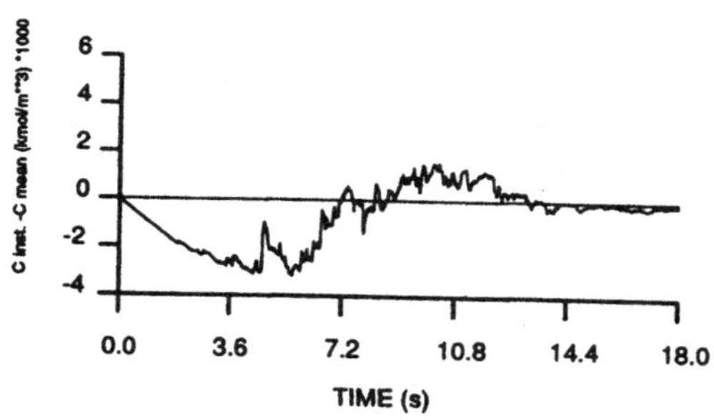

(d)

Figure 3 Recordings of the variation of tracer concentration with time.

around C2/T = 0.385 and then decreases as C2 is increased further. From the flow visualization results it is known that for C2 values lower than 0.385 merging flow is present, whereas for values larger than 0.4 parallel flow prevails and for C2 = 0.385 T the flow is unstable. The results indicate that the shortest mixing time is achieved with merging flow, while unstable flow yields the longest mixing times.

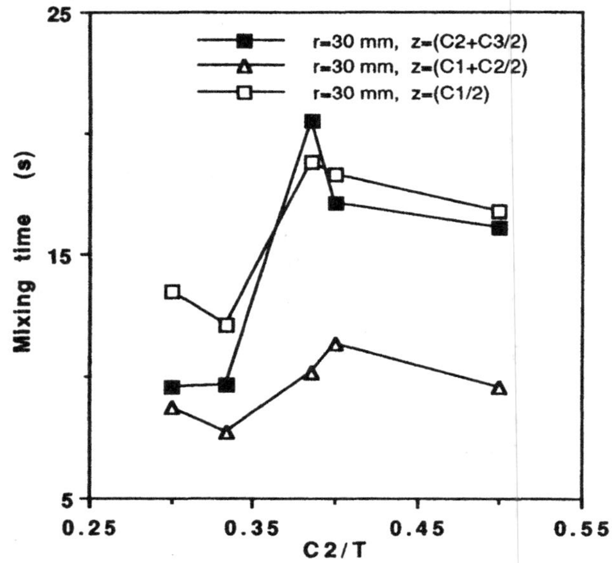

Figure 4 The effect of impeller spacing C2 on mixing time; N = 200 rev/min, H = T, C1 = C3.

However, it is evident that results obtained from a single probe may not be representative of the flow behaviour across the whole vessel. For this reason the average mixing times determined from measurements at eight locations throughout the vessel are shown in Fig. 5. It can be seen that the variation of mixing time with C2 is similar to those of the individual recordings of Fig. 4: more rapid mixing is achieved with the merging

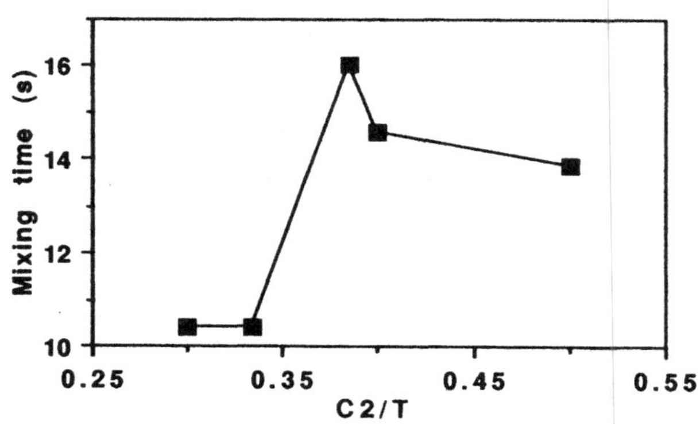

Figure 5 The effect of impeller spacing C2 on the average mixing time; N = 200 rev/min, H = T, C1 = C3.

flow (low C2 values), the unstable flow impeller spacings correspond to the longest mixing time and in the parallel flow region (high C2) mixing times are about 40% higher than for merging flow.

It must be noted that the flow visualization revealed that the impeller stream flow was significantly more turbulent when a merging pattern was formed, and therefore better mixing can be expected for this type of flow.

EFFECT OF IMPELLER SPEED ON MIXING TIME

Concentration measurements were carried out at different impeller rotational speeds after the variation of flow pattern with speed reported earlier was observed. It is generally accepted from results in single-turbine vessels that the mixing time is inversely proportional to the rotational speed of the impeller (15). However with the aforementioned flow pattern variation with two impellers, a simple relationship between mixing time and speed may not be anticipated.

The results presented in Fig. 6 show average mixing times from four recordings at one location and indicate a similar variation of mixing time with speed for all three insertion times. Mixing times decrease with increasing speed as might be expected from the above considerations, but the slope of the curves changes at around 110 rev/min. This may be partly a result of the laminar flow regions present at least in some locations in the vessel at low speeds, and partly a consequence of a change in flow pattern. Similar trends were observed with a larger C2 clearance , as indicated by the results shown as Fig. 7. However, both the results of Figs. 6 and 7 were obtained from measurements at one location and may not be representative of the variation of mixing time across the whole vessel. The results presented in Fig. 8 show average mixing times determined from recordings at four positions in the vessel. Although the data show some scatter, the variation of mixing time with impeller speed can be approximated as linear for the speed range investigated.

EFFECT OF TRACER INSERTION TIME

The ability to predict the variation of mixing time with insertion time is also very useful, but little is known about it. Insertion times varying from 2 to 27 s were used in the present work, and the results are summarised in Figure 9. For long insertion times, the mixing time increases almost in direct proportion. For short insertion times there is some scatter in the data, but if a

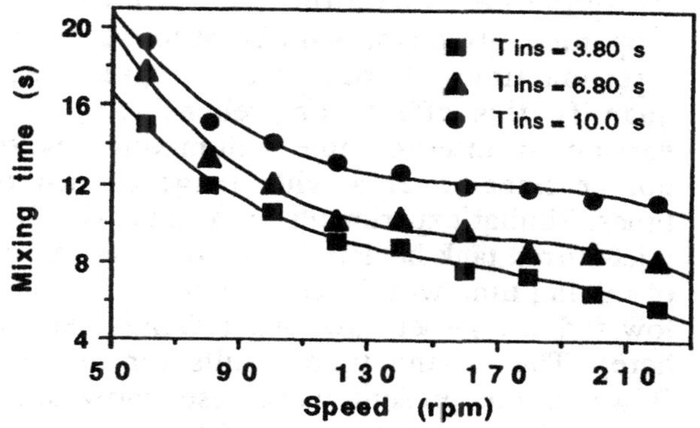

Figure 6. The effect of speed on mixing time for three insertion times; $C1 = C2 = C3 = T/3$. Probe located at $r = 30$ mm, $z = T/2$, $\theta = 90°$.

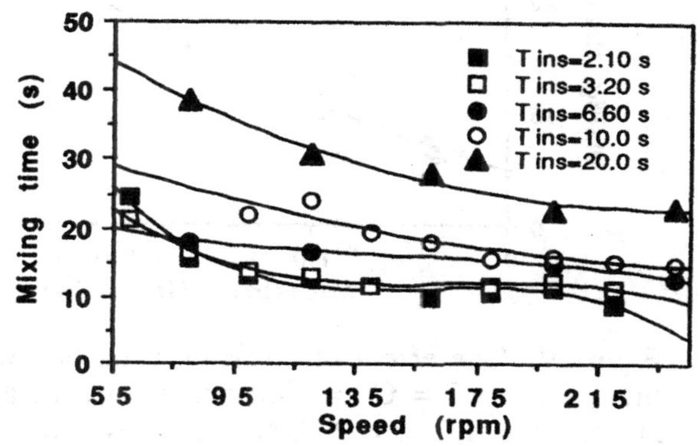

Figure 7. The effect of speed on mixing time for five insertion times; $C1 = 0.3$ T, $C2 = 0.4$ T. Probe located at $r = 60$ mm, $z = 0.7$ T, $\theta = 90°$.

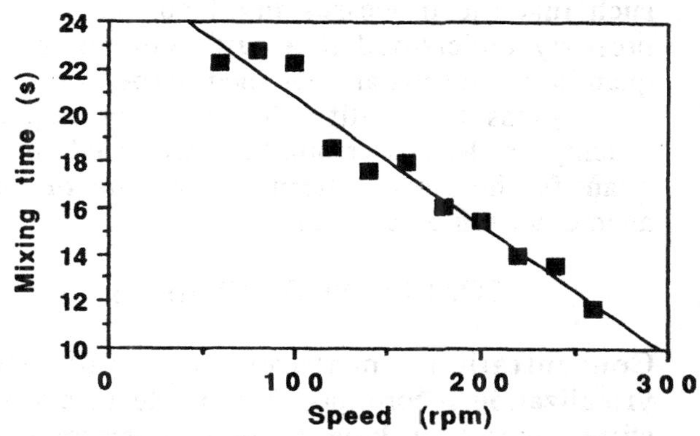

Figure 8. The effect of speed on the average mixing time; $C1 = 0.3$ T, $C2 = 0.4$ T.

straight-line variation is assumed, the proportionality factor is different to that for long insertion times. Further tests are required to quantify this effect comprehensively, as the variation of insertion time with mixing time may not be linear over a wide range of insertion times. Similar experiments by Yianneskis (16) in a jet-stirred tank have indicated that the variation of mixing time with insertion time is different at low and at high jet flow rates (cf. impeller speed here). The mixing time results for the stable flows in the present work also show similar trends with those reported in (17).

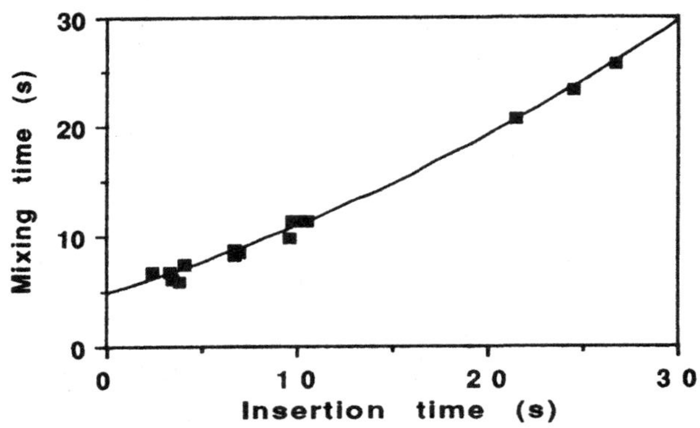

Figure 9. The effect of insertion time on mixing time; C1 = C2 = C3 = T/3; probe located at r = 30 mm, z = T/2, θ = 90°, N = 200 rev/min.

The results presented in this paper have thrown some new light in the processes in stirred vessels. However the complexity of the flow behaviour is such that the processes involved can only be properly understood if a sufficient number of quantitative results are obtained. Experiments are in progress to identify the effect of impeller spacing on the power numbers produced, and to quantify the flow patterns by means of laser anemometry measurements.

CONCLUDING REMARKS

Concentration measurements and flow visualization recordings were made in a vessel stirred by two Rushton turbines to determine the effect of impeller spacing on the flow patterns and mixing time. The results indicate that for different arrangements of impeller spacing and liquid level three stable flow patterns and four unstable combinations of these three patterns are produced in the vessel. The mixing time was shown to be affected by the variation of flow pattern. The results can therefore be expected to be of considerable significance for process design.

REFERENCES

1. Kuboi, R. and Nienow, A.W. (1982). The power drawn by dual impeller systems under gassed and ungassed conditions. Proc. 4th European Conference on Mixing, Noordwijkerhout, The Netherlands, 27-29 April.
2. Machon, V., Vlcek, J. and Skrivanek, J. (1985). Dual impeller systems for aeration of liquids: an experimental study. Proc. 5th European Conference on Mixing, Wurzburg, Germany, 10-12 June.
3. Roustan, M. (1985). Power consumed by Rushton turbines in non standard vessels under gassed conditions. Proc. 5th European Conference on Mixing, Wurzburg, Germany, 10-12 June.
4. Solomon, J., Nienow, A.W. and Pace, G.W. (1981). Flow paterns in agitated plastic and pseudoplastic viscoelastic fluids. Fluid Mixing, I.Chem.E. Symposium Series No. 64, pp A1-A13.
5. Crozier, D.B.A. (1989). A detailed study of the flow characteristics in a tall stirred tank. Proc. 3rd International Conference on Laser Anemometry - Advances and Applications, Swansea, Wales, 26-29 September.
6. Hudcova, V., Machon, V. and Nienow, A.W. (1989). Gas-liquid dispersion with dual Rushton turbine impellers. Biotechnology and Bioengineering, vol. 34, pp. 617-628.
7. Bader, F.G. (1986). Mass transfer in a multiturbine fermentor. I. Gassed and ungassed power inputs. International Conference on Bioreactor Fluid Dynamics, Cambridge, England, 15-17 April.
8. Cronin, D.G. (1988). The design of a pilot-scale protofermenter and its application to mixing, power consumption, heat-transfer and mass-transfer studies. Ph.D. Thesis, University of Birmingham.
9. Smith, J.M., Warmoerskerken, M.M.C.G. and Zeef, E. (1986). Paper 117F, A.I.Chem.E. Annual Meeting.
10. Komori, S. and Murakami, Y. (1988). Turbulent mixing in an unbaffled vessel with double impellers. Proc. 6th European Conference on Mixing, Pavia, Italy, 24-26 May.
11. Yianneskis, M., Popiolek, Z. and Whitelaw, J.H. (1987) An experimental study of the steady and unsteady flow characterisctics of stirred reactors. Journal of Fluid Mechanics, vol. 175, pp. 537-555.
12. Popiolek, Z., Whitelaw, J.H. and Yianneskis, M. (1984). Unsteady flow over disk turbine blades. Proc. 2nd International Symposium on Applications of Laser Anemometry to Fluid Mechanics, Lisbon, 2-4 July.
13. Nouri, J.M., Whitelaw, J.H. and Yianneskis, M. (1987). The scaling of the flow field with impeller size and rotational speed in a stirred reactor. Proc. 2nd International Conference on Laser Anemometry - Advances and Applications, Glasgow, UK, 21-23 September.
14. Khang, S.J. and Fitzerald, T.J. (1975). A new probe and circuit for measuring electrolyte conductivity. Ind. Eng. Chem. Fundam., vol. 14, pp. 208-213.
15. Ruszkowski, S.W. and Muskett, M.J. (1985). Comparative mixing times for stirred tank agitators. Proc. 5th European Conference on Mixing, Wurzburg, Germany, 10-12 June.
16. Yianneskis, M. (1991). The effect of flow rate and tracer insertion time on mixing times in jet-agitated vessels Proc. 7th European Conference on Mixing, Brugges, Belgium, 18-20 September.
17. Voncken, R.M., Holmes, D.B. and den Hartog, H.W. (1964). Fluid flow in turbine-stirred, baffled tanks. Chem. Engng. Sci., Vol. 19, pp. 209-213.

DEVELOPMENT OF TRANSITIONAL FLOW MIXING IMPELLER

R.J. WEETMAN, Lightnin, New York, USA

ABSTRACT

Many important mixing applications are performed in the transitional flow regime, (i.e. mixing Reynolds number from 20 - 2000, viscosities up to 100,000 cps). When the mixing impeller operates in this regime, some of the liquid volume may be in the laminar flow regime. In many processes, the viscosity changes during the process. Where blending is important, a flow efficient impeller is desired when operating in the turbulent regime as well as the transitional flow regime. This paper will discuss the development of a new mixing impeller (called the A410) which address these issues. The techniques used in its development will be shown.

The impeller performance characteristics were measured with a two-dimensional laser doppler velocimeter (LDV). The flow characteristics were studied in both water and glycerin solutions to cover the turbulent and transitional flow regimes. In order to analyze its process characteristics, blending studies were conducted to observe mixing in all areas of the tank. This is especially important since the laminar regions away from the impeller must also be blended.

Computational fluid dynamics (CFD) was an integral part of the impeller development. The experimental flow characteristics in the transitional flow regime were compared to the calculated velocities of the computational field. Blending calculations using computational fluid dynamics were performed and compared to blending experimental results.

The new A410 impeller produces high flow efficiency in the turbulent regime and possesses unique characteristics in the transitional flow regime. The flow field measured by the LDV agrees well with the computational fluid dynamics calculations. In addition, the blending studies demonstrate the value and insights gained by the use of computational fluid dynamics in mixing studies.

INTRODUCTION

Important mixing applications occur in the transitional flow regime, (i.e. mixing Reynolds number from 20 to 2,000, viscosities up to 100,000 cps). When the mixing impeller operates in this regime, some of the liquid volume may be in the laminar flow regime. In many processes, the viscosity (therefore the Reynolds number) changes during the process. This is especially important in small batch sizes (which account for a large percentage of all mixing applications). Small volumes, requiring small impeller diameters, are more likely to be in the transitional flow regime. If these small batches have a large product value, then it is extremely important that the products are processed efficiently under all conditions. A flow efficient impeller is desired when operating in the turbulent range, as well as, in the transitional flow regime. This paper will discuss the developments of a new mixing impeller (designated the LIGHTNIN A410 Impeller) which address these issues and it will show many of techniques used in its development.

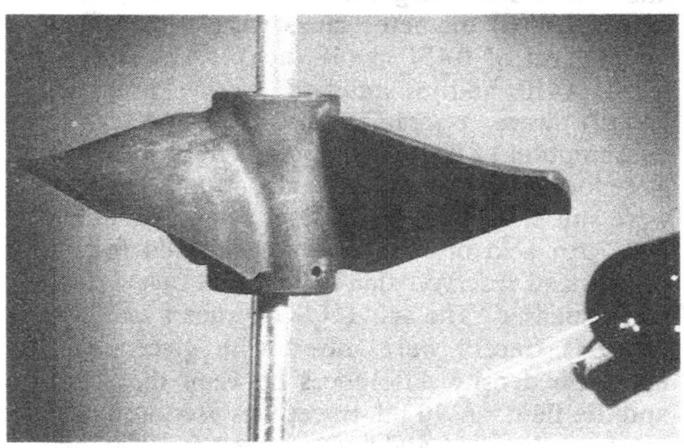

FIG.1 A410 IMPELLER

[A410] TRANSITIONAL FLOW IMPELLER REGIME

The A410 impeller as shown in Fig. 1 has been developed to operate in the laminar or transitional flow regime. As stated above, this is a common and extremely important operating range for small batch size mixers (often called portable mixers). Fig. 2 shows a portable drive mixer incorporating the A410 impeller. This new impeller is made of corrosion resistant, composite plastic materials. The impeller was designed with computer-aided design (CAD), finite element analysis (FEA) and computational fluid dynamics (CFD). In addition, blending studies were done to analyze the particular characteristics in this range.

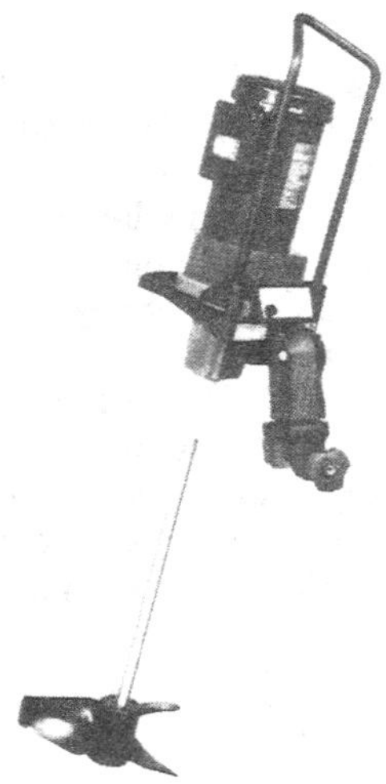

FIG.2 PORTABLE MIXER

BLENDING

Below we will discuss some of the blending attributes in the transitional flow regime. Fig. 3, 4 and 5 show a blending test. Two side-by-side 0.445 m (17.5 inch) diameter tanks were used with a liquid level of 0.457 m (18 inch). A .178 m (7 inch) A410 and a standard marine propeller (A110) were operating at an equal power consumption of 7.9 watts. Both .178 m impellers were located .178 m off the tank bottom. The tests were run at a Reynolds number of approximately 200. An acid or base tracer is injected from a pressurized reservoir that is activated by a Tridak 250 dispenser. The acid (H_2SO_4) tracers and base (NaOH) tracers were mixed with glycerine to minimize density differences between the tracer and the fluid. A 10 ml. tracer was used at a 30% solution. Fig. 3, 4 and 5 were taken at 20, 40 and 60 seconds after the acid tracer was injected. The reduction of color shows (with the phenolphthalein indicator) the areas of decreased pH levels. The photos show a much faster blend time for the A410 impeller (left) compared to the standard marine impeller (right). Also seen in Fig. 5 is a stratified, unmixed layer beneath the marine impeller.

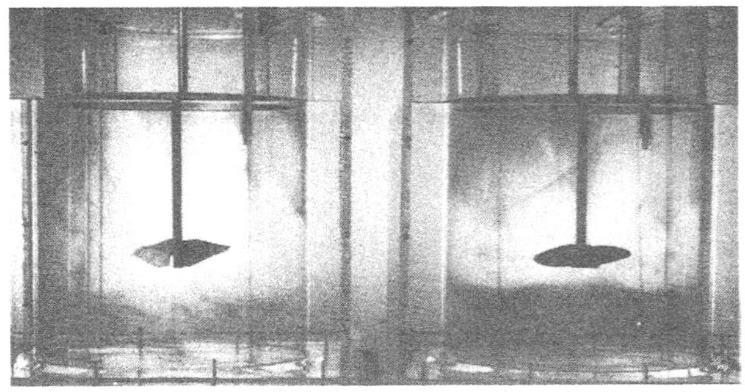

FIG.3 20 SEC.

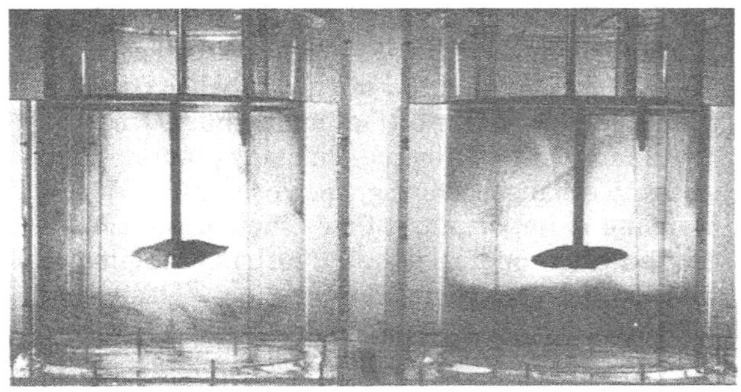

FIG.4 40 SEC.

FIG.5 60 SEC.

In the transitional flow regime an axial impeller looses its effectiveness to pump in the axial direction. This occurs initially at the hub. It will be described in more detail in the laser doppler velocimeter (LDV) measurements below. Therefore, it is important that the inside of the impeller near the hub pump as much as possible. The way to accomplish this is to have an extremely high twist on the blade, i.e. the difference in blade angle between the tip of the impeller and the blade angle at the hub. The angle of the blade is measured from the chord of the airfoil blade section relative to the horizontal. Fig. 6 shows results of blending experiments conducted as a function of twist angle from the tip to 0.4 radius. It shows a large reduction in blend time as the twist angle increases. Illustrated in Fig. 6 is a sharp reduction (improvement) in blend time from conventional axial flow impellers that have twist angles in the range of 15° compared to a twist angle greater than 30°. Thus, the A410 impeller, with a twist greater than 30°, is able to pump more near the hub than a standard marine impeller. As we'll see later, the A410 impeller still experiences a reduction in axial flow as it approach the laminar region. Even with this reduction, it has sufficient pumping action near the hub to prevent stratification as was seen in the above Fig. 3, 4 and 5.

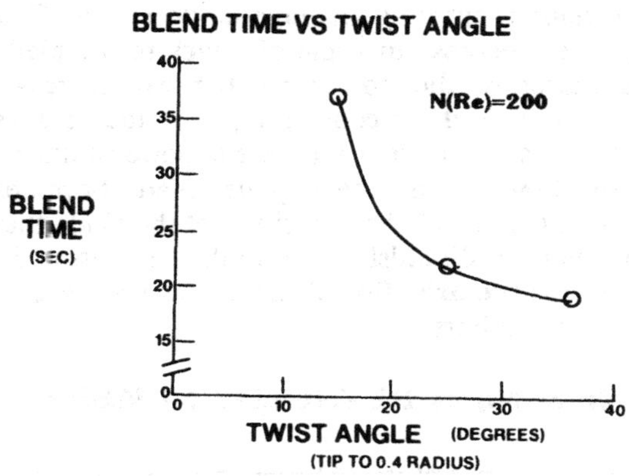

FIG.6

It is extremely important to have a contoured airfoil shaped impeller where complete freedom is available to change the attributes of the blade, i.e. twist, camber, thickness, etc.

In the development of the A410 impeller tufts were used to critically examine the flow and separation on the surfaces of the blade. The tufts or threads on the blades were examined with a Hi Band 8mm video with a 1/2000 shutter speed.

An example of an impeller with high twist that is not a true airfoil shape is called the A210 Impeller. The highly flow efficient A310 impeller with a twist angle of 16° can be compared to the A210 impeller. The A210 impeller has the same flow number as the A310 impeller (0.56) but it has a 0.53 turbulent power number compared to a 0.30 power number for the A310. The A210 impeller has a twist of 45° compared to the A310 twist of 16°. This high twist as well as other airfoil differences requires the A210 to consume 77% more power at the same flow. This example illustrates the need to use extreme care when increasing the twist of an impeller. This paper shows that the A410 has high twist for blending in the transitional flow regime as well as high flow efficiency in the turbulent regime. Characteristics of various impellers are described by Weetman, et. al.[1]. Additional discussions on impellers operating in the transition region is discussed in Weetman, et. al.[2].

LDV MEASUREMENTS

Measurements of fluid velocities have been performed using a dual channel laser doppler velocimeter (LDV). Three laser beams shown in Fig. 1 are generated by a Dantec 55X two-color laser doppler velocimeter. It measures two velocity components simultaneously and is mounted on a computer-controlled, three axis traversing mechanism that allows a complete scan of a mixing tank. Measurements are collected via back-scattering, with both receiving and transmitting optics in the same module. This eliminates the need to re-align the receiving optics at each measurement point and allows an automated computer-controlled operation. This laboratory is described in more detail by Weetman[3].

Fig. 7 shows velocity vectors in the RZ plane for the A410 composite impeller. These measurements were taken in a 0.445 m (17.5 inch) diameter tank with a 0.457 m. (18 inch) liquid level. The impeller diameter is 0.178 m (7 inch). The velocity vectors emanate from the horizontal and vertical scan lines. The axes in the figures are the tank centerline and the bottom half of the tank. The vertical line near the tank wall represents the width of the four baffles. The off-bottom height of the impellers is 0.178 m (7 inch). In Fig. 7 the flow diagrams were measured in the turbulent range, i.e. Reynolds number = 2.2 x 10^5. The Reynolds number, the ratio of inertia forces to viscous forces, for mixing impellers is calculated as:

VELOCITY VECTORS IN R-Z PLANE

N(Re)=2.2E5

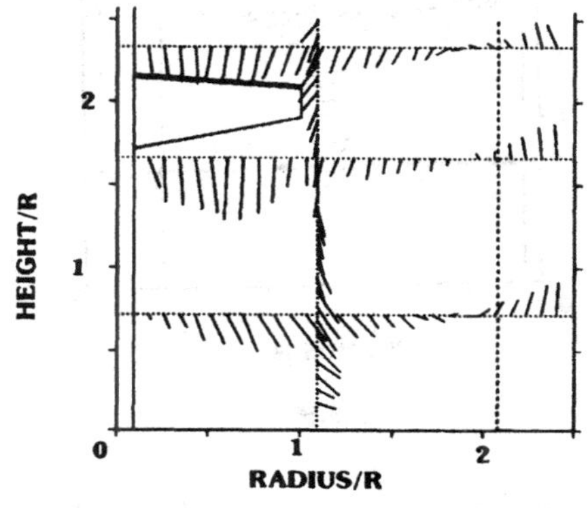

FIG.7

$$N_{Re} = \rho\ ND^2/\mu$$

D = Impeller Diameter (m)
N = Impeller Speed (rps)
ρ = Density (Kg/m^3)
μ = Dynamic viscosity (Ns/m^2)

As the Reynolds number decreases, the fluid resistance to flow increases causing the axial impellers' discharge flow to become more radial. In the limit of laminar flow, the discharge flow from an axial impeller is radial. Examining Fig. 8 and 9, the progression of axial to radial flow of axial flow impellers is shown. Fig. 8 shows the flow discharge angle at a Reynolds number of approximately 700 and Fig. 9 shows it at a Reynolds number of approximately 200.

VELOCITY VECTORS IN R-Z PLANE

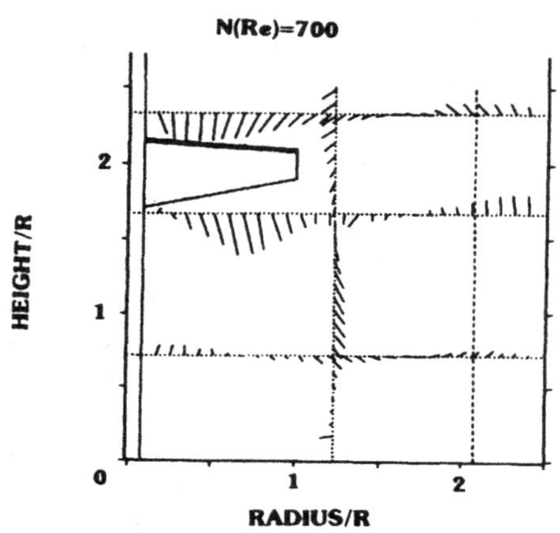

FIG.8

VELOCITY VECTORS IN R-Z PLANE

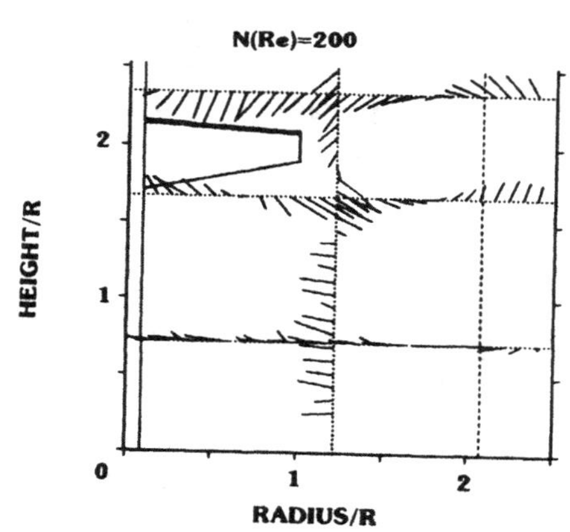

FIG.9

The test measurements at the lower Reynolds numbers (700 and 200) were conducted in glycerine, while tests at the higher Reynolds numbers were done in water. Care is needed to obtain good measurements in higher viscosity materials like glycerine. To obtain quality signals from the LDV, the glycerine needs to be kept free of contaminates. Obviously, with glycerine's viscosity sensitivity to temperature, a tight control of temperature also needs to be maintained. The power number and flow number versus Reynolds number for the A410 are shown in Fig. 10. The power number curve has the classical shape of a flat minimum in the turbulent range and then increases as the Reynolds number decreases (or with increasing viscosity). A change (increase) in power number in the transition region is also observed with a change (decrease) in flow number as shown in Fig. 10. As the load on the impeller increases due to the viscous nature of the fluid, the fluid velocity entering the impeller decreases. This causes an increase in angle of attack to the blade and therefore the power number will increase. Also, increased viscous drag on the blades contributes to a further increase in power number. As the flow around the impeller blade becomes laminar, the airfoil characteristics of the blade that produce lift is diminished. With the decreasing lift (or thrust) the axial flow decreases as observed in the flow numbers.

Np & Nq vs REYNOLDS NUMBER

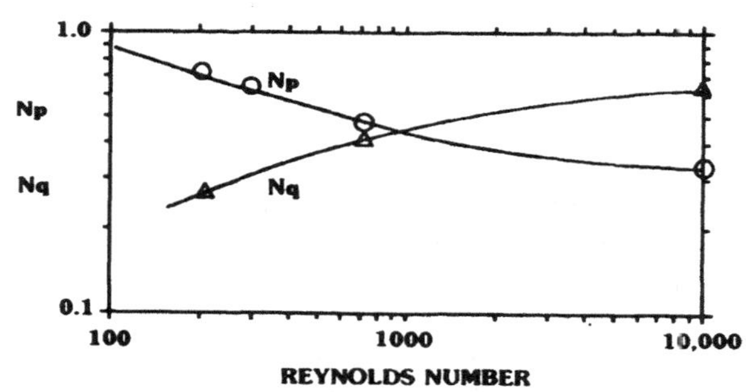

FIG.10

It is difficult to compare different impellers that have different shapes in the transition region. Fig. 9 shows that the outlet flow is beyond the impeller diameter. This is because the outlet flow has a radial and an axial component in the transitional flow regime. When comparing different impellers, care is required in measuring the outlet flow number. A larger (diameter and/or height) impeller occupies more effective volume in the tank. Therefore, the flow number alone may not accurately describe its process performance. It may be necessary to measure the flow at the same relative position on the tank. The flow numbers in Fig. 10 include the radial as well as the axial outlet flow.

CFD RESULTS AND LDV COMPARISONS

Computational fluid dynamics is one of the tools used in the development of the A410 transitional flow mixing impeller. Background on the commercial code used and on some of the techniques employed are described by Hutchings et. al.[4]. The program used is FLUENT by creare.x, Inc. Presented here are velocity fields from the CFD computations for the turbulent flow conditions and for a Reynolds number of 200.

The results obtained at a Reynolds number of 2 x 10^5 are presented in Fig. 11 through 14. These are for the same conditions that were discussed previously for a 0.445 m (17.5 inch) diameter tank. Fig. 11 shows the velocity vectors for the entire tank and these compare favorably with those of Fig. 9. Obviously, the advantage of the CFD is that one is able to obtain other comparisons after one has obtained the solution of the velocities. As shown in Fig. 12, 13 and 14, the streaklines, kinetic energy contours and turbulent energy dissipation are presented respectively. These results are based on the k-ε turbulent model for the Navier-Stokes Equations. These analyses are three-dimensional with the input for the impeller from the outlet velocity profile as measured by the LDV. The input using the maximum values of the velocity components in the flow field that occur at the impeller outlet is advantageous.

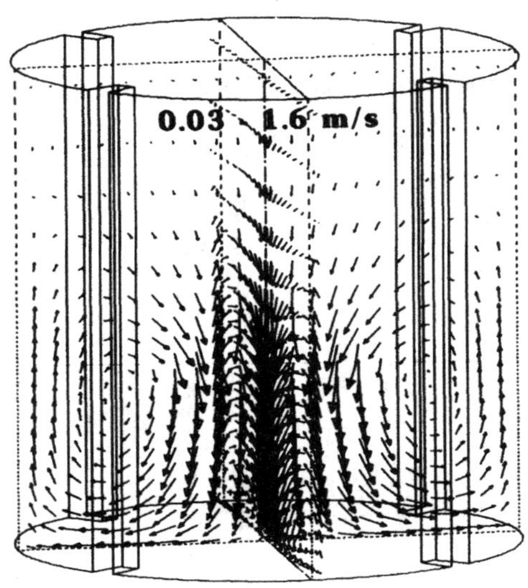

FIG.11 VELOCITY VECTORS NREY=2E5

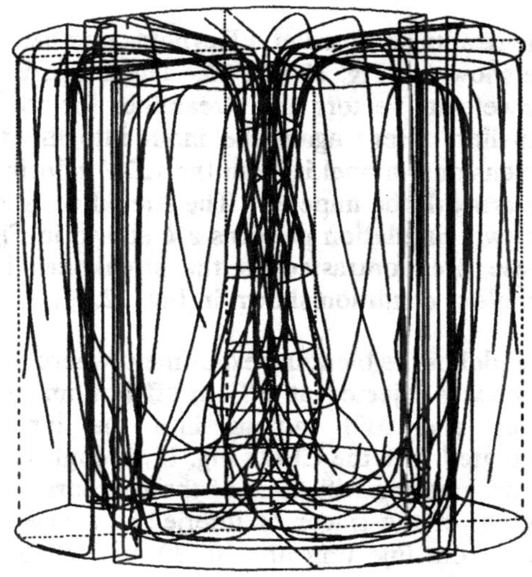

FIG.12 STREAKLINE NREY=2E5

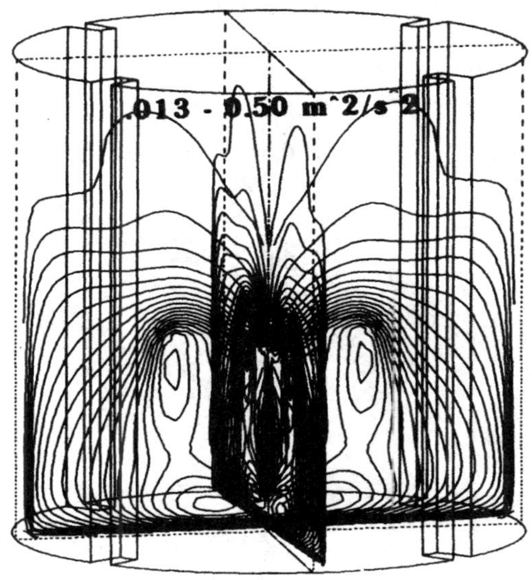

FIG.13 KINETIC ENERGY NREY=2E3

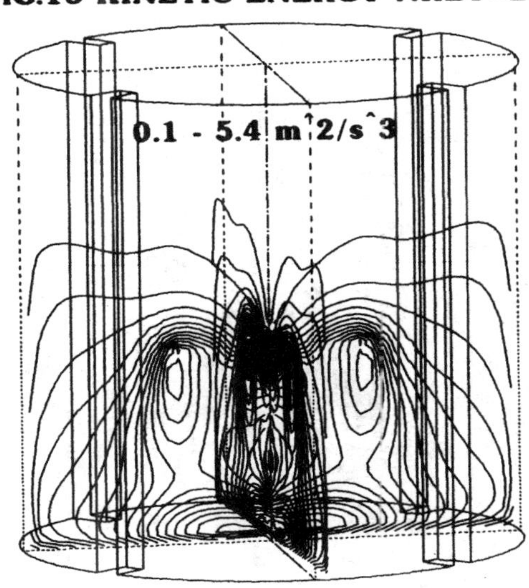

FIG.14 EDDY DISSIPATION NREY=2E5

The results obtained at a Reynolds number of 200 are shown in Fig. 15 and 16. These figures show the velocity vectors and streaklines for the A410 Impeller. Here again the input into this three-dimensional model is from the LDV velocity data underneath the impeller. The streaklines showing the two circulation patterns are shown in Fig. 16. These are contrasted to the streaklines in the turbulent condition shown in Fig. 12.

A model of the blending experiment where the dye tracer was injected at .383 m off bottom, .146 m radius and 0.68 radians into the tank was calculated with the CFD. Fig. 17 through 19 show the progression with time of the contours of mass fraction of the tracer or interjected specie. The times from injection are 10, 30 and 60 seconds.

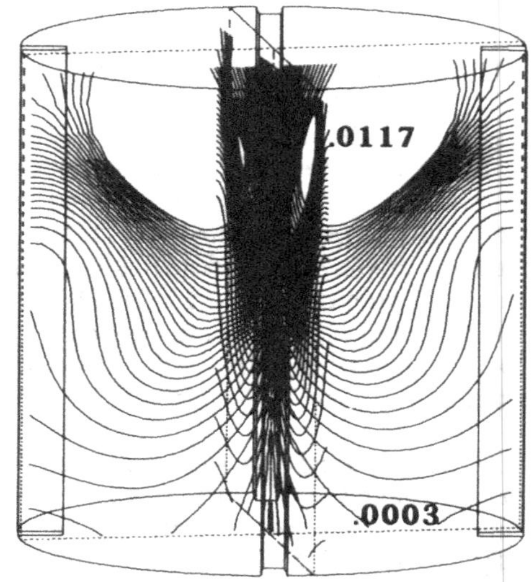

FIG.17 MASS FRACTION 10 SEC.

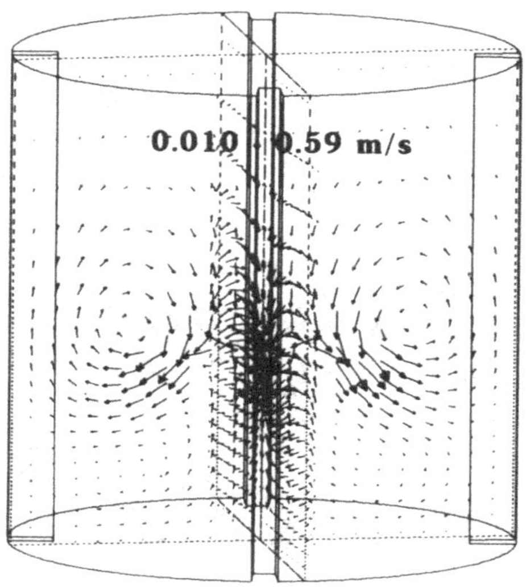

FIG.15 VELOCITY VECTORS NREY=200

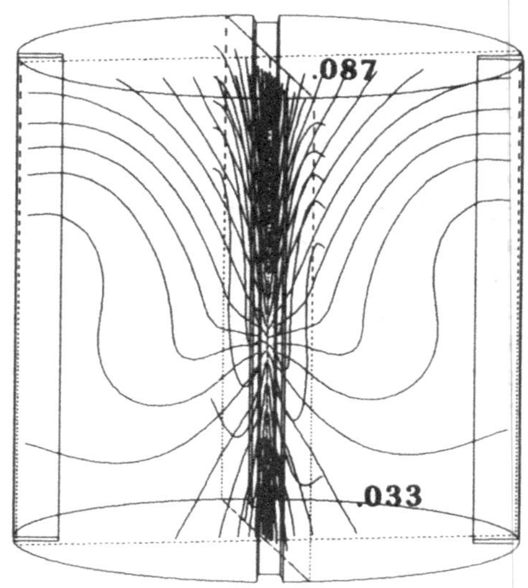

FIG.18 MASS FRACTION 30 SEC.

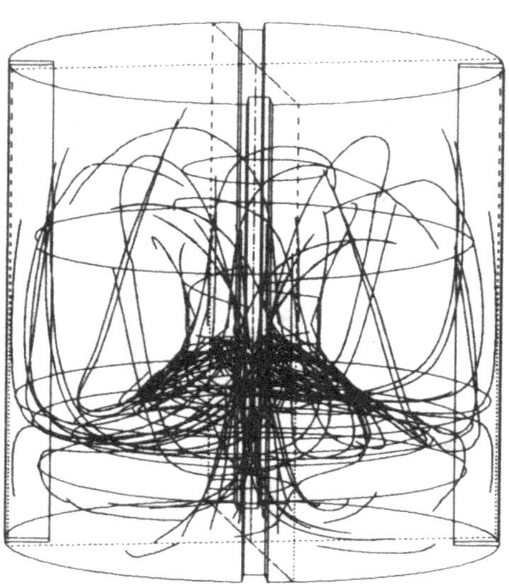

FIG.16 STREAKLINES NREY=200

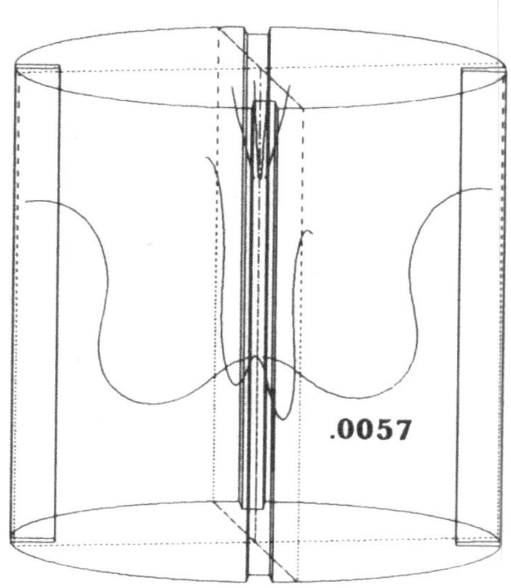

FIG.19 MASS FRACTION 60 SEC.

The progression shows over-saturation at the top of the tank followed by spreading out of the contours. The mass fraction contour lines are at 0.0006 intervals. Three locations in the tank were noted and the tracers for the mass fraction versus time was plotted in Fig. 20. The three positions were at 11, 70 and 383 mm from the bottom of the tank, at a radius of 15 mm and at an angle of .05 radians. Fig. 20 shows the overshoot of the highest position (383 mm) at the top of the tank compared to the other two positions. As seen in this figure, the three positions are almost at the same concentrations at 60 seconds. This is in good agreement with the blending study noted earlier that obtained uniformity in approximately 60 seconds.

CONCLUSIONS

This paper presents the effectiveness of the new A410 impeller in the transitional and turbulent flow regimes. Results show the blending time reduction using a high twist which enables the impeller to pump effectively near the hub. Care is required in incorporating a large twist on the impeller blade.

Results from LDV measurements showed the different flow characteristics of the A410 in the turbulent and transitional flow regimes. Also presented were blending studies showing the characteristics of blending in the transitional flow regime where laminar regions are required to be mixed. The advantages of using CFD computations to examine the entire flow field were received. Good agreement was obtained between the CFD computations and the LDV measurements for both the flow measurements and the blending experiments.

MASS FRACTION VS TIME

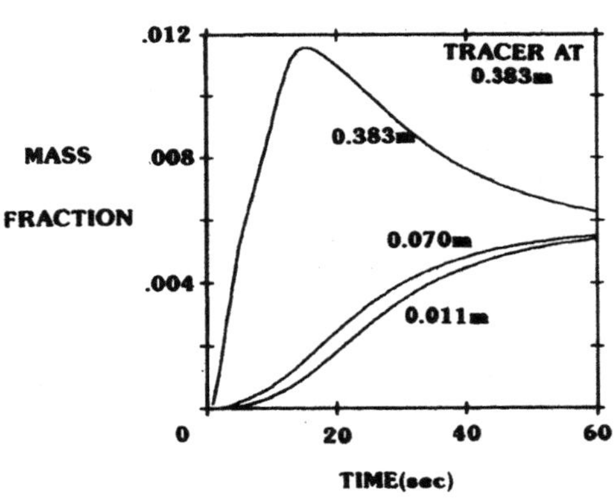

FIG.20

NOMENCLATURE

A110 = Marine 1.0 Pitch Impeller
A210 = High Twist Impeller
A310 = Fluidfoil Impeller
A410 = Fluidfoil Composite Impeller
C = Impeller Off-Bottom (m)
D = Swept Impeller Diameter (m)
N = Impeller Speed (rps)
N_Q = Q/ND^3
N_P = $P/\rho N^3D^5$
N_{Re} = $\rho N D^2/\mu$
Q = Outlet Flow (m^3/s)
P = Power (W)
T = Tank Diameter (m)
H = Liquid Height (m)
ρ = Density (kg/m^3)
μ = Viscosity (Ns/m^2

REFERENCES

1. Weetman, R.J., Oldshue, J.Y., "Power Flow and Shear Characteristics of Mixing Impellers", Proceedings of the 6th European Conference on Mixing. Pavia, Italy, 24-26 May, 1988.
2. Weetman, R.J., Coyle, C.K., "The Use of Fluidfoil Impellers in Viscous Mixing Applications", Presented at the AIChE Annual Meeting. San Francisco, CA, 5-10 November, 1989.
3. Weetman, R.J., Salzman, R.N., "Impact of Side Flow on Mixing Impeller", Chemical Engineering Progress, 71-75 (June, 1981).
4. Hutchings, B.J., Weetman, R.J., Patel, B.R., "Computation of Flow Fields in Mixing Tanks with Experimental Verification", Presented at the ASME Winter Annual Meeting. San Francisco, CA, 10-15 December, 1989.

REFEREES' QUESTIONS AND AUTHORS' ANSWERS

1. Why do you use a value of Z/T < 1? (HL/T < 1? in the official nomenclature).

   Z/T is U.S. AICHE offical nomenclature.

2. How was the power consumption of 34 watts measured?

   A torque cell w/speed sensor was used.

3. In test page 2, the A410 impeller is compared to the standard marine propeller. Are the geometry of both agitation systems (D/T and h/D (C/D for you), namely) optimized? They are not optimized for a Reynolds number of 200.

   They are the same for both and it is designed to show the differences in performances and not the optimized position.

4. For Fig. 6, what is the value of the Reynolds number?

   200.

5. On Fig. 5, could you also give the point corresponding to the pitched-blade turbine A200?

   I do not have the exact comparison, but it would be higher than the valleys on the graph for blend time and it would be plotted at zero twist angle.

6. It is difficult to use your Fig. 1. Could you also give the values of Np and Nqp (N_Q for you) at some Reynolds number (10, 50, 500, 100 and 10000, for instance). In that table, would you mind giving Np and Nqp for the A210 and A310 impellers?

   I assume you are referring to fig. 10 not fig. 1. This figure is being re-done as well as figures 1, 7, 8, 9 and 10 as stated in the letter with the original paper.

7. From Fig. 10, it seems that for the A410, the flow is already fully turbulent at Re = 10^4. Is it right?

   Yes.

8. Could you give the curve n.tm = f(Re) for the A410 and compare on a figure P.tm = f(Re), the homogenization efficiency of the standard marine propeller, the pitched-blade turbines, the A210, the A310 and the A410 in a set stirred tank?

   Those are good points and we do not figures for presently for those curves, but we will be doing those in the future.

DISTRIBUTION OF TURBULENCE CHARACTERISTICS IN AGITATED SYSTEM WITH AXIAL HIGH-SPEED IMPELLER AND BAFFLES

I. FOŘT, M. VOZNÝ, B. FORSTOVÁ, Faculty of Mechanical Engineering, Czech Technical University, Prague, CS

ABSTRACT

The paper deals with the description of distribution of turbulence characteristics (Reynolds stress and eddy viscosity) in a system with high-speed impeller (impeller with four asymmetrically profiled blades at the angle α of 67° - 26°) placed axisymmetrically in cylindrical vessel with four baffles. The turbulent characteristics mentioned are calculated from the measured radial profiles of axial and radial components of the mean velocity in the spaces below and above the body of rotating impeller (laser-doppler anemometry was used) according to von Karman's hypothesis of similarity of turbulent flow applied to axially symmetrical sytem with cylindrical co-ordinates.

INTRODUCTION

Axial high-speed impellers (impellers with inclined plane blades or propellers) are currently used for homogenisation of miscible liquids, for suspension of solid particles in liquids (crystallisers) or for intensification of heat transfer in jacketed mixing vessels. Recently, these types of mixers have been introduced in mechanically agitated bioreactors[1,2] because of their ability to form an intensive circulation of agitated charge at rather low level of the turbulent (Reynolds) stress affecting the rate of production activity of microorganisms present.

THEORETICAL

The study deals with the description of turbulence characteristics (Reynolds stress and eddy viscosity) in a system with axial high-speed impeller (see Fig. 1) placed axisymmetrically in cylindrical system with four baffles (see Fig. 2).

The turbulent characteristics mentioned can be calculated from the measured radial profiles of axial and radial components of the mean velocity below and above the body of rotating impeller - see Fig. 3 and 4 (laser-doppler anemometry (LDA) of the back scatter mode was used[3)]) according to von Karman's hypothesis applied to axially symmetrical system with cylindrical co-ordinates[4,5)]. While the radial component of the mean velocity exhibits rather flat radial profiles (between axis of symmetry and the vessel wall) in both measured spaces of agitated system, the axial component of the mean velocity exhibits significant slopes of radial profiles, especially in the stream streaking from the impeller body, and, further, at the vessel wall (see Figs 5 and 6). Therefore, we can write[4] for axially-radial component of the Reynolds stress

$$\bar{p}_{ax,rad} = -\rho k^2 \left| \frac{\left(\frac{d\bar{w}_{ax}}{dr}\right)^3}{\left(\frac{d^2\bar{w}_{ax}}{dr^2} - \frac{1}{r}\frac{d\bar{w}_{ax}}{dr}\right)^2} \right| \times \left(\frac{d\bar{w}_{ax}}{dr}\right) \qquad (1)$$

and for corresponding components of eddy viscosity and fluctuating velocity correlation relation

$$\bar{p}_{ax,rad} = -\mu^t_{ax,rad}\left(\frac{d\bar{w}_{ax}}{dr}\right) = -\rho\,\overline{w'_{ax}\,w'_{rad}} \quad , \qquad (2)$$

where quantity k is a universal von Karman's similarity constant and quantity ρ is a density of agitated charge.

For scaling up reasons we can transform the above mentioned turbulence characteristics with respect to the impeller tip speed πDn. Then we have dimensionless quantities

$$P_{ax,rad} = \frac{\bar{p}_{ax,rad}}{\rho\, n^2 D^2} = -\pi^2 k^2 \left| \frac{\left(\frac{dW_{ax}}{dR}\right)^3}{\left(\frac{d^2W_{ax}}{dR^2} - \frac{1}{R}\frac{dW_{ax}}{dR}\right)^2} \right| \times \left(\frac{dW_{ax}}{dR}\right) \quad , \qquad (3)$$

$$\frac{\mu^t_{ax,rad}}{\rho\, n^2 D^2} = \frac{\nu^t_{ax,rad}}{nD^2} = \pi k^2 \left(\frac{T}{D}\right) \left| \frac{\left(\frac{dW_{ax}}{dR}\right)^3}{\left(\frac{d^2W_{ax}}{dR^2} - \frac{1}{R}\frac{dW_{ax}}{dR}\right)^2} \right| \qquad (4)$$

and

$$W'_{ax,rad} = \frac{\overline{w'_{ax}\,w'_{rad}}}{(\pi Dn)^2} = k^2 \left| \frac{\left(\frac{dW_{ax}}{dR}\right)^3}{\left(\frac{d^2W_{ax}}{dR^2} - \frac{1}{R}\frac{dW_{ax}}{dR}\right)^2} \right| \left(\frac{dW_{ax}}{dR}\right) , \qquad (5)$$

where dimensionless co-ordinate and velocity component were introduced

$$R = r/T, \quad W_{ax} = \frac{\bar{w}_{ax}}{\pi Dn} \quad . \qquad (6a,b)$$

Further, we suppose that all radial profiles of the axial component of the mean velocity can be expressed as straight lines, i.e. a set of linear dependences in the given cross-sections

$$W_{ax} = A_i\, R_i + B_i \;,\; i = 1,2,3 \ldots \qquad (7)$$

Then we can transform the dimensionless turbulence characteristics (see Eqs. 3,4,5) into the forms

$$P_{ax,rad,i}\,(R) = -\pi^2 k^2 \,|R^2 A_i|\, A_i \;, \quad i = 1,2,3 \ldots , \qquad (8)$$

$$\frac{\mu^t_{ax,rad}}{\rho\, n^2 D^2}\, i\,(R) = k^2 \left(\frac{T}{D}\right) R^2 A_i \;, \quad i = 1,2,3 \ldots , \qquad (9)$$

$$W'_{ax,rad,i}\,(R) = k^2 \,|R^2 A_i|\, A_i \quad i = 1,2,3 \ldots \qquad (10)$$

A universal von Karman's similarity constant is considered to be[5)]

$$k = 0.4 . \qquad (11)$$

RESULTS AND DISCUSSION

All dimensionless radial profiles of the axial component of the mean velocity (see Fig. 5) were transformed in to sets of their slopes A_i, $i = 1,2,3, \ldots$ and corresponding turbulent characteristics were calculated (see Fig. 7). The eddy viscosity was then expressed in dimensional form at the impeller Reynolds number $Re_m = 1.0.10^5$ for the investigated pilot plant equipment (see Figs. 1 and 2) - see Fig. 8.

The axially - radial components of the Reynolds stress and of eddy viscosity in the regions below impeller exhibit considerably high values, much higher than for laminar regime of flow. On the other hand both turbulent characteristics mentioned appear very low close above the body of rotating impeller where the radial profiles of both axial and radial components of the mean velocity can be described as piston flows.

REFERENCES

1. Reuss,M.: Influence of mechanical stress in the growth of Rhisopus nigrificants in stirred bioreactors. Chem. Eng. Technol. 11, 178 - 187, 1988.
2. Oldshue,J.Y.: Fluid mixing 1989. Chem. Eng. Prog., May 1989, 33 - 42.
3. Klaboch,L., Ptáčník,M., Lamka,J., Fořt,I.: Application of the laser-doppler anemometry for determination of velocity field in the cylindrical baffled system with pitched - blade impellers. Proceedings of the 6th European Conference on Mixing, Pavia, 1988, pp. 35 - 42.
4. Bird,R.B., Stewart,W.E., Lightfoot,E.N.: Transport phenomena. John Wiley and Sons, Inc., New York 1965.
5. Brodkey,R.S.: The phenomena of fluid motions. Addison-Wesley, Inc., Reading (MA) 1967.

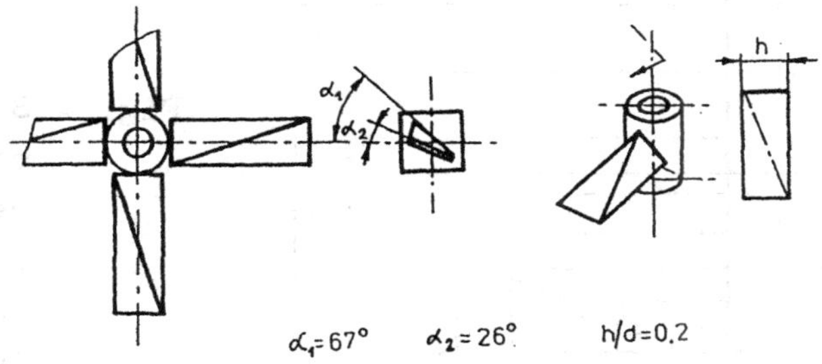

Fig. 1 Investigated impeller with four asymmetrically profiled inclined blades.

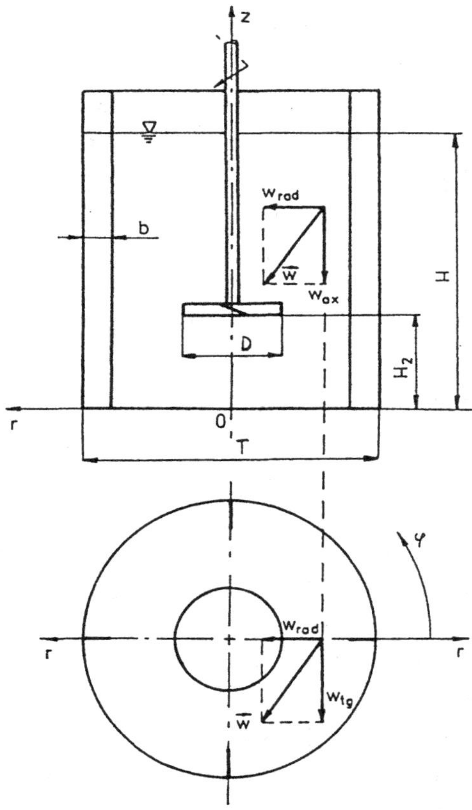

Fig. 2 Sketch of agitated system:
$T = 0.297$ m, $H/T = 1$,
$D/T = 1/3$, $H_2/D = 1$,
$b/T = 1/10$.

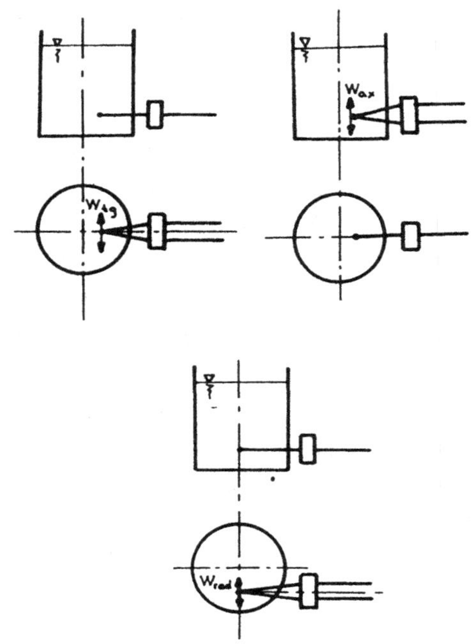

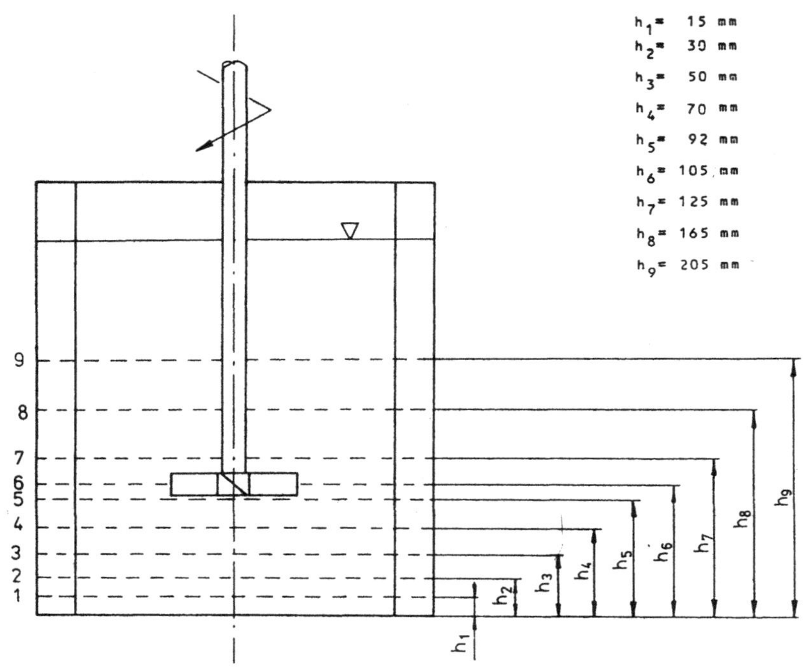

Fig. 4 Sketch of the LDA measurement of axial, radial and tangential components of the instaneous velocity in agitated system.

Fig. 3 Distribution of cross-sections for measurement of axial and radial components of the mean velocity in agitated system.

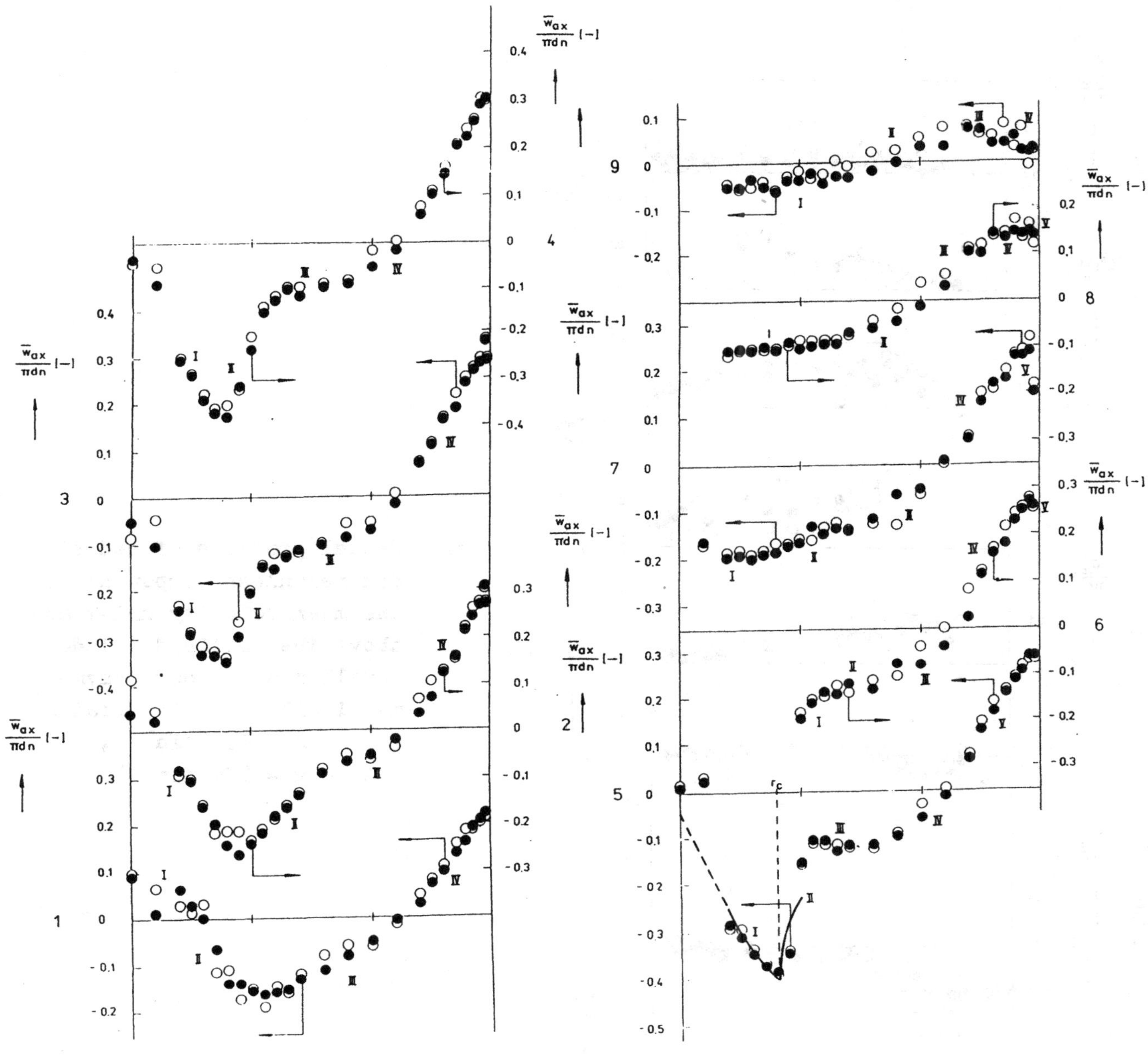

Fig. 5 Radial profiles of axial dimensionless component of the mean velocity under and above the inclined blade impeller with four asymmetrically profiled blades (○ - n = 450 min^{-1}, ● - n = 700 min^{-1}).

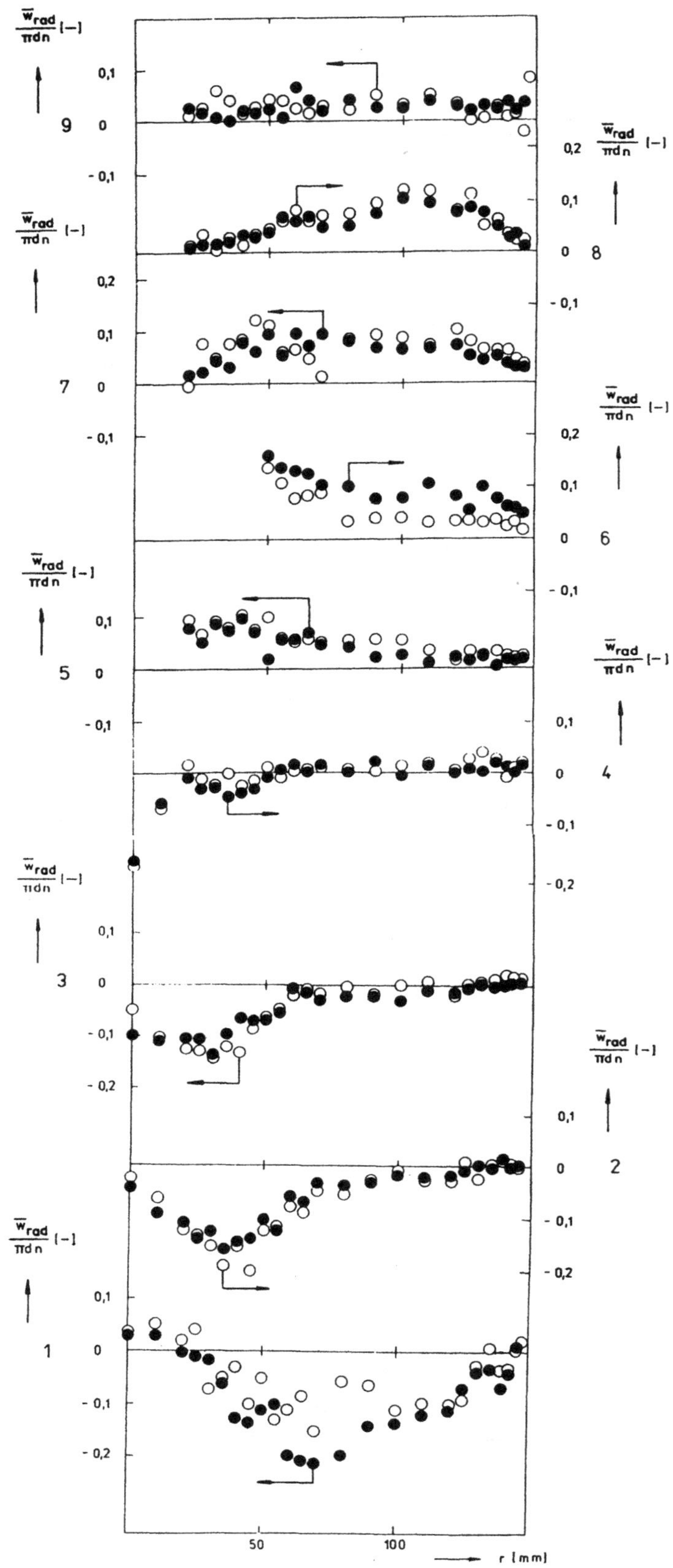

Fig. 6 Radial profiles of radial dimensionless component of the mean velocity under and above the inclined blade impeller with four asymmetrically profiled blades (○ − n = 450 min^{-1} , ● − n = 700 min^{-1}).

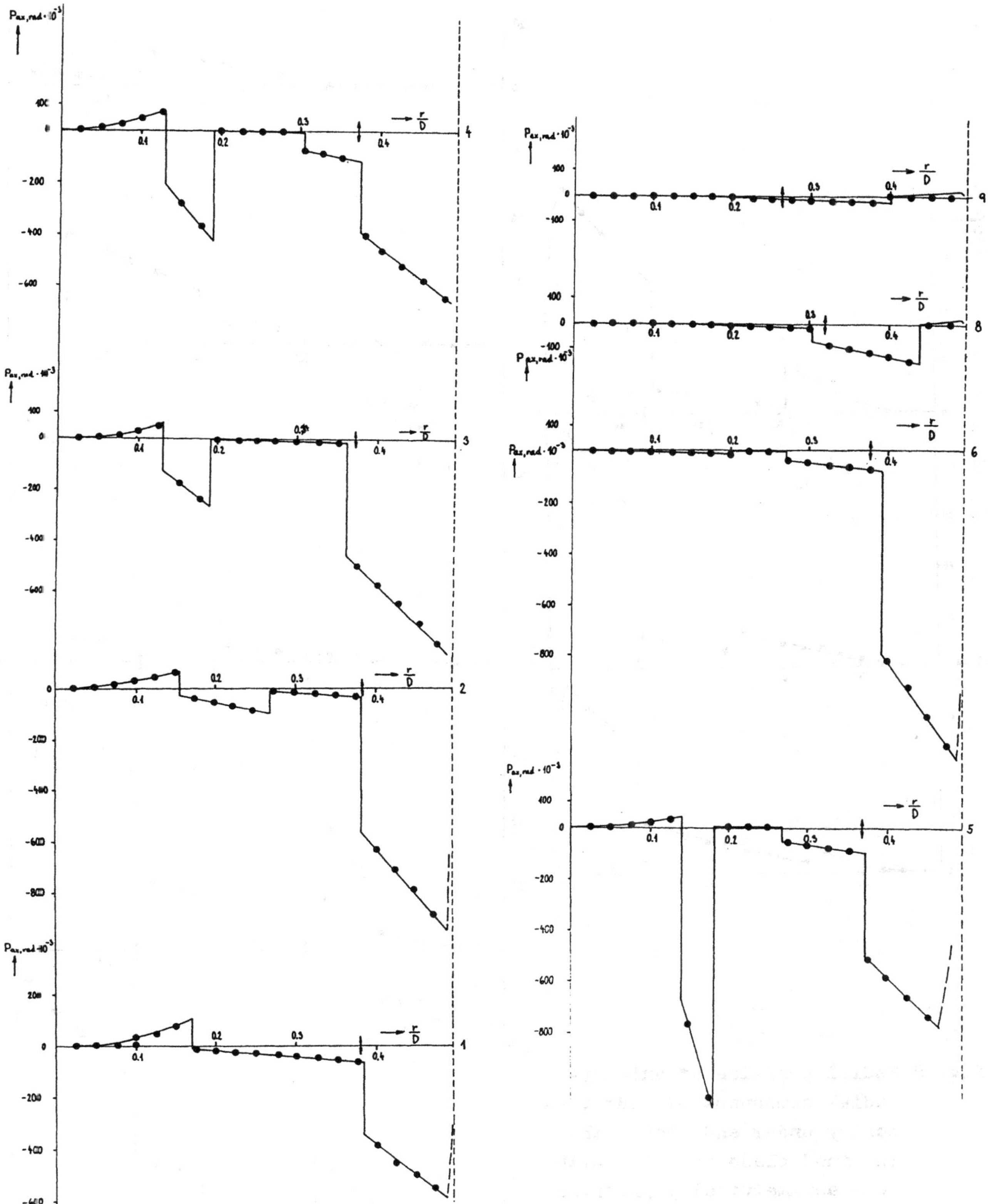

Fig. 7 Radial profiles of axially - radial dimensionless component of the Reynolds stress under and above the inclined blade impeller with four asymmetrically profiled blades.

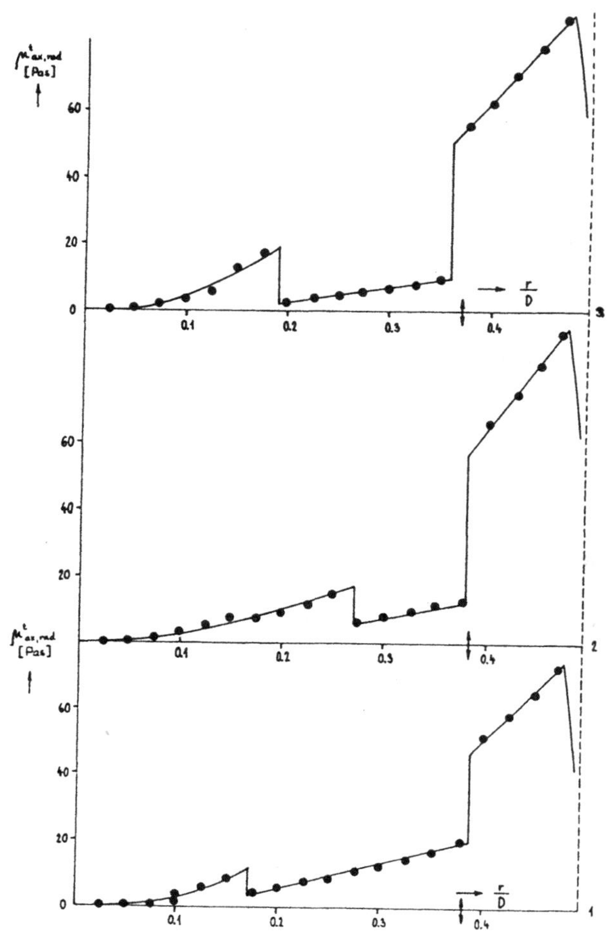

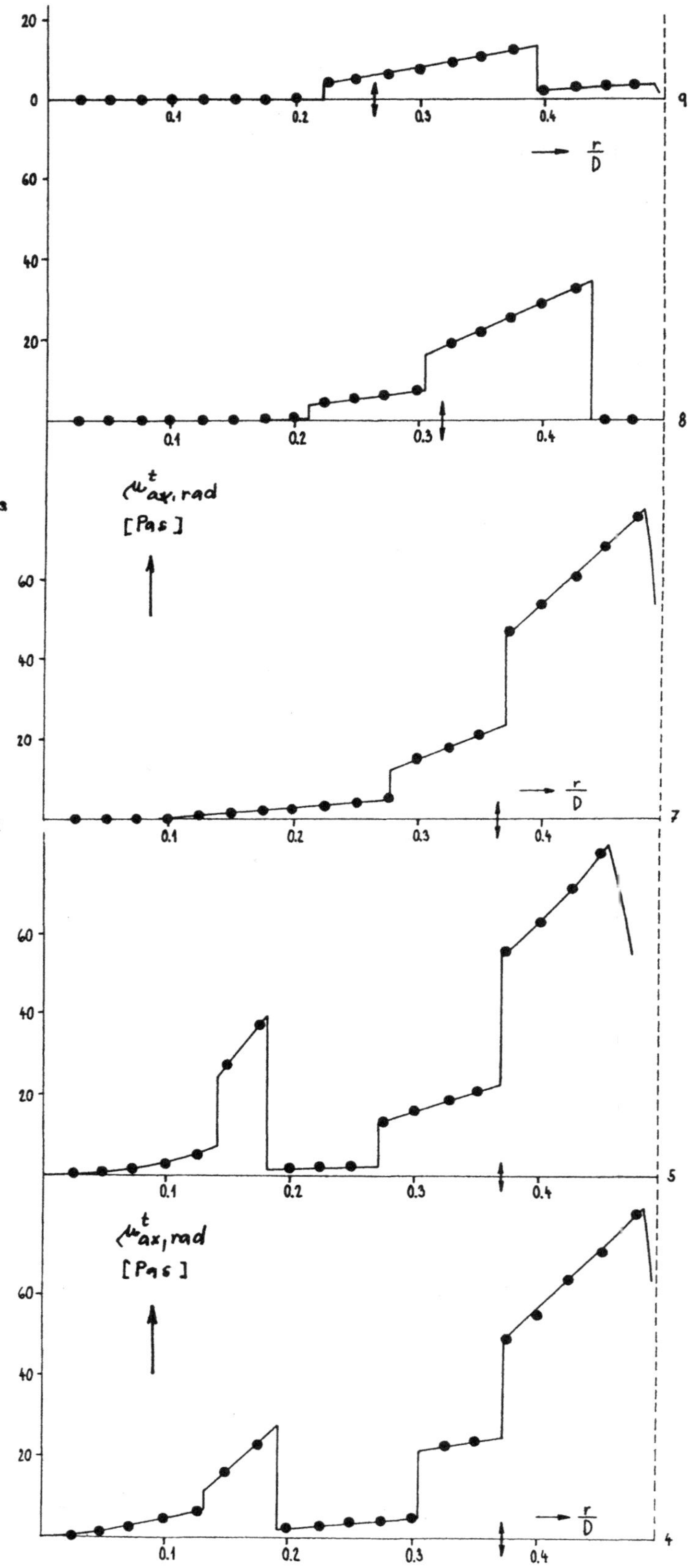

Fig. 8 Radial profiles of axially-radial component of eddy viscosity under and above the inclined blade impeller with four asymmetrically profiled blades (impeller Reynolds number $Re_m = 1.0 \cdot 10^5$).

NOMENCLATURE

A	Parameter in Eq. (7)
B	Parameter in Eq. (7)
b	Baffle width (m)
D	Overall agitator diameter (m)
H	Height of liquid from bottom of vessel (m)
H_2	Distance of the lower edge of impeller blades from bottom of vessel (m)
k	Universal von Karman's similarity constant
n	Frequency of impeller revolution (Agitator speed) (s^{-1})
$\overline{p}$	Reynolds stress (Pa)
R	Dimensionless radial co-ordinate
r	Radial co-ordinate (m)
Re_m	Agitator Reynolds number
I	Inside diameter of stirred tank (m)
w	Mean velocity (m/s)
w'	Turbulence fluctuating velocity (m/s)
W	Dimensionless mean velocity
z	Axial co-ordinate (m)
α	Blade pitch angle to the horizontal plane (°)
ρ	Density (kg/m^3)
μ^t	Eddy (turbulent) viscosity (Pas)

Subscripts

A_{ax}	axial
A_i	i-th value
A_{rad}	radial

A COMPUTATIONAL STUDY ON DISPERSING GAS IN A STIRRED REACTOR

A. BAKKER, H.E.A. VAN DEN AKKER, Kramers Laboratorium voor Fysische Technologie, Delft University of Technology, NL

INTRODUCTION

The numerical computation of the flow field in stirred tank reactors has received attention since the beginning of the 1980's, starting with Harvey (1980) and Harvey and Greaves (1982). Most times computations are limited to single-phase flow. Reviews of the work done on these single-phase computations have been given by Ranade and Joshi (1990) and Ranade et al. (1989). Stirred vessels are generally used for multiphase mixing but full computation of the flow field of a liquid phase in which a gas is dispersed is not yet possible. This is partly due to a lack of knowledge of bubble dynamics and the influence of bubbles on the turbulence structure, and partly due to computational limitations.

In spite of these difficulties, several attempts have been made to model the flow in a gassed stirred tank in a less rigorous way. Issa and Gosman (1981) calculated the flow in a gassed stirred vessel equipped with a Rushton turbine. In their calculations they assumed a very small (0.5 mm), constant bubble diameter. Further they used very coarse grids and it was not possible to verify the results of their simulations because of a lack of experimental data. Looney at al. (1985) presented a model for the turbulent flow of solid/liquid suspensions in stirred vessels. This model incorporated mass balance and momentum equations for both phases, together with a two-phase turbulence model. Pericleous and Patel (1987) calculated both the single-phase flow and the two-phase flow in a stirred tank. Their calculations were done for various impeller types and combinations of impellers. Due to the use of a simple one-equation k-l type turbulence model, their velocity predictions had a limited accuracy only. Their two-phase calculations were done assuming a constant bubble size and a constant bubble rise velocity, and the results could not be verified with experimental data. Mann (1988) followed a different approach. He modeled the flow created by a disc-turbine by a network of zones, and calculated the gas transport in this network by solving the continuity equation for the gas phase. Although his results were interesting and had a qualitative appeal, the simplified flow pattern makes extension of the model to other geometries difficult. An extensive model has been proposed by Trägårdh (1988). Not only did he manage to incorporate the momentum exchange between the gas phase and the liquid phase, but he also incorporated a model for local mass transfer and for the growth of microorganisms. However, his code was not capable of calculating the local bubble size and, like the other authors mentioned above, he used a constant value of the slip velocity between the gas bubbles and the liquid phase, without taking the influence of turbulence on bubble rise velocity into account.

It will be clear that there is a strong need for more extensive models, capable of calculating local gas holdup, local bubble size and local mass transfer rate. In the current research project, the general purpose code FLUENT is used for calculating the single-phase flow in a stirred vessel equipped with either an axial flow impeller or a disc-turbine. Provided that at low void fractions this overall flow pattern is not affected by the gas, this flow pattern is used as input for an in-house code named *GHOST!* (Gas Holdup Simulation Tool!) that calculates the distribution of the gas in the vessel on the basis of conservation equations. A mathematical model for bubble break-up and bubble coalescence, based on local turbulence intensity and local energy dissipation rate as calculated by FLUENT, is incorporated in this code. *GHOST!* is capable of calculating local values of void fraction, bubble size, interfacial area and mass transfer. Further details regarding the several modeling steps are given below.

SINGLE-PHASE FLOW PATTERN COMPUTATION

The single-phase flow pattern is calculated by solving the momentum equations and mass conservation equation with the aid of the general purpose code FLUENT. An Algebraic Stress Model, as described by Boysan (1984), is used for calculating the turbulent Reynolds stresses. To investigate possible grid dependence of the flow fields computations were done with several grid sizes. Two-dimensional calculations, assuming rotational symmetry, were done with 50x25, 42x27 and 28x14 grid nodes. Three-dimensional computations were done with

50x25x20, 52x27x17 and 56x35x11 grid nodes (in (z, r, ϕ) coordinates). The 3D calculations were done for a 90 degree segment of the vessel. In these computations the impellers have been treated as black boxes. For the disc turbine the u, v and w-velocities, together with the turbulent kinetic energy k and dissipation rate ε have been prescribed on the vertical swept boundary of the impeller, using symmetric, parabolic profiles according to the data presented by Ranade and Joshi (1990). In their case the impeller was located midway in the vessel at C/T = 0.5. In our case the impeller is located closer to the vessel bottom at C/T = 0.3 and the velocity profiles might well be asymmetrical, but no clear literature data was found supporting this suspicion. For the computations with a six-bladed downwards pumping pitched blade turbine (PBT) the velocity and turbulence profiles were prescribed at the bottom surface of the impeller (fig. 1) according to own experimental LDV data.

Further, since the calculated flow patterns are used for calculating the gas transport through the vessel it is necessary to correct the prescribed liquid velocities for the decrease in the dimensionless pumping number N_{qp} of the impeller when the impeller is gassed. This is done assuming that the decrease in pumping capacity of the impeller is proportional to the decrease in power number (Joshi et al, 1982, Trägårdh, 1988):

$$N_{qp,g} = N_{qp,u} \left(\frac{Po_g}{Po_u} \right) \tag{1}$$

All further information on the geometries investigated, together with the values for the impeller pumping numbers are listed in Table 1.

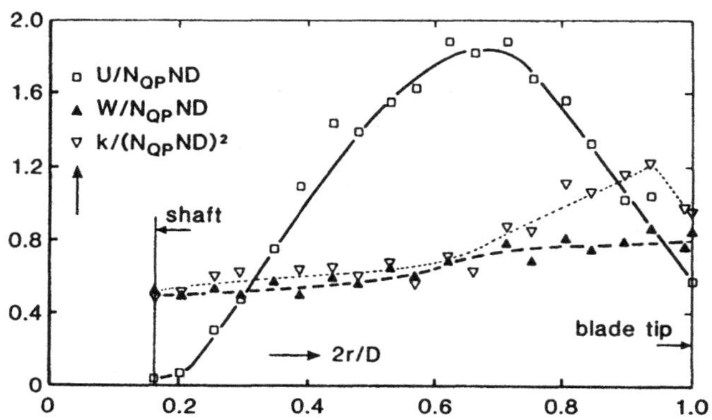

FIGURE 1 Experimental liquid velocities and turbulent kinetic energy in the outflow of a pitched blade turbine.

Table 1, Geometrical information.

	DT	PBT	General
Po_u	5.00	1.55	P/V = 1520W/m^3
Po_g	3.63	1.21	v_{sg} = .0036m/s
$N_{qp,u}$	0.75	0.80	C/T = 0.30
$N_{qp,g}$	0.55	0.63	D/T = 0.40
n	5.55	8.00	D_s/D = 0.75
Fl	.019	.013	S/D = 0.60
			T = 0.444 m

THE GAS-LIQUID MODEL

The flow pattern calculated following the procedure described in the preceding section is used for calculating the transport of the gas through the vessel using the model equations given below.

The continuity equation for the gas-phase reads:

$$\frac{\partial \alpha}{\partial t} + \vec{\nabla}.(\alpha\, \vec{u}_g) = S_g \tag{2}$$

S_g denotes the gas source. On performing the Reynolds decomposition and averaging the equation we get for a quasi-steady-state situation:

$$\vec{\nabla}.(\bar{\alpha}\, \bar{\vec{u}}_g) + \vec{\nabla}.(\overline{\alpha'\, \vec{u}'_g}) = \bar{S}_g \tag{3}$$

The second term on the left-hand side describes the transport of gas due to turbulent motion of the gas-phase and is not known in general. This leads to a so-called closure problem. This turbulent transport could be modeled by an analogy with the kinetic gas theory. According to this theory the number of collisions per unit time and per unit area between molecules exhibiting random motion and the boundary of a volume element is (1/4) $n\,\bar{c}$, where $\bar{c}$ is the average velocity of the molecules and n their number density. Since the bubbles in a turbulent gas-liquid dispersion also exhibit random motion we can calculate the turbulent gas flux from one cell into an other by (see also figure 2:

$$\phi_{g,turb} = \frac{1}{4}\, \bar{\alpha}\, \overline{|\vec{u}'_g|} \tag{4}$$

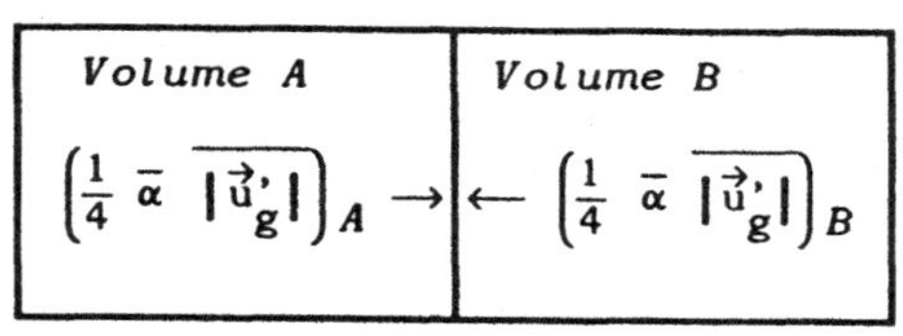

FIGURE 2 Illustration of the turbulent exchange between cells A and B.

Further we assume that the average random fluctuating velocity of the gas phase is proportional to that in the liquid phase:

$$\overline{|\vec{u}'_g|} \sim \overline{|\vec{u}'_l|} = \sqrt{2k} \tag{5}$$

The linear gas velocity is calculated as the sum of the linear liquid velocity in a single-phase system as calculated with FLUENT and the slip velocity between the two phases:

$$\vec{u}_g = \vec{u}_l + \vec{u}_s \tag{6}$$

The slip velocity $\vec{u}_s$, in (r, ϕ, z) coordinates can be calculated by a force balance on the bubbles:

$$\vec{F}_s = \left(\rho_l \, v_b \frac{w^2}{r} \, , \, 0 \, , \, \rho_l \, g \, v_b \right) \tag{7}$$

$$\vec{F}_s = C_D \frac{1}{2} \rho_l \, |\vec{u}_s| \, \vec{u}_s \frac{\pi}{4} d_b^{\,2} \tag{8}$$

The force in the radial direction equals the centripetal force on the bubbles, the force in the vertical direction is of course the buoyancy. The bubbles are assumed to follow the main liquid flow in the circumferential direction, and thus there is no net tangential force on the bubbles.

The constant C_D is a function of bubble Reynolds number and can be calculated from the correlations given by Morsi and Alexander (1972). The bubble Reynolds number is calculated from:

$$Re_b = \frac{\rho_l \, |\vec{u}_s| \, d_b}{\eta_\bullet} \tag{9}$$

Here $\eta_\bullet$ is the sum of the liquid viscosity and a term proportional to the turbulent viscosity:

$$\eta_\bullet = \eta_l + C_\bullet \, \rho_l \frac{k^2}{\varepsilon} \tag{10}$$

This extra term is introduced to account for the decrease in slip velocity when a bubble is moving in a turbulent flow field instead of in a stagnant liquid. A similar approach has already been followed by Barnea and Mizrahi (1975) for the flow in bubble swarms, but in contrast to their work our correction reflects the influence of turbulence in the liquid phase.

A first-order model is used to describe the process of bubble formation in the bulk:

$$\frac{\partial n_b}{\partial t} + \vec{\nabla}.(n_b \, \vec{u}_g) = \omega(n_{b\infty} - n_b) + S_g / v_{b,in} \tag{11}$$

The number of bubbles per dispersion volume n_b is related to the gas-holdup and the average bubble volume in the volume element investigated:

$$n_b = \alpha / \langle v_b \rangle \tag{12}$$

Further $n_{b\infty}$ is the number density of bubbles in which the process of bubble break-up and coalescence would result in case no new bubbles would be added, in other words when the source and convection terms in eq. (12) would be zero. Such a process would result in an average bubble size equal to the maximum stable bubble size according to Hinze (1955):

$$d_{b\infty} = C_{b\infty} \left(12 \frac{\sigma}{\rho}\right)^{3/5} \varepsilon^{-2/5} \tag{13}$$

The effective coalescence/breakup frequency ω may be proportional to the number of collisions between the bubbles. Another analogy with kinetic gas theory may result in:

$$\omega = C_\omega \frac{3}{2} \sqrt{2} \, \frac{\alpha}{d_b} \overline{|\vec{u}'_g|} \tag{14}$$

Equations (11) to (14) are used for calculating the local bubble size in the bulk of the dispersion volume only. For the impeller it is assumed that the impeller breaks up the gas bubbles to the maximum stable bubble size as calculated by eq. (13), on basis of the average energy dissipation rate inside the impeller. Therefore $C_{b\infty}$ could be estimated to be 0.75 from bubble size data reported by Greaves and Barigou (1988).

When the local gas holdup and bubble size are known, the mass transfer coefficient $k_L a_d$ can be calculated by using the equation for k_L given by Kawase and Moo-Young (1990):

$$k_L = 0.301 \, (\varepsilon \nu)^{1/4} \, Sc^{-1/2} \tag{15}$$

The k and ε values as calculated for a single-phase flow by FLUENT (ε_{1p} and k_{1p}) are corrected for the energy input by the gas-phase. The energy input per second for a single bubble is given by:

$$P_b = |\vec{F}_s| \cdot |\vec{u}_s| \tag{16}$$

Therefore the total energy dissipation rate will be:

$$\varepsilon = \varepsilon_{lp} + \frac{n_b P_b}{\rho_l (1-\alpha)} \tag{17}$$

Provided that the turbulence as generated by the bubbles has a length scale of the order of the bubble size, and when the turbulent length scale is calculated by:

$$L_t = \frac{k^{3/2}}{\varepsilon} \tag{18}$$

the total kinetic energy is given by:

$$k = k_{lp} + \left[d_b \frac{n_b P_b}{\rho_l (1-\alpha)} \right]^{2/3} \tag{19}$$

All the equations above are incorporated in *GHOST!* (Gas Holdup Simulation Tool!). *GHOST!* is capable of calculating local values of void fraction, bubble size, interfacial area and mass transfer. It should be stressed that the computational algorithms are set up in a very flexible way and are not limited to any particular geometry, which means that a quick assessment of the influence of geometrical changes with respect to e.g. sparger position and impeller placement on gas holdup and mass transfer can be made. Further the code allows for prescribing the shape of the gas-holdup profile in the outflow of the impeller. The calculations can be performed on any type of 2D or 3D, uniform or non-uniform, cylindrical finite-difference grid.

EXPERIMENTAL

To verify the calculated single-phase flow patterns several velocity measurements were performed using a TSI Laser Doppler velocimeter, incorporating a 4W Spectra-Physics Argon laser, a fiber optics measurement probe, and a colorburst frequency shifter.

For verification of the calculated gas-holdup profiles, local gas-holdup measurements were performed using a single point optical fiber probe. This measurement method was described by Frijlink (1987).

SINGLE PHASE RESULTS

Figure 3 shows the calculated vector components in the plane midway between two baffles (ϕ = 45°) for a disc-turbine. The computations were done on a 3D grid with 50x25x20 nodes. Due to the radial pumping direction of this impeller the flow is characterized by two main circulation loops, one on each side of the impeller. Such a flow pattern has already been described by numerous authors. This gives confidence in the results.

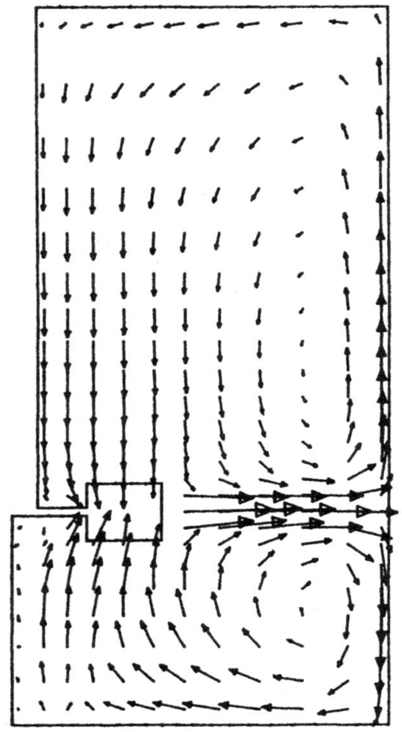

Figure 3 Calculated velocity vectors for a Disc Turbine, midway between the baffles (ϕ = 45°).

Figures 4a,b,c show the flow pattern for a pitched blade turbine as calculated on a 3D grid with 52x27x17 nodes. Figure 4a shows the flow pattern in the plane midway between to baffles (ϕ = 45°). In this plane the flow pattern consists of one large circulation loop, and a small recirculation loop below the impeller. Just in front of the baffle (ϕ = -4°) the flow pattern is approximately the same, although it can be seen that there are differences to be found in the upper half of the vessel. The flow a small distance behind the baffles (ϕ = 15°) is very different from the flow pattern in front of the baffle. It can be seen that a second recirculation loop has formed in the upper half of the vessel, and that the flow along the wall in the upper part of the vessel is directed downwards instead of upwards. The existence of this second recirculation loop has not been reported before in the literature. The fact that such a secondary recirculation loops really exists is shown in figure 5. This figure shows experimental axial liquid velocities as measured in the plane midway between two baffles, near the vessel wall. It can be seen that the axial velocities change sign at a distance of approximately Z/H = 1/3 from the liquid surface, thus proving the existence of this second recirculation loop. The experimental data show that this loop extends at least 45° away from the baffle, whereas the simulations say that the loop extends only approximately 33° from the baffle.

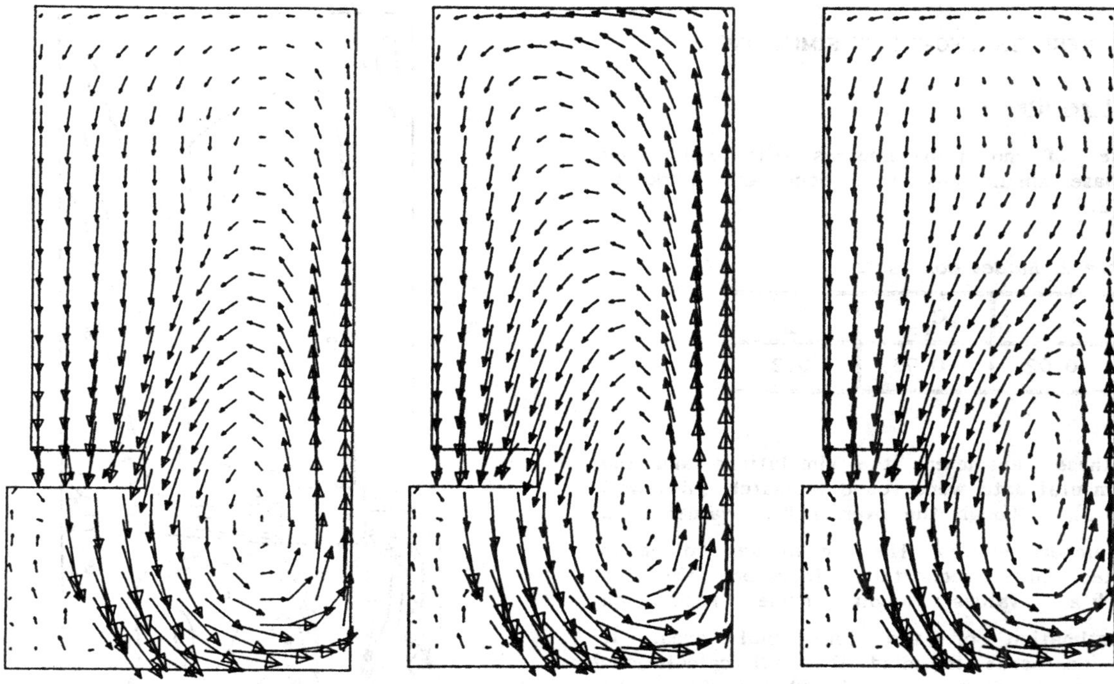

Figure 4a,b,c. Calculated velocity vectors for the PBT, at $\phi = 45^o$, $\phi = -4^o$ and $\phi = 15^o$.

This means that the simulated flow patterns are qualitatively correct, showing details not reported before in the literature, but that the quantitative accuracy should be improved even further. It should be mentioned that the recirculation loop was neither found when doing 2D, axisymmetric simulations nor when doing full 3-dimensional simulations with 11 grid nodes in the tangential direction. This might mean that for a correct quantitative prediction of the flow pattern the number of grid nodes in the tangential direction should even be larger than the 17 nodes used in this calculations but due to computational restrictions this could not yet be tested. There may be two reasons for the fact that the existence of such a recirculation loop has not been reported before in the literature. The first reason is that such a loop may not be formed when the impeller is mounted at a larger impeller to bottom clearance. The second reason might be that until now no 3D computations with such a large number of grid nodes were performed.

In general the predicted overall energy dissipation rates compare very well with the total energy input delivered by the impeller, provided that the ASM turbulence model is used. For example with the disc-turbine the calculated total energy dissipation rate in the bulk

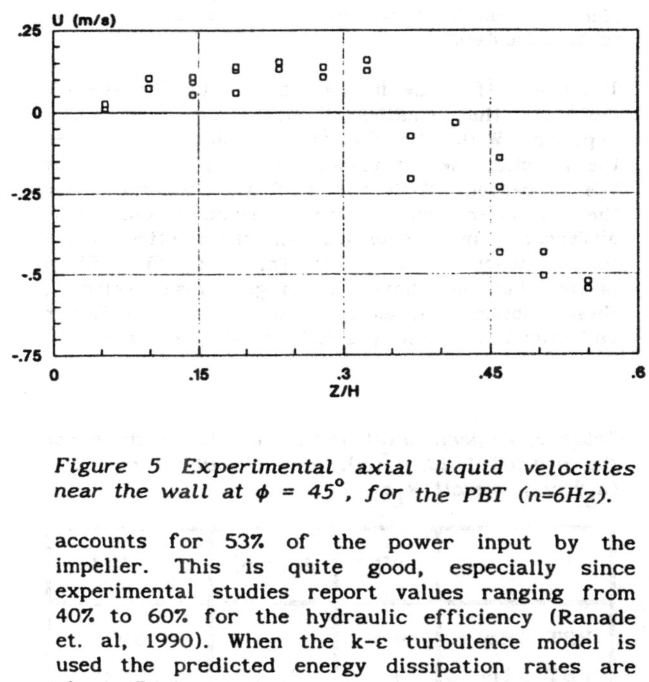

Figure 5 Experimental axial liquid velocities near the wall at $\phi = 45^o$, for the PBT (n=6Hz).

accounts for 53% of the power input by the impeller. This is quite good, especially since experimental studies report values ranging from 40% to 60% for the hydraulic efficiency (Ranade et. al, 1990). When the k-ε turbulence model is used the predicted energy dissipation rates are about 15% lower than when the ASM model is used.

RESULTS TWO-PHASE SIMULATIONS

Overall results

The set of model parameters with which the two-phase simulations were done are listed in Table 2.

Table 2, Model constants

$C_\bullet$	$C_{b\infty}$	C_ω
0.02	0.75	0.2

With these parameters the simulations and the experimental data were found to match quite well (Table 3), although the overall k_La_d values tend to be underpredicted. The overall gas holdup is predicted quite accurately. Therefore the too low k_La_d values might arise from an underestimation of the interfacial area. It should be borne in mind that all calculations are done assuming spherical bubbles. Since the bubbles are possibly ellipsoidal, the surface area will be underpredicted. Further, the number of experimental data is limited and it might turn out that when more experimental data are available, especially with respect to bubble size and bubble shape, the model parameters have to be adjusted.

The reason for the higher gas holdup for the DT lies in the smaller bubble size for this impeller. With the DT the bubbles first enter the impeller, which breaks them up, before they are dispersed. With the PBT the bubbles enter the impeller only after recirculation. This difference in dispersing characteristics leads to a larger bubble size for the PBT. Since larger bubbles have a larger rise velocity, these bubbles will escape from the vessel faster and thus the overall gas holdup will be lower.

Table 3, Experimental and simulation results for the overall holdup (α), overall bubble diameter (d_b) and overall k_La_d.

	α(%)	d_b(mm)	k_La_d (1/s)
PBT exp.	4.0	-	0.0036
PBT sim.	4.1	3.77	0.0025
DT exp.	4.7	-	0.0038
DT sim.	4.6	3.38	0.0029

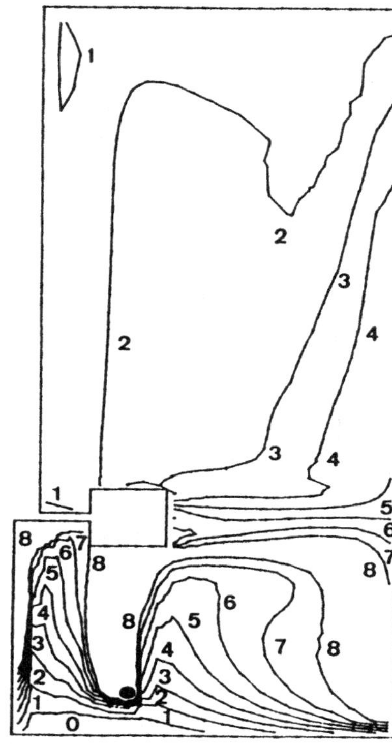

Figure 6. Calculated contours of constant α(%) for the DT at ϕ = 45°

Results Disc Turbine

Figure 6 shows contours of equal gas holdup for the disc-turbine in the plane midway between the baffles. It can be seen that the gas holdup is high near the vessel wall, particularly in the lower part of the recirculation loop where the motion of the rising gas is opposed by the downwards liquid flow. Further the gas holdup shows a peak in the gas flow, issuing from the sparger and rising up into the impeller. For verification of the modeling results several local holdup measurements were performed. Figure 7 shows a comparison between experimental holdup data and simulated holdup data as a function of the radial coordinate 2r/T, at a distance Z/H = 0.92 from the unaerated liquid surface, just above the sparger. It is clear that the rising gas results in a locally high holdup and in a very steep gradient. The holdup rises again near the vessel wall. Figure 8 shows a comparison between experiment and simulation at a constant radius of 2r/T = 0.9, close to the vessel wall. The holdup data have been plotted as a function of the axial coordinate. Both the experimental data and the simulation results show low holdup values near the top surface. The holdup gradually increases with increasing distance to the liquid surface. There is a steep increase in holdup just below the centerplane of the impeller, and a steep decrease again near

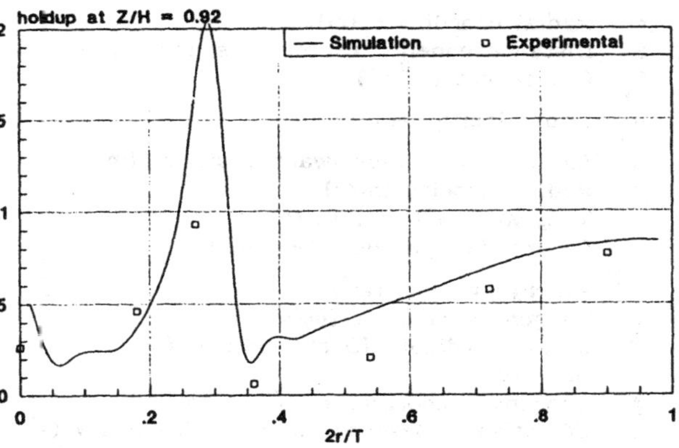

Figure 7 Comparison between experimental holdup data and simulated holdup data (DT).

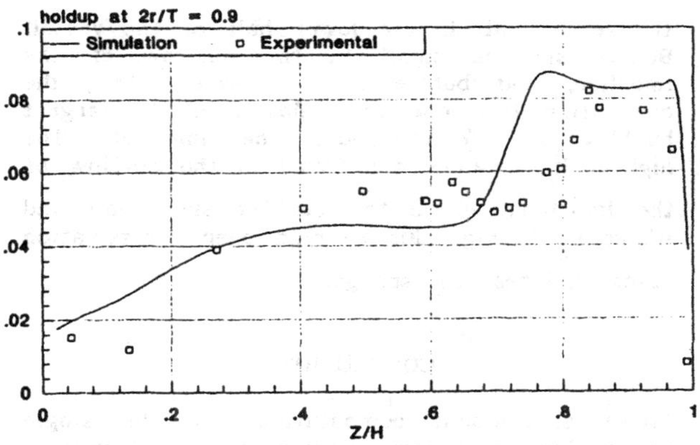

Figure 8 Comparison between experimental holdup data and simulated holdup data (DT).

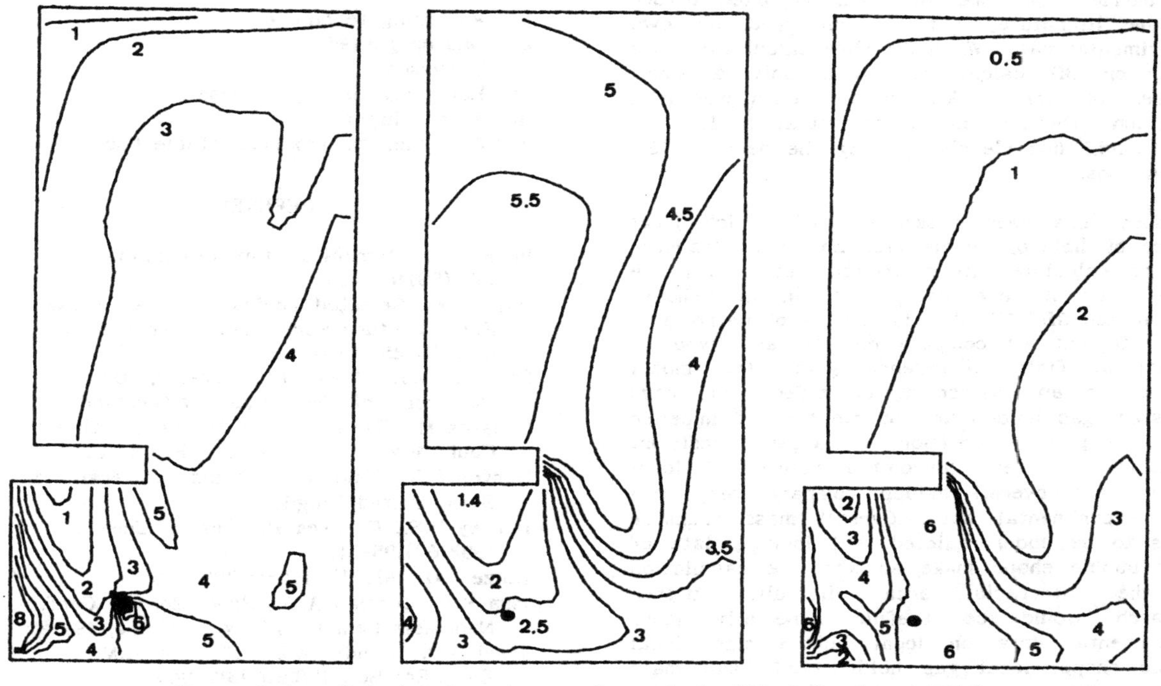

Figure 9a,b,c. Calculated contours of constant α(%), d_b(mm) and $100{*}k_La_d$ for the PBT at ϕ = 45°. The sparger is marked by a black dot. The numbers denote the minimum values in the areas enclosed by the contours.

the vessel bottom. The latter indicates that hardly any gas reaches the bottom. In general the predicted hold-up profiles match the experimental data quite well, thus giving confidence in the modeling method.

GHOST! also predicts the formation of a gas filled vortex in the center of the vessel, below the impeller. In reality such a vortex was not seen. Instead the gas rising from the sparger precesses around the vessel axis with a period of several seconds. Such periodic motions can not be calculated. The fact, however, that it was found that the *GHOST!* solution converged particularly slowly in this region may be an indication of flow instabilities.

Results Pitched Blade Impeller

Figures 9a,b,c show calculated contours of constant holdup, constant bubble size and constant k_La_d respectively, for the PBT. Peaks in the gas holdup are found near the impeller blade tip, near the sparger and, as with the disc turbine, in the center of the vessel, below the impeller. A high gas holdup is found near the vessel wall in the lower half of the vessel. Bubble size is small in the outflow of the impeller, and bubble size increases along the circulation loop due to coalescence: the largest bubbles are found above the impeller. The highest k_La_d values are found in the outflow of the impeller, where the bubbles are small and where k_L is high due to high energy dissipation rates, and near the sparger.

CONCLUSIONS

Three dimensional computations of the single phase flow in a stirred vessel equipped with a PBT, as done with the general purpose code FLUENT, show that a second recirculation loop is formed in the upper part of the vessel, behind the baffles. This has not been reported before in the literature, but is confirmed by own experimental data. When 2D flow calculations are done, or 3D calculations with only a small number of grid nodes in the circumferential direction, this secondary recirculation loop is not found, thus clearly showing the need for 3D simulations.

A model has been presented with which local values of holdup, bubble size and mass transfer can be calculated in a stirred vessel. All the model equations are incorporated in an in-house code named *GHOST!*, that is capable of performing both 2D and 3D computations on any type of cylindrical finite difference grid. The model allows for an assessment of effects on mass transfer, gas holdup and bubble size of impeller type, impeller position, sparger position, aeration rate, etc. Predicted values of local holdup, and overall holdup compare very well with experimental data. Overall mass transfer tends to be underpredicted, but lacking data of local bubble shape make an accurate calculation of the interfacial area difficult. Further research would be useful. Especially more experimental data on local bubble size, local bubble shape, local gas holdup and local mass transfer rate is required, to make an extensive verification of the modeling results possible.

ADDITIONAL NOMENCLATURE

General

- C Distance impeller center to bottom (m)
- C_∞, C_ω, $C_\bullet$ Model constants
- d Diameter (m)
- D_s Sparger diameter (m)
- $\vec{F}_s$ Slip force vector (N)
- k Turbulent kinetic energy (m^2s^{-2})
- L_t Turbulent length scale (m)
- P Power input (W)
- Po Power number
- r Radial coordinate (m)
- S Distance impeller center to sparger (m)
- S_g Gas source (m^3s^{-1})
- $\vec{u}_s$ Slip velocity vector
- u Axial velocity (downward positive) (m/s)
- v Radial velocity (m/s)
- w Tangential velocity (m/s)
- v_{sg} Superficial gas velocity (m/s)
- <v> Average volume (m^3)
- w Tangential velocity (m/s)
- Z Axial coordinate (0 at surface) (m)
- α Gas holdup
- ϕ Tangential coordinate (0° at baffle)
- ω Effective coalescence/breakup frequency (s^{-1})
- $\eta_\bullet$ Effective viscosity defined by eq. 10

Subscripts

- b Referring to bubble
- g Gas or gassed
- u Ungassed
- 1p Referring to single phase
- in At gas input
- ∞ Referring to maximum stable bubble size

REFERENCES

Barnea E., Mizrahi J., Can.J.Chem.Eng. 1975(53)461-469

Boysan F., Selected Topics in Two-Phase Flow, Lecture Series no. 9, Trondheim, may 25-25, 1984

Frijlink J.J., 1987, PhD Thesis, Delft University of Technology, Netherlands

Greaves M., Barigou M., 1988, Proc. 6th Eur. Conf. Mixing, pp 313-320, BHRA, Pavia

Harvey P.S., 1980, PhD Thesis, University of Bath, United Kingdom

Harvey P.S., Greaves M., Trans.I.Chem.E. 1982(60)195-210

Hinze J.O., AIChEJ 1955(1)289-295

Issa R.I., Gosman A.D., Proc. 2nd Int.Conf. Num.Meth.Lam.Turb.Flows, Venice, 1981

Joshi J.B., Pandit A.B., Sharma M.M., Chem.Eng.Sci. 1982(37)813-844

Looney M.K., Issa R.I., Gosman A.D. Politis S. Fifth Int.Conf. on Mathematical Modelling, Berkeley, July 29-31st, 1985

Kawase Y., Moo-Young M., Chem.Eng.J. 1990(43)B19-B41

Mann R., Hackett L.A., 1988, Proc. 6th Eur.Conf. Mixing, pp 321-328, BHRA, Pavia, Italy

Morsi S.A., Alexander A.J., J. Fluid Mechanics 1972(2)193-208

Pericleous K.A., Patel M.K., Physico Chemico Hydrodynamics 1987(8)105-123

Ranade V.V., Joshi J.B., Marathe A.G., Chem.Eng.Comm, 1989(81)225-248

Ranade V.V., Joshi J.B., Trans.I.Chem.E. 1990(68)34-50

Trägårdh Ch., Proc. 2nd Int.Conf.Bioreactor Fluid Dynamics, pp 117-134, BHRA, Cambridge U.K., 1988

REFEREE'S QUESTIONS

1) For your table 1, please give the rotational speed and peripheral speed of the impellers.

	DT	PBT
Rotational speed $n(s^{-1})$	5.55	8.00
Tip velocity (m/s)	3.07	4.42

2) Is the arrow length in figs. 3 & 4 proportional to the local (average) velocity?

The arrow length is indeed proportional to the local average liquid velocity.

3) In fig. 6, is the black dot the gas sparger?

Both in figs. 6 and 9, the gas sparger is marked by a black dot.

4) What is the fluid-dynamic regime for the cases examined in figs. 6 & 9? (At least loading in the former case, probably indirect loading for the latter).

The disc turbine is operated in the loading regime, with vortex/clinging cavities.

The pitched blade turbine is operated in the indirect loading regime and, as far as it could be seen, with clinging/growing cavities.

5) In your opinion, would it be interesting to use a sparger diameter larger than the impeller diameter?

Bakker and Van den Akker (1990) reported for an A315 impeller that the use of a large sparger lead to a decrease in gas-holdup below the impeller and consequently to a more stable gas-handling, but also to a decrease in mass transfer.

Nienow et al. (1988) reported for Rushton turbines with a diameter of D/T = 1/3, that the use of ringspargers with a diameter larger than the impeller lead to an improvement in the performance. Recently, in Delft, we started a case study, in collaboration with prof. A. Lübbert of the University of Hannover, to the gas dispersion performance of a D/T = 1/3 Rushton turbine in combination with a ringsparger with a diameter of 1.9 times the impeller diameter. Preliminary calculations with *GHOST!* showed that this leads to a decrease in the gas loading of the impeller at a slightly larger gas holdup, as compared to a 0.5 D ringsparger mounted close to the impeller. This confirms the results of Nienow et al. However, this work is not finished yet and will be reported later.

REFERENCES

Bakker A., Van den Akker H.E.A., "Gas-liquid contacting with the Lightnin A315 impeller, Effects of flow-pattern", AIChE Annual Meeting, Session on Scale-Up and Industrial Mixing II, November 11 to 16, 1990, Chicago, U.S.A., Unpublished.

Nienow A.W., Allsford K.V., Cronin D., Liu Huoxing, Wang Haozhung, Hudcova V., "The use of large ringspargers to improve the performance of fermenters agitated by single and multiple standard Rushton turbines", 2nd Int.Conf. on Bioreactor Fluid Dynamics, September 21-23, 1988, Cambridge, U.K.

MEASUREMENTS AND MODELLING OF FLOW IN GAS SPARGED, AGITATED VESSELS

G.K. PATTERSON, University of Missouri-Rolla, USA

ABSTRACT

Effects of poor gas distribution, liquid mixing, and rates of interphase mass transfer continue to be important issues in the design and scaleup of gas-liquid, gas-liquid-solid, slurry catalyzed gas-liquid, and biological-growth reactors. Most studies of mass transfer rates, which directly affect reaction rates and yield, have been done on a volume-averaged basis, although some measurements of bubble size and bubble surface area have been done to determine local variations. Preliminary work is reported in this paper on measurements and simulation of local bubble dynamics and effects on liquid hydrodynamics (flow patterns). Laser-Doppler anemometry (LDA) was used to measure liquid velocities with and without sparged gas. It was necessary to maintain low relative gas rates and high impeller rates in order to maintain a regime where LDA gave useful results. Air was sparged into tap water giving bubble sizes which were estimated to be 1 to 5mm. Simulations done with the code FLUENT gave results with close similarity to the experimental measurements and showed aspects of bubble dynamics that are very difficult to see visually. Importantly, most of the gas, even under highly dispersed conditions, channels upward along the upstream sides of the baffles. Also the higher the gas loading at a given impeller rotation rate, the greater is the channelling tendency and the poorer is gas distribution. Preliminary mass transfer rate computations gave promising results and will lead to more studies.

STATUS OF MASS TRANSFER IN GAS SPARGED, AGITATED VESSELS

Since almost all research on mass transfer in agitated, gas sparged vessels has been on a volume-averaged basis, existing correlations and scaleup relationships are on that basis. That means that changes in impeller type, vessel geometry (such as baffle design), location of gas spargers, liquid feed location, solids particle size or density, and many other aspects cannot usually be accounted for without extensive experimental work and the making of more correlations. If a method for accounting for effects of changes such as those mentioned can be developed based on local conditions, making use of well established hydrodynamic modelling capabilities, more flexibility and reliability can be achieved for modelling and scaleup.

There are several steps in developing such a model. First, the effects of gas bubble loading for specified bubble sizes on the liquid flow patterns must be incorporated into a functioning hydrodynamic modelling code. Second, the effect of bubble loading on liquid-phase pumping rate must be incorporated, so that the impeller-induced flows may be properly specified. Third, the relationship between bubble size distribution (and consequently bubble surface area) and the local hydrodynamic conditions must be incorporated. Fourth, local hydrodynamic effects on mass transfer rate per unit area must be incorporated. In step one, the motions of the bubbles must be simulated to determine their flow patterns, when they escape from the liquid surface, and to allow computation of momentum exchange between the liquid and gas phases. In the last step, as mass transfer occurs, the bubble size and composition changes must be simulated so that effects of bubble depletion on mass transfer rates will be accurately accounted for. This paper concentrates on the first two steps, hydrodynamics for a specified bubble size and effects on liquid velocity patterns.

Previous studies have provided the information necessary to properly chose conditions that can most easily be simulated and where measurements of local liquid velocities are most likely to be successful. Such conditions are necessary to provide a starting point from which to develop methods for more difficult (and usually practical) conditions. Joshi *et al.* (1982) reviewed correlations for effects of gas sparging on impeller power consumption, mixing times, volume-average mass transfer rates and bubble surface areas, and heat transfer rates, and effects of multistage contacting, impeller gas induction, and surface aeration methods. Smith (1985) and Nienow *et al.* (1985a) have reviewed in depth how sparged gas interacts with impellers in stirred vessels and how this affects the power consumed by the agitation. Unfortunately, no information on how the bubbles affected liquid pumping rates and velocity patterns was available.

For air-water systems the regimes where the sparged air affects hydrodynamics only slightly and where effects are much stronger can be determined from data such as that of Nienow *et al.* (1978). For a given gas-liquid system and impeller type, the ratio of gassed power to ungassed power (P_G/P_0) is approximately correlated by the gas flow number ($F_l = G/nD^3$). [Some, i.e. Hughmark (1970) have suggested that the gas flow number should be G/nV to give better correlation at various vessel sizes.] For instance, at F_l = 0.015, $P_G/P_0 \approx 0.85$. Therefore, little effect of the gas bubbles on flow patterns would be expected as long as the vessel is not gas flooded, i.e. impeller rotation rate is high enough. The minimum impeller rotation rate for dispersed conditions was treated by Nienow *et al.* (1985b), and their work shows that the 300 rpm used for this work was well above the minimum.

Bubble sizes produced by the method of sparging and by the impeller have been studied by Calderbank (1958), Shumpe and Deckwer (1980), Takahoshi *et al.* (1986), and Greaves and Barigou (1988). Such studies have generally shown that for air-water systems, the bubbles are mainly in the range of 1-5mm with the peak of the distribution varying between 2mm and 4mm, depending on agitation conditions and location in the vessel. Calderbank's correlation for surface area (Sauter) average bubble diameter, which applies to many different liquids is as follows:

$$d_b=4.15\left[\frac{\sigma^{0.6}}{(P/V)^{0.4}\rho_l^{0.2}}\right]H^{0.5}+0.09$$

where σ is surface tension in dynes/cm^3, (P/V) is in hp/cu. ft., ρ_l is density in gm/cm^3, and H is volume fraction gas in the vessel. For modelling and simulation a bubble size distribution in the 2mm to 5mm diameter range should give good results.

One of the most important aspects of modelling the effects of sparged gas on mixer hydrodynamics is the momentum exchange between the liquid and the bubbles. Since the bubbles have little mass, acceleration effects will be minimal and they will tend to move with the liquid; although buoyancy affects large bubbles strongly. That means that low concentrations of small bubbles should have only small effects on velocity distributions in a stirred vessel. That remains to be proven experimentally and is addressed by this paper. At high bubble concentrations, the combined drag effects of bubbles moving upward because of gravitational displacement will cause significant modifications in liquid velocities. It is under those conditions that the power/volume ratio decreases below the ungassed value at constant impeller rotation rate. The mechanism for the decreased power draw is related to the effective shape change of the impeller blades caused by gas cavities (see Smith, 1985) and to the upward drag of gas bubbles near the vessel wall, which "pulls" liquid through the normal circulation pattern at a higher rate. This phenomenon is illustrated by the simulations discussed below.

HYDRODYNAMIC MODELLING OF TURBULENT FLOWS WITH SUSPENDED PARTICLES AND BUBBLES

Numerical modelling of single phase (gas or liquid) turbulent flows in complex, three-dimensional flow fields is now a well established technique (Patterson, 1991). Although refinements continue to be made and choices exist as to what modelling accuracy to seek based on model complexity, engineering use of turbulent flow simulation is accepted. Several commercial computer codes are now available so that the work of solving the sets of partial differential equations generated for each flow problem is eased. Most codes allow for considerable flexibility as to turbulence model used (k-ϵ, Reynolds stress, etc.), values of constants, representation of differentials (upwind, quadratic, etc.), solution method (single direction sweep, alternating direction sweep), pressure computation (SIMPLE, SIMPLEC, and PISO) and replacement of source terms for custom model construction. Among the commercial codes available are FLUENT (Creare Corp., Hanover, N.H.), PHOENICS (Cham Corp., Huntsville, AL), FIDAP (Fluid Dynamics Internl., Evanston, IL). FLUENT and PHOENICS are both finite volume based codes developed for the combustion and chemical processing industries, whereas FIDAP is a finite element based code developed for the aerospace industry. Body fitted coordinate systems, which require coordinate transformation before solution, are available for the finite volume codes. Finite element codes can easily fit any geometry.

Modelling of bubble and particle dynamics and effects of bubbles and particles on the fluid flow patterns required some method for accounting for the motions of the bubbles and particles. Two approaches are common. One way is to treat the bubbles or particles as a second continuous phase with its own fluid dynamics and variations in density (see Soo, 1967). The other way is to track individual particles in a Lagrangian way as they move through the fluid, accounting for gravitational and drag forces at each location. The code FLUENT has the tracking computation built-in and also computes the momentum exchange that occurs between fluid and particle or bubble. By iteration between particle track computations and modified fluid velocity computations, effects of particles or bubbles on fluid flow patterns may be obtained. Unfortunately, no correction for very high bubble concentrations (hindered settling effect) is built into the code. It is also possible to track particles or bubbles which have a Rosin-Rammler size distribution

where $M_d=\exp[-(d_p/\overline{dp})^n]$ gives the fraction M_d, of particles with diameter, d_p, greater than $\bar{d}_p$.

The momentum exchange rates between bubbles and liquid are given in FLUENT by the following relationships:

$$\frac{du_i,p}{dt}=-F_i(u_{i,p}-u_i)+g_i$$

$$F_i=\left(\frac{18\mu}{\rho_p d_p^2}\right)\left(\frac{C_D R_e}{24}\right)$$

$$R_e=\frac{\rho_f d_p|v_p-v_f|}{\mu}$$

$$C_D=\frac{K_1}{Re}+\frac{K_2}{Re^2}+K_3$$

where K_1, K_2 and K_3 are coefficients which vary with Reynolds number range, Re (see Morsi and Alexander, 1972). Other variables above are u_i, fluid velocity in i-direction, $u_{i,p}$, particle velocity in i-direction; g_i, acceleration of gravity in i-direction, μ, fluid viscosity; ρ_p, particle density, ρ_f, fluid density; v_p and v_f, vector velocities.

EXPERIMENTAL CONDITIONS

The velocity measurements were made by use of a Dantec one-directional laser-Doppler anemometer with a counter-processor using the forward scatter mode. The water was seeded with a very low concentration of 5 micron polystyrene scattering particles in order to provide Doppler-shifted scattered light for the measurements. The counter was interfaced to a computer to give a variety of output modes: average velocity, mean square velocity, and velocity probability distributions among them. Even though in many cases the light travelled through more bubbly fluid in forward scatter, more reliable results were obtained because the scattered light from the particles was much stronger. Direct, real-time velocity measurements were not possible. Instead the peaks of the distributions of occurrence frequency versus Doppler shift (velocity probability distributions) were used. The peak for velocity from small particles in fluid not at interfaces could generally be distinguished from the peaks caused by bubble passages and by particles being carried by bubble-fluid interfaces.

The general experimental setup is shown in Figure 1. The stirred vessel was a flat-bottom, fully-baffled round tank of 27cm diameter and 27cm liquid height. The impeller was a disk turbine 9.3cm in diameter with 1.8cm blade height. It was positioned 9cm above the tank bottom.

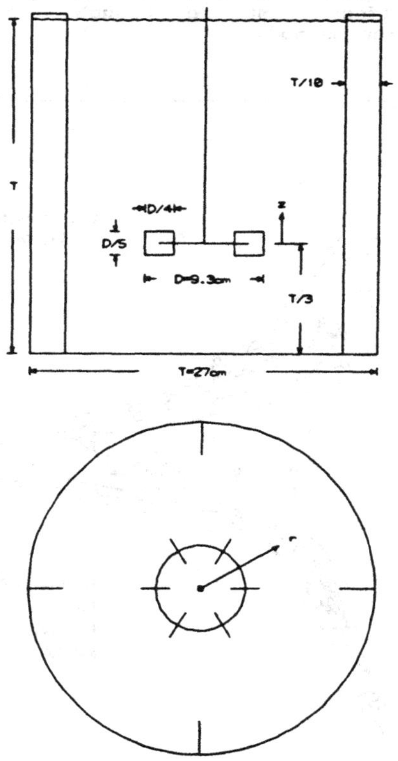

Figure 1. Tank and agitator simulated and used in experiments.

The fluid stirred was water, and air was the sparged gas. The sparger was a ring 6cm in diameter with 46 holes, each of which was 0.05cm in diameter, equally spaced. It was positioned 4.5cm above the tank bottom. Gas was sparged at 0.062 l/s and the impeller rotation rate was 300 rpm, giving a gasflow number, F_l, of 0.015.

During some experiments, partially deoxygenated water (13.5% saturated) was pumped into the vessel at the bottom center at a rate of 1.64 l/min. An oxygen probe was used to measure the oxygen concentrations at several points in the vessel, showing nearly uniform distribution of oxygen concentration at 81.5% of saturation at 25°C. Under these conditions the volume average value of K_la was $0.0049S^{-1}$.

BOUNDARY CONDITIONS FOR SIMULATION

The vessel size simulated was as described in the previous section. Also impeller rotation rate and gas flow rate were the same, except for the cases where gas flow rate simulated was 10 times the experimental rate, giving F_l =0.15. The gas inlet position in the simulations was exactly the same as in the experiments, except only 40 sparger holes were simulated. The flow produced by the impeller was fixed by prescribing the velocities and level of turbulence at the impeller blade tip as given in Table I. The rotating disk was modelled by specifying the disk angular velocities at each node point on the disk to correspond to 300rpm. The water surface was modelled as a flat boundary with no momentum transport (zero viscosity air above it) and totally porous to bubbles, so that they could escape.

Table I

Fixed Velocities and Turbulence Levels at Blade Tip * (r=5 cm, 300 rpm)

2z/w	Axial Velocity m/s	Radial Velocity m/s	Tangent. Velocity m/s	Turbul. Energy m2/s2	Turbul. Dissip. m2/s3
0.75	0	0.3	0.4	0.10	2.5
0.25	0	1.2	1.1	0.08	1.0
-0.25	0	1.2	1.1	0.08	1.0
-0.75	0	0.3	0.4	0.10	2.5

*Based tip velocity is 1.46 m/s; Fluid velocities and turbulence based on measurements by Wu and Patterson (1989)

Figure 2 shows the grid for x-z plane number 2 within the quarter section of the vessel which was modelled. Fifteen x-z planes spaced evenly except near the baffle were used. The cyclic nature of the vessel with four baffles and four identical regions was used to reduce the volume necessary to include in the simulation. The x-z planes between the quarters were so-called cyclic planes which assure continuity of all values and gradients. The impeller disk is shown but the blades are not, since they move through the space and their effect is given by the fixed velocities and turbulence levels. The simulation would be classed as a three-dimensional, axisymmetric simulation. The k-ϵ model was used and the grid dependence has not yet been checked since high accuracy was not sought in these exploratory computations.

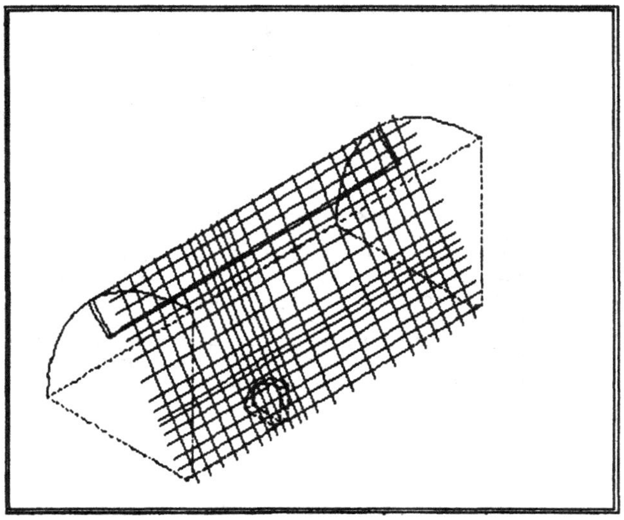

Figure 2. Grid used for computations. Symmetry allows simulation of a quarter section. There are 15 grids distributed angularly.

RESULTS OF MODELLING AND EXPERIMENTAL MEASUREMENTS

The experimental measurements of velocities in the presence of low concentrations of air bubbles in water (F_l = 0.015) showed only small differences from the velocities obtained in plain water. The bubbles did increase the upward water velocities near the wall from the impeller level up to the water surface. That effect is caused by the higher concentration of bubbles near the wall because they are carried there by the impeller stream. Very little effect was seen on the radial velocities produced by the impeller, although some compensating effects may have been brought about by the bubbles' reduction of the average density and by the shape effect of bubbles caught in the wake of the impeller blades. The former would increase the radial velocities but the latter would tend to decrease them. Neither effect should be large since the effect of the sparged gas on impeller power is small at F_l = 0.015.

The effects discussed above are illustrated by Figures 3 and 4, which show comparisons of experimentally measured axial (vertical) and radial velocities for plain water and sparged gas at F_l = 0.015. Figures 5 and 6 show the axial and radial velocity results of the computer simulation with FLUENT for 1mm bubbles compared to the experimental results at F_l = 0.015. The simulation shows essentially the same effects as compared to plain water, except that a larger effect on the radial velocities was computed. This is probably because the bubble shape effect on the pumping capacity of the impeller blades is not included. As this research progresses that effect must be brought into the simulation method.

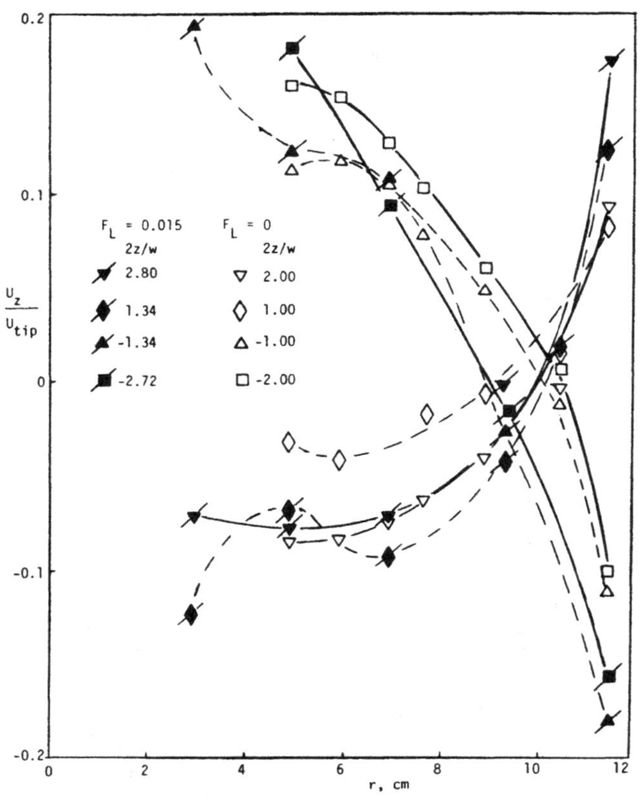

Figure 3. Measured axial velocities for gassed and ungassed stirring. (z is height above impeller disk, w is blade height, r is radial distance from tank axis)

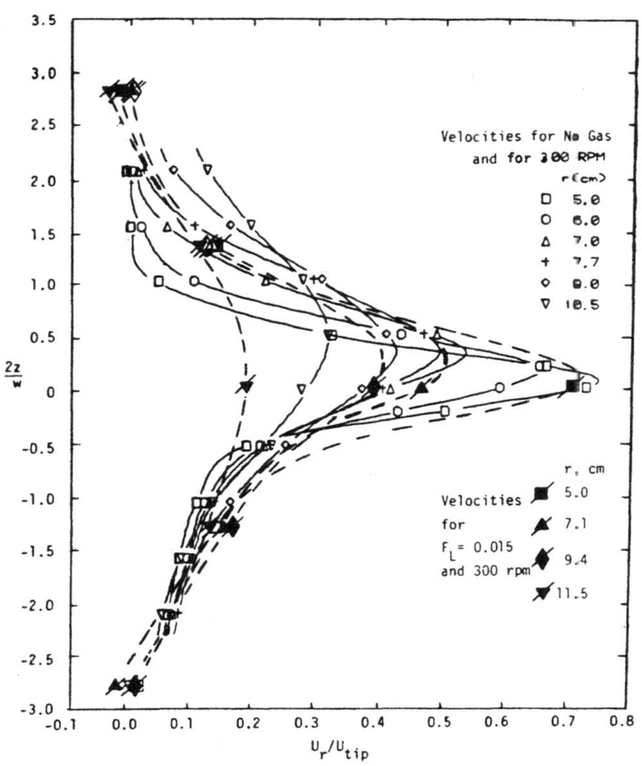

Figure 4. Measured radial velocities for gassed and ungassed stirring.

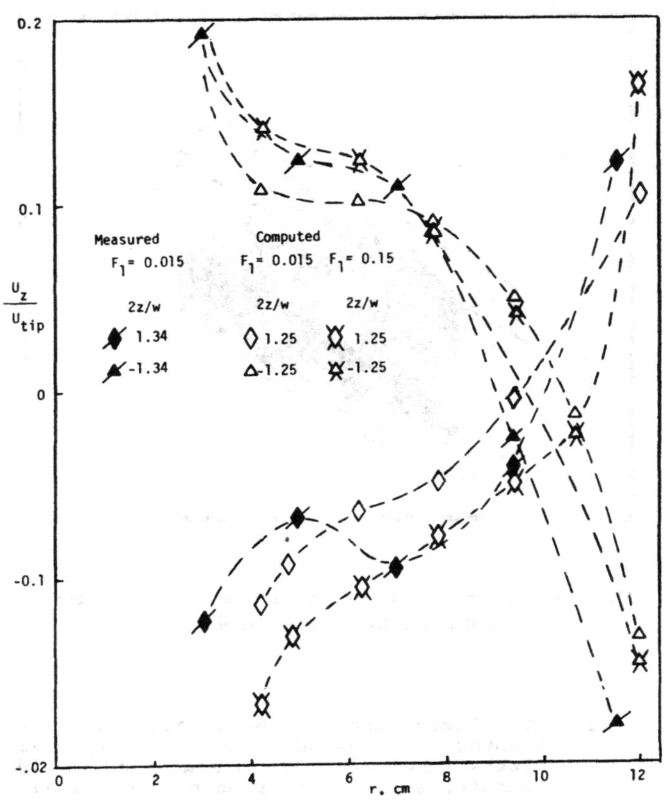

Figure 5. Comparison of measured and computed axial velocities.

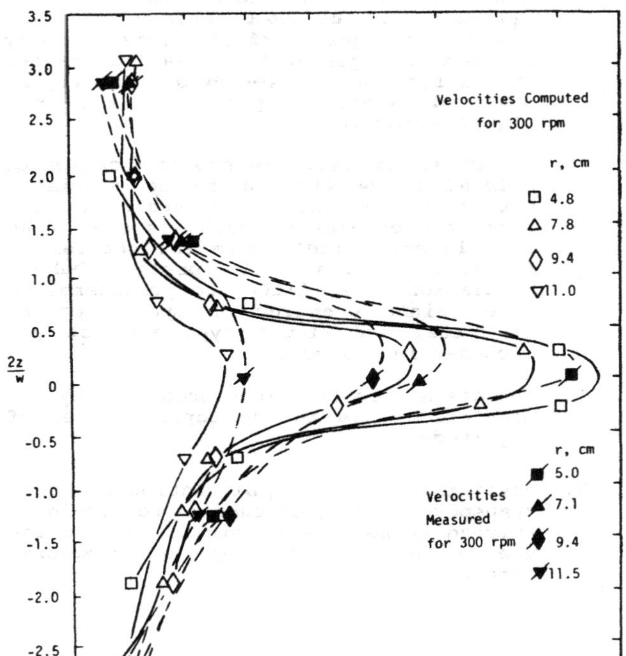

Figure 6. Comparison of measured and computed radial velocities for F_L = 0.015.

Illustrations of the bubble tracks and of the gas holdup for 2mm bubbles are shown in Figures 7 and 8. It is clearly shown that the bubbles prefer to rise along the upstream sides of the baffles. The main differences in simulations with 1mm and 6mm bubbles was that the 2mm bubbles did not recirculate to the lower region of the vessel below the impeller and they escaped more rapidly from the liquid surface. Differences in effects on liquid velocities were slight.

Effects of bubbles on liquid velocities at F_l = 0.15 (10 times the former) were substantial. No experimental measurements have yet been successful at that gassing rate, but the simulations show much stronger effects of the bubbles on vertical velocities near the wall and movement downward and toward the axis of the vessel of the center of rotation of the upper circulation pattern. The strength of the upward velocities is shown in Figures 5 and 9. Figure 5 shows the substantial increase in upward velocities near the wall and Figure 9 shows the effect by the region enclosed by a surface of constant axial velocity of 0.5 m/s (blade tip velocity is 1.46 m/s) near the wall and also from the sparge ring up to the impeller. The case with no gas bubbles showed almost no region with vertical velocities above 0.5 m/s. As illustrated by comparison of Figures 8 and 10, the volume fraction of bubbles at F_l = 0.15 is approximately times that at the lower gas rate of F_l = 0.015. The average gas bubble holdup fractions for these simulations are approximately 0.016 and 0.065. This relationship nearly agrees with previous average gas holdup data (see de Figueiredo and Calderbank, 1979), which yielded holdup approximately proportional to the superficial gas velocity to the power 0.75. de Figueiredo and Calderbank's correlation is H = $0.34(P/V)^{0.25}\ (V_s)^{0.75}$, which yields 0.012 and 0.056 gas holdup fractions at F_l = 0.015 and 0.15, respectively, and 300 rpm.

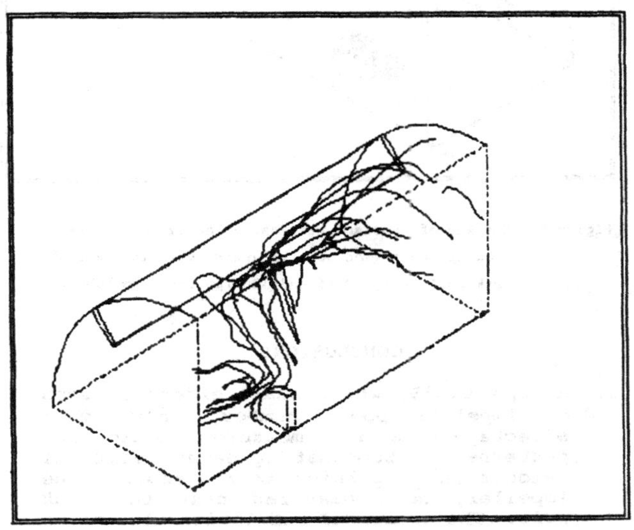

Figure 7. Bubble tracks at F_1 = 0.015 and 300 rpm for 2 mm bubbles. Bubbles escape at the far end.

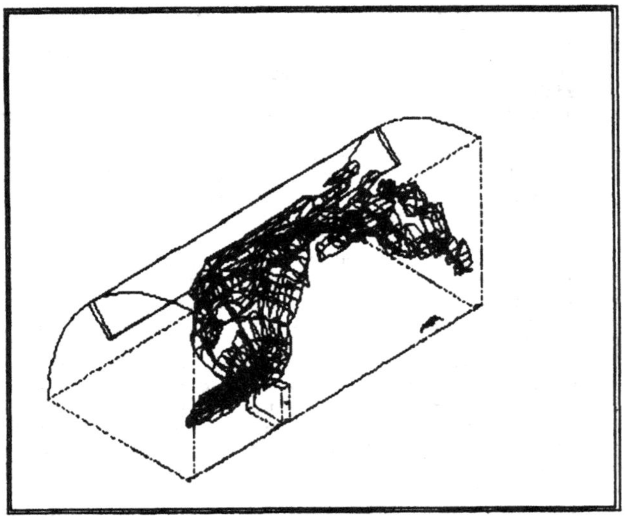

Figure 8. Surface of constant gas fraction of 1.6% by volume for F_1= 0.015 and 300 rpm for 2 mm bubbles.

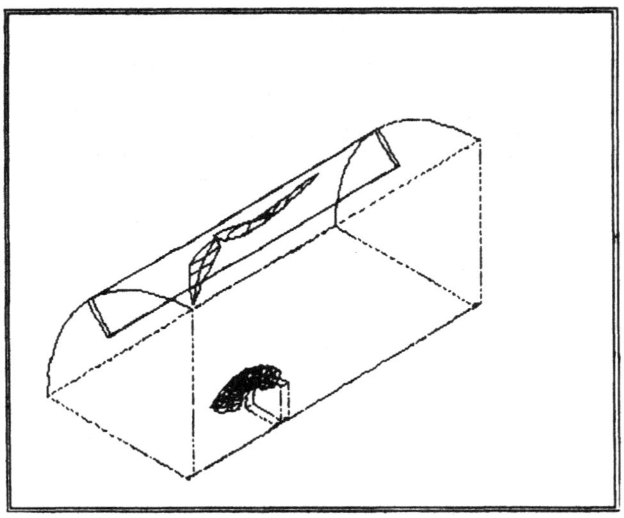

Figure 9. Surface of constant vertical velocity of 0.5 m/s, showing the region of high upward velocity near the tank wall at F_1= 0.15 and 300 rpm for 2 mm bubbles.

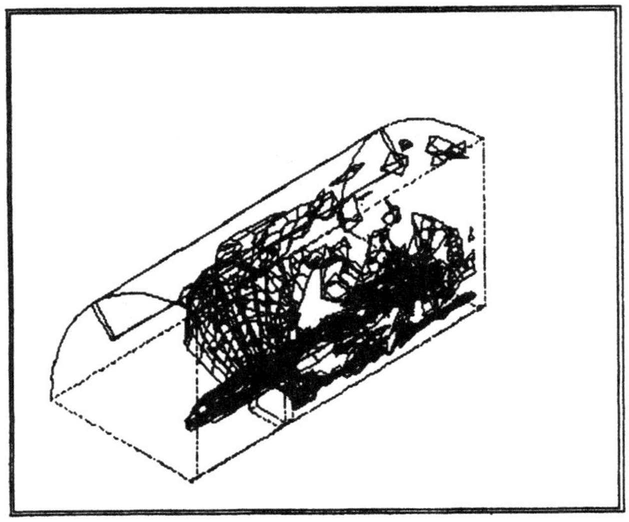

Figure 10. Surface of constant gas fraction of 6.5% by volume for F_1 = 0.15 and 300 rpm for 2 mm bubbles.

CONCLUSIONS

1. At F_1 = 0.015, where small effect is seen on impeller power number, only small effects are seen on measured liquid flow patterns. Somewhat greater vertical velocities, particularly above the impeller, were measured near the tank wall. The numerical simulation also shows these vertical velocity changes near the wall.

2. Significant flow pattern differences are shown by the numerical simulation for F_1 = 0.15 as compared to the flow pattern with no gas injection. This would be expected since the power number is significantly reduced.

3. The laser Doppler anemometer with a counter processor and no special alterations is capable of measuring liquid velocities in the presence of small bubbles at F_1 = 0.015, but at F_1 = 0.15 too much scattering from the bubbles occurs. Improved methods must be developed for higher, more practical gas rates.

4. The simulations clearly show the strong tendency for bubbles to channel along the upstream sides of the baffles as they rise toward the liquid surface. This tendency becomes stronger at high gas flow rates. This effect can be seen experimentally if observed carefully, particularly at the liquid surface.

5. Numerical simulation of bubble dynamics in stirred tanks appears to be a viable option in the study of scale up and geometric effects on such vessels. The same is most likely true for studies of suspended solids. At present bubble simulations are limited to predetermined sizes, since dependence on impeller power, gas rate, and liquid type are not yet included in the model.

6. The gas holdup fractions predicted by the simulations are in the correct order of magnitude.

7. Preliminary tests on simulation of local transport rates from bubbles to liquid in stirred tanks show promising results for use in studying scaleup and geometric effects.

REFERENCES

Calderbank, P.H., Trans. Instn. Chem. Eng., 36, 443 (1958).

de Figueiredo, M.M.L., and P.H. Calderbank, Chem. Eng. Sci., 34, 1333 (1979).

Greaves, M., and M. Barigou, in Proc. 6th Eur. Conf. on Mixing, Pavia, Italy, May 1988, BHRA, Cranfield, England, p. 313 (1988).

Hughmark, G., Ind. Eng. Chem., Proc. Des. Dev., 19 638 (1980).

Joshi, J.B., A.B. Pandit, and M.M. Sharma, Chem. Eng. Sci., 37, 813 (1982).

Morsi, S.A., and A.J. Alexander, J. Fluid Mech., 55, 193 (1972).

Nienow, A.W., D.J. Wisdom, and J.C. Middleton, in Proc. 2nd Eur. Conf. on Mixing, Cambridge, U.K., April 1977, BHRA, Cranfield, England, p. F1-1 (1978).

Nienow, A.W., M. Konno, and W. Bujalski, Paper 2 in Proc. 5th Eur. on Mixing, Würzburg, Germany, June 1985, BHRA, Cranfield, England, p.1 (1985a).

Nienow, A.W., M.M.C.G. Warmeskerken, J.M. Smith, and M. Konno, in Proc. 5th Eur. Conf. on Mixing, Würzburg, Germany, June 1985, BHRA, Cranfield, England, p. 143 (1985b).

Patterson, G.K., Chapter 21 in Turbulent Flows: The Practical Use of Theory, by S.W. Churchill, Butterworths (1991).

Shumpe, A., and W.D. Deckwer, Chem. Eng. Sci., 35, 2221 (1980).

Smith, J.M., Chapter 5 in Mixing of Liquids by Mechanical Agitation, ed. by J.J. Ulbrecht and G.K. Patterson, Gordon and Breach, p. 139 (1985).

Soo, S.L., Fluid Dynamics of Multiphase Systems, Chapter 6, Blaisdell, p. 248 (1967).

Takahashi, K., W.J. McNanamey, and A.W. Nienow, Actes Colloque "Agitation Mecanique", Toulouse, France, June 1986.

Wu, H., and G.K. Patterson, Chem. Eng. Sci., 44, 2207 (1989).

NOMENCLATURE

g_i	acceleration of gravity in i-direction
d_b	bubble diameter
$\rho_l = \rho_f$	density of liquid
C_D	drag coefficient
K_1, K_2, K_3	empirical constants
$K_l a$	mass transfer rate per unit volume
F_l	gas flow number (G/nD^3)
H	gas holdup fraction
G	gas volumetric flow rate
z	height above impeller disk plane
w	impeller blade width
D	impeller diameter
$P_o = P$	impeller power with no gas
P_G	impeller power with sparged gas
n	impeller rotation rate (revolutions/time)
μ	liquid velocity
u_i	liquid velocity in i-direction
ρ_p	particle density
d_p	particle diameter
$u_{i,p}$	particle velocity in i-direction
v_f	resultant liquid velocity
v_p	resultant particle velocity
R_e	Reynolds number
σ	surface tension
V_s	velocity
V	vessel volume

SIMPLE PERFORMANCE CORRELATIONS FOR AGITATED VESSELS

J.M. SMITH, Dept. of Chemical and Process Engineering, University of Surrey, Guildford, GB

ABSTRACT

This paper considers simplified correlating equations for preliminary design of process equipment. The uncertainties associated with process change are much greater than inaccuracies resulting from the streamlining of the original equations.

Predictions are presented for gas hold-up and mass transfer rates in aqueous aerated mixtures which are independent of process scale. Experimental data shows that these simple dimensionless correlations can accommodate a wide range of process scales and many impeller and tank geometries.

INTRODUCTION.

The spectrum of gas liquid processes sets daunting requirements on those concerned with design specification and performance prediction. During recent years our understanding of these systems has increased very greatly, almost to the extent that the complexities threaten to mask any possible underlying simplicity. The infinite variety of process conditions has tended to make designers and operators suspicious of general rules - the every case is different syndrome - to the extent that work carried out say with air and water - is dismissed as generally irrelevant to process needs. At the same time those concerned with computational fluid mechanics are quite pleased to assume that their programs will solve flow fields - for air-water or any other species. It must be allowed that some processes - notably those associated with coalescence phenomena and hence a whole group of transfer performance indices - are not yet at the point of rigorous solution. Nevertheless, outside this group of phenomena there is real progress to be made.

The search for reliable predictive and scaling rules for process equipment has led to the development of a multitude of correlating relationships. Sometimes these have been dimensionally inconsistent whilst in other cases an unrealised commonality in the correlated variables has misleadingly enhanced apparent value of the method presented. (Michel and Miller (1962), Nienow (1977)) The present paper demonstrates that dimensionless correlating equations can be used to establish the hydrodynamic conditions and on this basis to predict useful performance data. Knowledge of the relative importance of any differences between conditions used in establishing the correlation and those in the operating plant allows sensible assessment of both the value and limitations of the proposed equations.

Much of the literature relating to gas-liquid operations has been established in academic studies carried out in "standard" geometry equipment at small scale. This limitation is perceived by industry as being significant and severe. Unfortunately it is both expensive and time consuming to carry out scientific investigations in process scale equipment: there is a need to establish guidelines for the use of small scale data to predict performance in large scale applications, including some assessment of validity and reliability. Ideally the extrapolation procedures should allow extension to somewhat modified conditions, e.g. to reactors using multiple impellers on a single shaft.

Six blade Rushton impellers have been the usual choice for gas dispersion and mass transfer, while pitched blade impellers have featured for solids suspension. In recent years many publications have dealt with a succession of profiled impellers - hydrofoils and their derivatives - which have been particularly successful in applications requiring efficient pumping of large volumes of liquid, e.g. heat transfer, large scale liquid blending and solids suspension. The widely distributed and low power dissipation of these impellers makes them rather suitable for liquid- liquid dispersion; one would expect a rather more uniform drop size distribution than can be achieved with a disc turbine for example.

Although this paper is presented as relevant to Rushton turbines, the principles are more general.

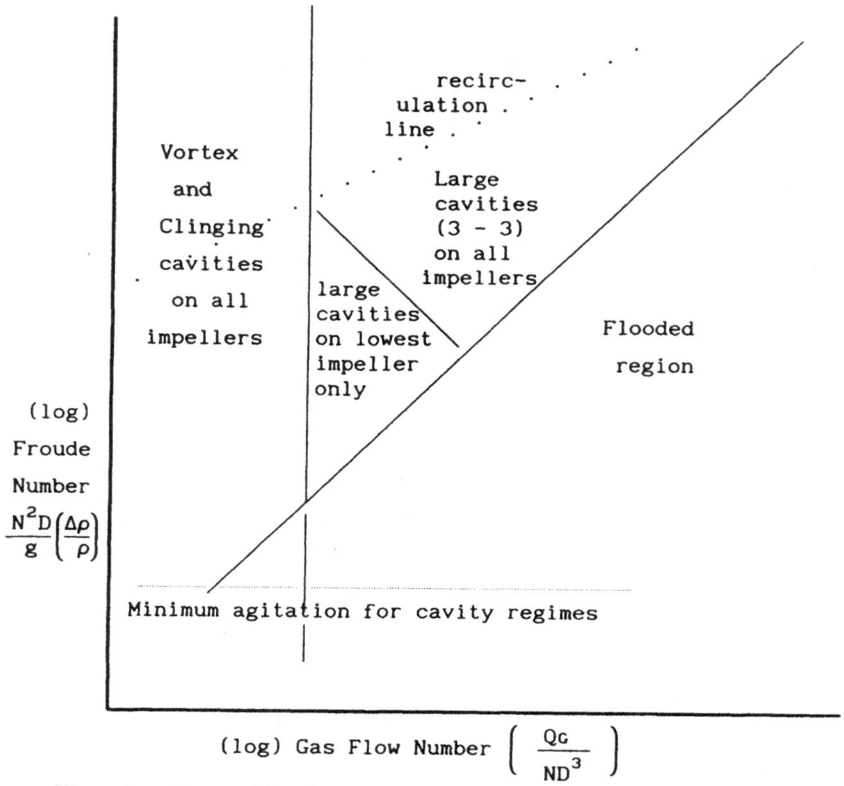

Fig. 1. Generalised Flow Map for Rushton Turbines.

Note. As will be evident from the equations presented by Smith et al. the precise positions of these lines depends on the geometry of the vessel. The minimum agitation line is at a Froude number of about 0.03, the cavity formation line at a gas flow number of about 0.03, the flooding line has a slope of exactly 1.0, corresponding approximately to Fl = 1.2*Fr. The complete dispersion line has a slope of exactly 0.50 and crosses the cavity development line at a Froude number of about 0.5.

THE ROLE OF FLOW MAPS

It is recognised that impellers operating in gas liquid systems develop local flow regimes which depend on the conditions in ways which can be quantified in terms of key dimensionless groups, Smith and Warmoeskerken (1985). Conditions in the immediate vicinity of the impeller control the energy transfer from the shaft to the fluid, both in terms of the (directional) pumping and its (pseudo) random turbulent motion. The near field processes include the regions of most rapid micromixing, and local phase dispersion. The bulk convective flows generated by the impeller determine the conditions further away, notably possible recirculation of dispersed phase to the impeller, macroscopic mixing time, solid suspension or drawdown, and heat transfer to containing surfaces or internals.

Second phases do not always affect the flows significantly: moderate amounts of solid or a second dispersed liquid phase are of minor influence whereas the effects of dispersed gases cannot be ignored.

Flow Maps.

Gas liquid flow maps provide a very useful tool for establishing the effects. With a Rushton turbine, the impeller flow regime may involve vortex-clinging cavities, large cavities or be flooded. The vast majority of industrial scale operations take place in the large cavity regime, in which alternate larger and smaller cavities are held captive after successive impeller blades. The flow map, Fig 1, which allows prediction of hydrodynamic conditions in vessels agitated with either single or multiple Rushton impeller, delineates the main areas of operation, (Smith et al., 1987). The transitions are defined by dimensionless criteria.
The straight lines on this graph correspond to:
vertical - the constant flow number at which large cavites develop,
horizontal - the minimum agitation at which cavities adhere to the impeller blades,
slope 1.0 - the flooding to loading transition,

slope 0.50 - the recirculation line above which dispersion is complete and virtually uniform throughout the vessel.

Comparisons between correlating equations used for single and multiple impeller systems give information about differing conditions at various locations within the vessel. The flow map shows that conditions in the upper part of the vessel differ significantly from those near the base.

with great accuracy. The sporadic overflows occasioned by waves in the free liquid surface were easily accommodated by this system. The measurements differ slightly from those conventionally obtained from changes in the free liquid surface level in that the volume of liquid in the vessel is reduced slightly under the highest conditions of gas retention: this will only be significant at very large void fractions, say above 10%.

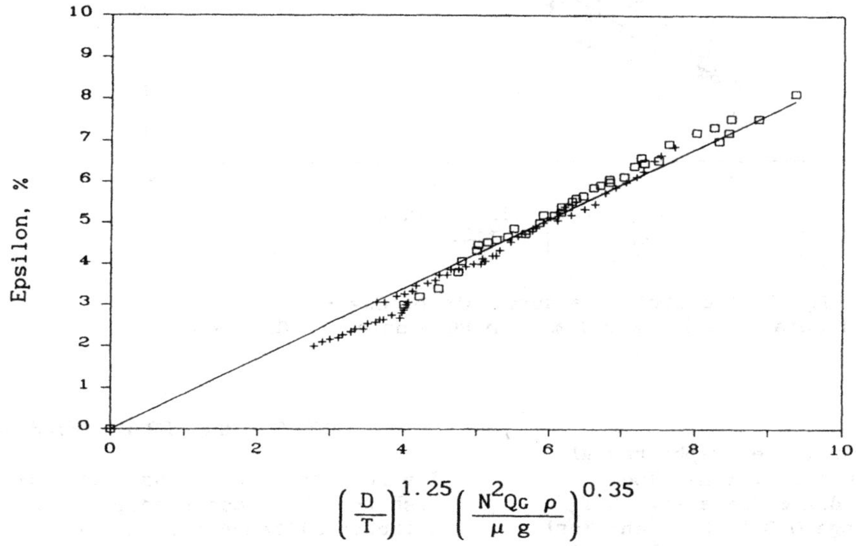

Fig. 2. Void fraction in standard vessels, Delft data.
Tank diameter + T = 0.44 m □ T = 1.20 m tank.

CORRELATION OF GAS FRACTION RESULTS.

Experimental determination of the gas retention in agitated vessels is complicated by the inherent instability of the surface. The simplest method is to measure the change in liquid level during operation though this may be difficult because of the constantly fluctuating free liquid surface. Subjectiveness can be reduced by detecting the position of the free liquid surface electronically. Various methods have been used for this, including ultrasonic, capacitance and conductivity probes, with a variety of traversing and remote sensing arrangements. Unfortunately these methods are often affected by the presence of even small quantities of foam on the liquid surface and there is a real and persistent difficulty in deciding the most representative position(s) on the liquid surface for the determination of the change in level. The measurements that are reported here utilise a slightly different method; the liquid in the vessel was allowed to overflow through a funnel to a measuring cylinder from which it was pumped back at a steady rate to the vessel. The change in the level in the measuring vessel could be very accurately determined. The steady flow conditions give an unvarying holdup in the return pump system and so that the volume of gas retained in the vessel could be measured

Void Fraction: Single Impellers in Standard Tanks.

In good agreement with Hassan and Robinson (1977), it was found that the gas fraction correlates linearly with the volumetric gas rate and the square of the impeller speed. This group occurs in the product (Flc•Fr•Re), which, like the Flow Map, is independent of scale. Results for two scales of equipment (T = 0.44 m and T = 0.61 m) are shown in Fig. 2.

Almost all the data points represent conditions in which the single Rushton impeller is operating in the large cavity regime; the condition of interest for most industrial operations. The lack of scatter in these data is remarkable. The correlating equation has the form

$$\varepsilon_G = \text{const} \bullet (\text{Re}\bullet\text{Fr}\bullet\text{Fl}_G)^{0.35} \qquad (1)$$

The independence of scale is confirmed in data which has kindly been made available by BHRA and ICI. The ICI data, (measured by Middleton), and BHRG (Fluid Mixing Processes consortium, measured by Muskett) are shown in Fig. 3. against the same correlating line used in Fig. 2.

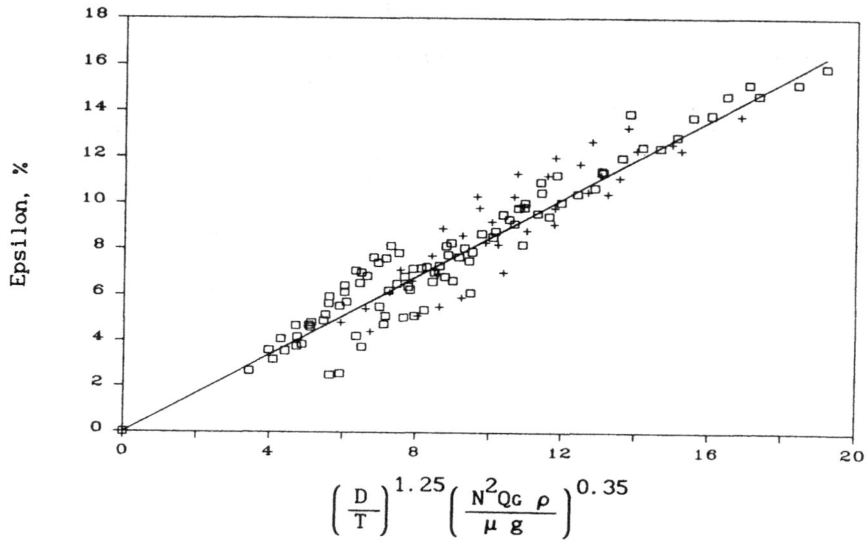

Fig. 3. Industrial measurements of Gas Fraction
\+ ICI data T = 1.8 & 2.1 m □ BHRG data T = 0.6, 1.6 & 2.7 m

The data represented in these graphs has all been obtained in air-water systems. The largest equipment was 4.2 m diameter. D/T ratios were in the range 0.3 to 0.5, and most experiments were carried out with $H/T \simeq 1$ in standard tanks. In the multiple impeller results referred to below, the vessels up to 1m in diameter, with $H/T \simeq 3$ and impeller separations $\geq$ D. Bottom clearances, C/T, were usually 0.25 or 0.33.

TANK TO IMPELLER DIAMETER RATIO.

The industrial data has been used to establish the (D/T) dependence to the power 1.25. Although although the data show rather greater scatter than that from the Delft laboratory study, the good fit is extended to much higher gas fractions. It is likely that in a coalescing system a transition at a gas fraction above perhaps 25% there will be a transition in flow regime to something corresponding to the churn turbulent regime in a vertical pipeline.

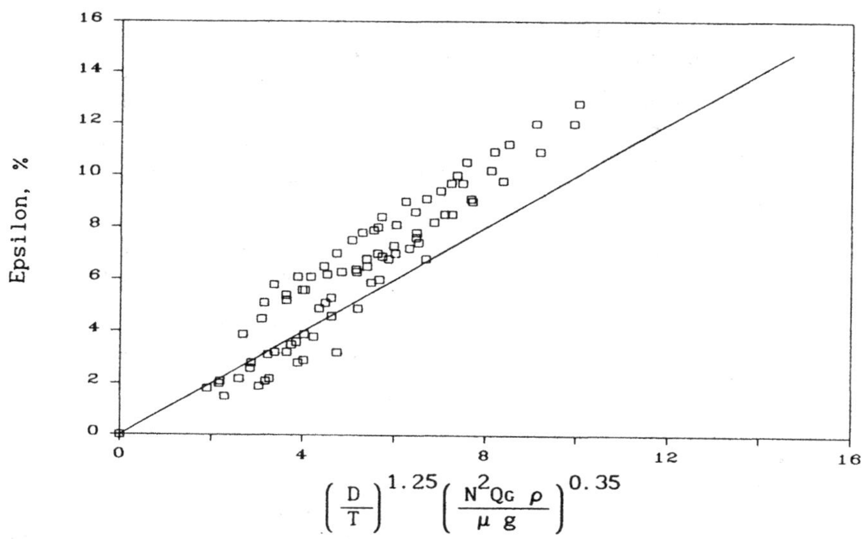

Fig. 4. Data from Greaves and Barigou
T = 1 m; D = 0.25, 0.33 & 0.5 m

Although the sensitivity to this exponent is low, and the trends are consistent, the data of Greaves and Barigou (1987), obtained with various impellers in a 1m diameter tank, do not fit well with the scale independent generalised result. There is no reason to doubt the accuracy of their extensive experimental results which were obtained with rapidly coalescing distilled water.

All their data were obtained in a 1 m diameter flat bottomed tank with a base to impeller clearance of 0.25T. For all three impeller diameter ratios used, their results show good initial agreement with the proposed correlation which is shown in Fig. 4. It is only as gas fraction increases that the differences become large, amounting to an additional 3% or so voidage. Although the largest impeller (D/T ratio = 0.50) seems to show a somewhat greater divergence than the other two sizes, the performance is reasonably consistent. Statistical analysis of their results, and imposing the requirement of an independence of scale, gives an exponent of almost exactly 2 on (D/T). However this worsens the correlation using (N^2Q) which has been justified by the large mass of other data, leaving 1.25 as the best compromise.

The resulting correlation then brings together

$$\varepsilon_G = \text{Const} \bullet \left(\frac{N^2 Q_G}{g\mu} \right)^{0.35} (D/T)^{1.25} \qquad (2)$$

In dimensionless terms this can be written as

$$\varepsilon_G = 0.85\ (\text{Re}\bullet\text{Fr}\bullet\text{Flc})^{0.35}\ (D/T)^{1.25} \qquad (3)$$

MULTIPLE IMPELLER GEOMETRIES.

The recycled overflow technique was has been used to measure the holdup in a vessel with triple Rushton turbines (Zeef (1985) and Hogervorst (1987)). Fig. 5. shows their results plotted against the same dimensionless grouping.

It is evident that it is not possible to fit a straight line through the origin to these multiple impeller data, with lower measured gas fractions at low gas rates than a linear correlation would lead us to expect. A line fitted through the origin to the higher gas fraction data has a significantly greater slope (1.25) than the standard tank results (0.85).

The flow regimes in multiple impeller vessels are significantly different from those in standard vessels. The flow map reflects the reduced gas loading on the upper impellers. The data which deviate at low gas fractions correspond to regimes in which the upper impellers are in the vortex cavity regime.

Again additional industrial results are available to confirm and extend the suggested correlation. Fig. 6 includes data from BHRg and ICI. One remarkable point about these data is that the configurations used were not all the same: some equipment consisted of triple disc turbines while in other experiments combinations of a dispersing disc turbine with profiled impellers above.

The full set of data available is shown in Fig. 7.

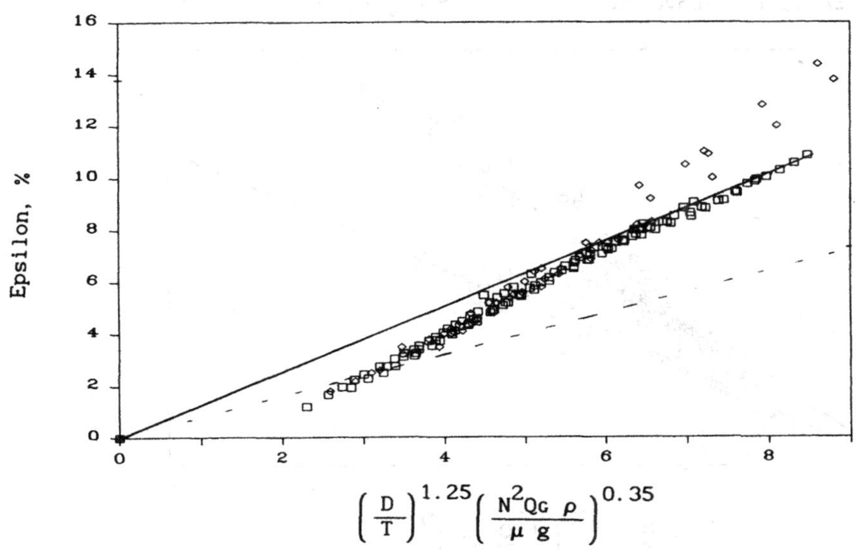

Fig. 5. Multiple Impeller Gas Void Fraction, Delft Data.
Triple Rushtons, T = 0.64 m D = 0.25 m.
◇ Zeef □ Hogervorst

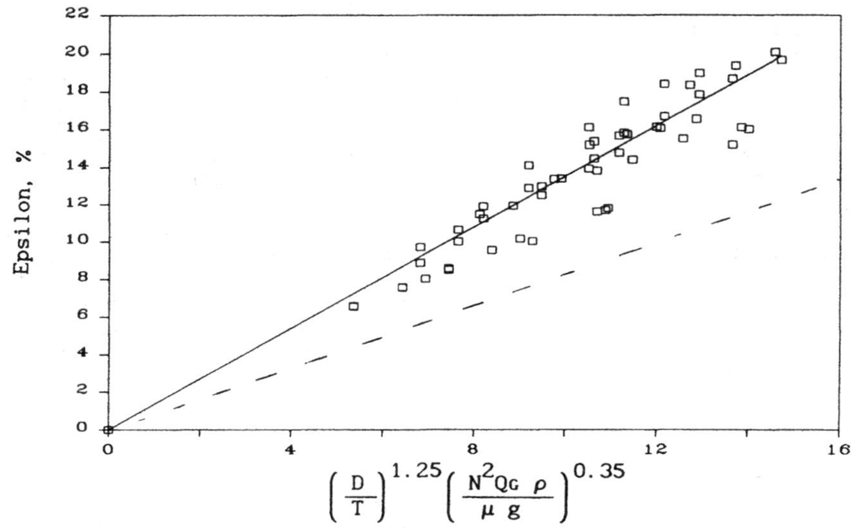

Fig. 6. Industrial Void Fraction Data,
Various Combinations of Triple Impellers, T = 0.64 m and 0.95 m.

Implications of higher voidage in Multiple Impeller vessels.

The ratio between the constants of the correlating equations (0.85 versus 1.25) implies that there is a very considerable deference in the local holdup above and below the impeller plane, at the very least in a single impeller vessel and possibly also above and below upper impellers, though this is less likely in view of the very different way in which gas reaches the upper impellers.

If we assume the simplest possible configuration of uniform gas distribution above the plane of the lowest impeller with three pairs of compartments of equal volume, centered on the impellers, a simple volume balance implies that the gas fraction below the lowest impeller is only about 15% of that in the upper reaches of the tank.

PHYSICAL PROPERTIES.

Although the analysis presented in favour of (Re•Fr•Flc), may appear convincing, important questions remain. The dependence on $N^2Q\bullet\left(\frac{\rho_L}{g \bullet \mu_L}\right)$ implies a mutual relationship

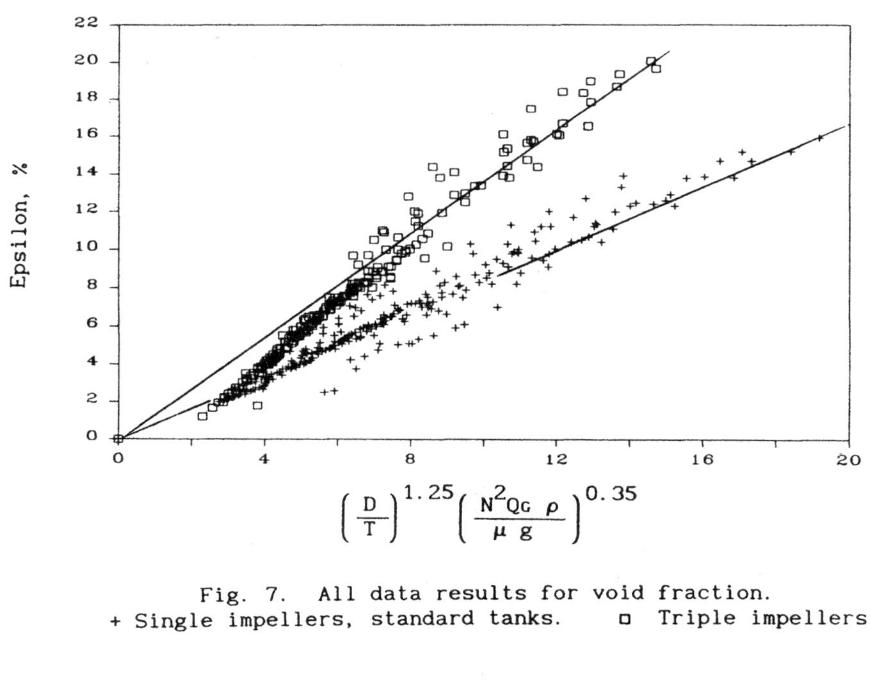

Fig. 7. All data results for void fraction.
+ Single impellers, standard tanks. □ Triple impellers

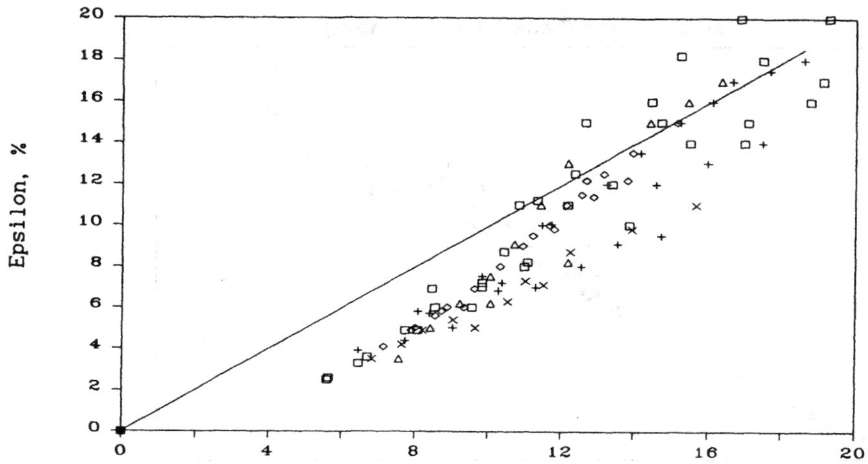

$$\left(\frac{D}{T}\right)^{1.25}\left(\frac{N^2Q_G}{\sigma}\right)^{0.35}\left(\frac{\rho^2\mu}{g^2}\right)^{\frac{1}{3}}$$

Fig. 8. Data of Hassan and Robinson in water, glycol and glycerol
□ Water, T = 0.15 m + Water, T = 0.29 m ◆ 8% Glycerol, T = 0.29 m
Δ 20% Glycerol, T = 0.29 m × 40 % Glycerol, T = 0.29 m

between density and viscosity which has not been justified by experiment. Useful guidance is provided by the data reported by Hassan and Robinson, (1977). Their paper studied coalescence affecting properties, notably the surface tension of the liquid phase. They found that the (dimensional) group (N^2Q/σ) was effective in correlating their results. If the surface tension is to be included in the correlation then dimensional considerations lead us to look for a correlation involving the terms :

$$\varepsilon_G \propto \left(\frac{N^2Q}{\sigma^x}\right)\left(\rho^{(1-x/3)}\mu^{(4x/3-1)}g^{(x/3-1)}\right) \quad (4)$$

With a value on x of 1.0 this reduces to:

$$\varepsilon_G \propto \left(\frac{N^2Q}{\sigma}\right)\cdot\left(\frac{\rho^2\mu}{g^2}\right)^{\frac{1}{3}} \quad (5)$$

which implies quite different interdependence of the variables in the last group to that which follws from Equ. (3). Unfortunately the data of Hassan and Robinson, while following the same general trends, diverge significantly from the other data presented in this paper. Many of their results were obtained in a very small vessel and involved extreme void fractions - in excess of 20 %. However using the other physical properties in the examination of their data does improve the consistency of their results, Fig. 8.

Even though the individual experimental data do not fit the proposed correlation quantitatively, the trends are unambiguous and the differences amount to perhaps 30% in the appropriate constants.

Henzler and Kauling (1985) included surface tension and other fluid properties in their consideration of mass transfer performance in viscous liquids. The real role of surface tension is as a coalescence rate parameter, and in that light the dynamic surface tension and surface elasticity are certainly as important as the static values used by Hassan and Robinson.

Given the sensitivity of the coalescence mechanism to small degrees of contamination it is worthwhile considering the present approach as being as likely as any other to offer the key to the useful description of the physical process.

MASS TRANSFER.

The correlation of mass transfer performance is much more complex than that of gas fraction. Apart from anything else, k_La is not dimensionless and k_La/N has an unfortunate dependence on the particular operating conditions. It is suggested that the group $k_La(D/g)^{0.5}$, derived by multiplying k_La/N by the square root of the Froude Number, would be useful, since this will be characteristic of the equipment itself rather than its mode of operation.
Using various mass transfer results from Delft, BHRg and ICI in vessels in the range 0.44 m to 2.4 m diameter, and basing the mass transfer calculation on well mixed liquid with a plug flow for the gas phase, a usable correlation emerges, Fig. 9. The line, which is within an rms deviation of 8% of the data points, has the equation:

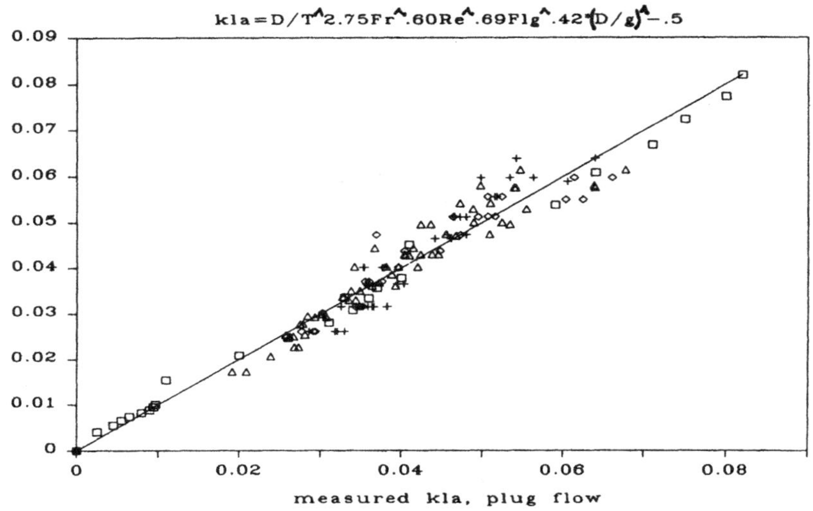

Fig. 9. Mass transfer performance in dimensionless terms.
□ Delft, (hollow blade) T = 0.44 m △ BHRA (Rushton) T = 0.6
\+ BHRG (imp type 1) T = 2.6 ◇ BHRA (imp. type 2) T = 2.6

$$k_L a = 1.25 \cdot 10^{-4} \left(\frac{D}{T}\right)^{2.8} Fr^{0.6} Re^{0.7} Flc^{0.45} \left(\frac{D}{g}\right)^{-0.5} \qquad (6)$$

Although the exponents on the correlating equation are not very sensitive, it has not been possible to reduce this grouping to a diameter independent set of variables as described the void fraction. This data was all derived from physical oxygen mass transfer experiments in clean air-water systems. It is appropriate to repeat the warning associated with the void fraction results above. Almost certainly surface tension (or some physical property related to it), ought to be included. If this is done then the interdependence of density and viscosity will, on the physical grounds of dimensionality, be different from that implied in equation.

CONCLUSIONS.

This paper has considered the possibilities of basing agitated vessel performance correlations on dimensionless groups. It has pointed out that in common with many multi variable optimizations the sensitivity to the parameter exponents is rather low, and that simplified rules are often worthwhile. For the correlation of gas holdup in agitated vessels operating with aqueous aerated mixtures the equation

$$\varepsilon_G = 0.85\ (Re \cdot Fr \cdot Flc)^{0.35}\ (D/T)^{1.25}$$

is sufficiently accurate for many purposes. The constant for multiple (triple) impeller equipment is 1.25. This leads us to the conclusion that the relative gas retention below the lowest impeller is probably not more than 20% of that in the upper regions.

Acknowledgements.

The author is particularly grateful to S. Ruszkowski of BHR Group, Cranfield, and J. Middleton of ICI P&C, Runcorn, for making it possible for results obtained in their organisations to be quoted in this paper. Acknowledgement is also due to R. Hogervorst, E. Zeef, and J Speur for their contributions to the experimental programme at T.U. Delft.

References.

Greaves, M. and Barigou, M.,
Gas Holdup and Impeller Power in a stirred vessel reactor
Fluid Mixing III., I.Chem.E. Symp. Ser. 108, 235 - 255, (1987)

Hassan, I.T.M. and Robinson, C.W.,
Stirred tank power requirement and gas holdup in aerated aqueous phases.
A.I.Ch.E.Jnl, 23, 48 - 56, (1977)

Henzler, H.J. and Kauling, J.,
Scale up of mass transfer in highly viscous liquids.
Fifth Eur. Conf. Mixing, 303 - 312 (1985) BHRA.

Hogervorst, R.
Engineer's Degree Report.
Kramers Laboratory for Physical Technology, T.U. Delft, 1987

Michel, B.J and Miller, S.A.,
Power requirements of gas-liquid agitated systems, A.I.Ch.E.Jnl., 8, 262 - 266, 1962.

Nienow A.W., Wisdom, D.J. and Middleton, J.C.
Effect of Scale and Geometry on Flooding, Recirculation and Power in Gassed Vessels.
Second. Eur Conf. Mixing, F1-1 -11, (1977) BHRA.

Smith, J.M. and Warmoeskerken, M.M.C.G.,
The dispersion of gases in liquids with turbines,
Fifth. Eur Conf. Mixing, 115 - 126. (1985) BHRA.

Smith, J.M., Warmoeskerken, M.M.C.G., and Zeef, E.
Flow conditions in vessels dispersing gases in liquids with multiple impellers.
Biotechnology Processes
eds Ho, C.S. and Oldshue, J.Y.
A.I.Ch.E., pp 107 - 115, (1987)

Zeef, E.
Engineer's Degree Report.
Kramers Laboratory for Physical Technology,
T.U. Delft, 1985

Notation

D	Impeller diameter	m
g	Gravitational acceleration	$m\ s^{-2}$
$k_L a$	Mass transfer factor	s^{-1}
N	Impeller Speed	s^{-1}
Q_G	Volumetric gas flow rate	$m^3\ s^{-1}$
T	Tank diameter	m
ε_G	Gas void fraction, percent	-
μ	Liquid viscosity	Pa s
ρ	Liquid density	$kg\ m^{-3}$
σ	Surface tension	$N\ m^{-1}$

Dimensionless Groups

Re $\left(\frac{N D^2 \rho}{\mu}\right)$ Fr $\left(\frac{N^2 D}{g}\right)$

Flg $\left(\frac{Q_G}{N D^3}\right)$ We $\left(\frac{\sigma}{\rho N^2 D^3}\right)$

GRADIENTS IN PRODUCTION SCALE BIOREACTORS

J. FEIJEN, J.J. HOFMEESTER, Gist-Brocades, Delft, NL

Abstract

In industrial fermenters micro-organisms are continuously exposed to a changing environment. These environmental changes may lead to a loss of viability, a reduction of the biomass and product formation and an increase in by-product formation.
In this study the critical gradients in an industrial bakers' yeast fermentation are determined. The results show that substrate mixing, oxygen transfer and oxygen consumption are the most critical factors in the process. Both substrate mixing and oxygen transfer were studied in a 150 m^3 fermentor. Advantages and disadvantages of the various measuring methods are discussed. The results of the experiments show that the physiological condition of the micro-organisms is dependent on their location in the fermentor.

Introduction

In most chemical processes mixing is the main cause of scale-up problems. Some researchers even state that in scaling-up there are no problems other than mixing problems.
In general in fermentation processes this statement is even more valid than in chemical processes. This is because in this type of process, environmental changes can lead to an irreversible loss of viability of the organisms.

In this paper some general scale-up aspects of fermentation processes are dealt with. The influence of these scale up phenomena on an actual large scale bakers' yeast process is also studied.

In fig. 1 the production process of bakers' yeast is given. The process consists of a fermentation step, a separation step and a drying step. Molasses and nutrients are fed into the fermentor. The mixing and aeration of the fermentor is done by sparging air into the broth. In some fermentors aeration and mixing is done by a combination of air supply and an impeller. In yeast fermentations both agitated and non-agitated fermentors are applied.

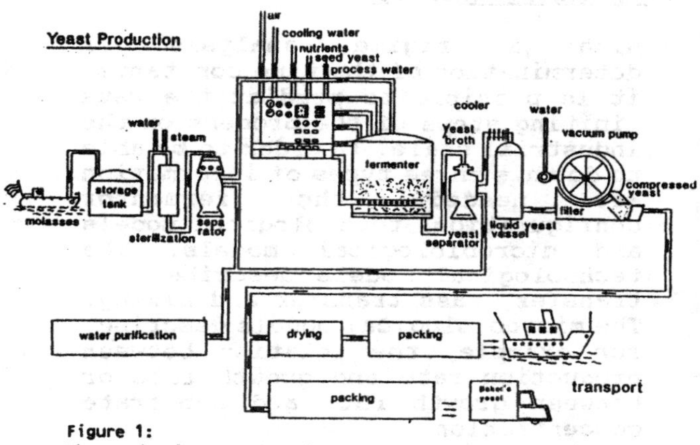

Figure 1:
The production process of yeast.

In general the quality of the final product (fresh yeast) is influenced by the process conditions in all the unit operations of the process.

The development and optimization of production scale processes can be done in two ways:
- the first is optimization through experiments in representative laboratory or pilot plant scale equipment,
- the second is (empirical) optimization in the production plant.

The problem with process optimization on production scale is that if a process change has a negative effect on the product quality, the experiments can be very expensive.
In a bakers' yeast production process it can be possible that the quality of the yeast is such that the product cannot be sold anymore. As this is the case in most fermentation processes many researchers have looked for ways to scale down production scale processes. The scale down - scale up approach has been schematized by Oosterhuis and Kossen (fig. 2).

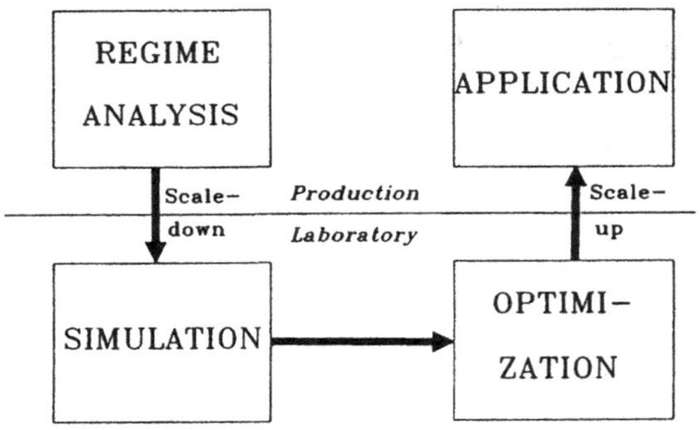

Figure 2:
Schematic view on a scale down - scale up approach after Oosterhuis and Kossen [4].

- 1 Regime analysis

With a regime analysis (a determination of the time constants) it is possible to predict the rate limiting steps of the process on the industrial scale. In order to be able to do this three types of information are needed: the fermentor configuration, technological models and microbiological models. The technological models describe heat transfer, mass transfer and mixing. The microbiological models describe, for example, the relation between production rate and growth rate or between growth rate and substrate concentration.
As most of the technological and microbiological models are empirical, they are often not well understood. Therefore one should only use these models in order to determine the order of magnitude of the characteristic times of the various process phenomena only. In addition most technological models have the disadvantage that they are derived from laboratory scale experiments.
In most cases this makes them unsuitable for use on industrial scale.

Mechanism	Definition	Magnitude
Mixing		
- Liquid	L^2/D_l	$10-10^3$
- Gas	L^2/D_g	$.1-10^3$
- Liquid circulation	$2L/v$	10-100
O_2 transfer	$1/k_la$	1 -100
O_2 consumption	C_o/r_o	1 - 10
substr. consumption	C_s/r_s	10-100
substr. addition	$vC_s/(F_sC_{so})$	10-100
Growth	$1/\mu$	10^4
heat transfer.		10^3

Table 1:
Some characteristic times for a large scale fermentation.

In table 1 some characteristic times for a fermentation are given.
The rate limiting step(s) of the process can be determined by comparing the characteristic time for transport with the characteristic time for consumption.

When e.g. the characteristic time for substrate consumption is much larger than that of the transport of substrate, no large substrate gradients or substrate limitations are to be expected. Since in most fermentations the characteristic time for oxygen consumption is of the same order of magnitude as for oxygen transfer, oxygen limitations are possible in fermentations.

- 2 Simulation

Scale down of a production scale process can be done by simulating the rate limiting step(s) on laboratory scale. It must be stressed that it is not possible to scale down all phenomena of a production scale fermentation process.

One should also keep in mind that by scaling down the rate limiting step of a production scale process, other physical and microbiological phenomena of the process can be changed in such a way that the simulation is no longer correct.

- 3 Optimization

When the experimental results indicate that it is possible to simulate the production scale process on laboratory scale the next step is to study the influence of changes in the transport/consumption ratio of the limiting step of the process.

- 4 Application

When the results from the optimization phase indicate a possible improvement of the process the process, change has to be implemented and tested on production scale.

In the past many researchers have applied this scale up-scale down approach. The major part of the work that is published in literature is focused on the second and third step of this approach (Larsson [2], Sweere [7], Vardar [8]). Little or no quantitative information is available on the process phenomena in production scale fermentation processes (step 1 and 4).

The work reported here is focused on step 1. The work deals with the quantification of transport phenomena in a non-agitated production scale bakers' yeast fermentation (gross volume: 150 m^3).

Theory.

Various models are used to describe mixing in large scale fermentors. The three basic models are:
- A circulation model (mostly used for agitated fermentors).
- A compartment model (mostly used for multi-impeller systems).
- A dispersion model (mostly used for non-agitated fermentors).

The models are given in figure 3.

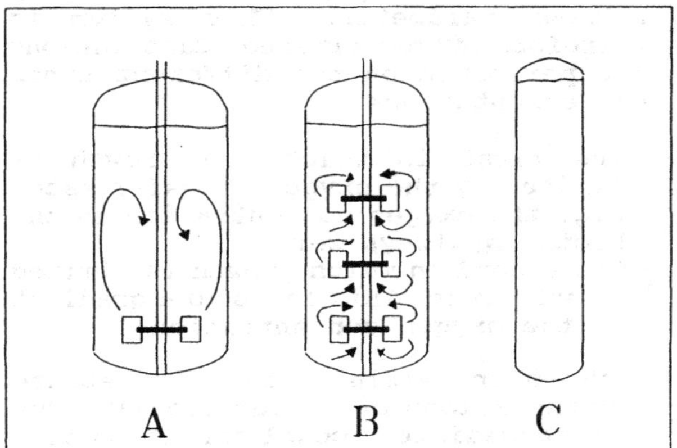

Figure 3:
Schematic drawing of the most common types of fermentors.

Many yeast fermentations at Gist-brocades are done in bubble column reactors. This research describes this type of bioreactor. Fermentations under both oxygen limited and substrate limited conditions are described.

The description of the models.

There are two kinds of models used in this paper. The first kind are technological models which describe the mixing in the fermentor. With this model and the balance equation it is possible to calculate the concentrations of different compounds in the various parts of the fermentor. A general representation of a balance equations is:

Change = flow + mixing + transfer + production. (1)

As already stated it is obvious to describe the hydrodynamics of non-agitated fermentors by a so called axial dispersion model. This is done by dividing the reactor in a number of small perfectly mixed horizontal slices.

The exchange of liquid between the slices is described by a linear flow term and a dispersion term (figure 4).

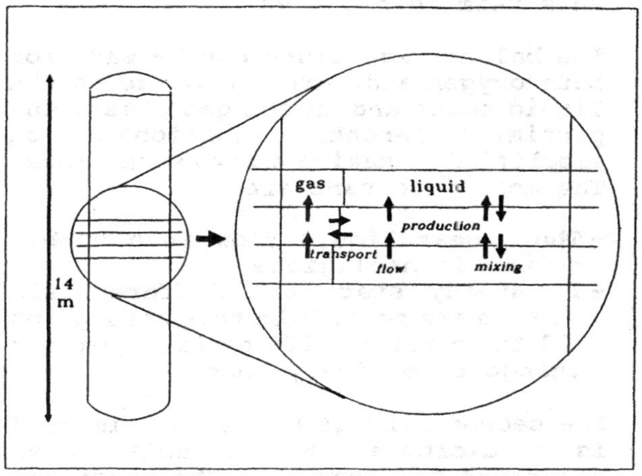

Figure 4:
Schematic representation of the axial dispersion model used in this paper.

This gives the following equation for the oxygen concentration in the gas phase in slice number i:

$$\frac{\partial C_{lo}}{\partial t} = \frac{\partial}{\partial h}(v_l(1-\epsilon)C_l) - \frac{\partial}{\partial h}\mathbb{D}_l\frac{\partial C_{lo}}{\partial h} + k_l a_o(C_{lo}^* - C_{lo}) + r_{lo}. \quad (2)$$

The first term of the right hand side of this equation gives the changes in the concentration as a result of the linear flow in the fermentor. In circulation loops this is a very important term. In the type of fermentor described here this term is negligible.

The second term in equation 2 gives the changes in concentration as a result of mixing. In this model mixing is described as an exchange of mass between two well mixed zones. The dispersion coefficient $\mathbb{D}_l$ gives the amount of volume which is exchanged over a unit of length per unit of time.

The third term gives the change of concentration due to transfer of mass between the gas and the liquid phase. In this term the factor $k_l a$ is called the transfer coefficient.
The factor $(C_{lo}^* - C_{lo})$ is the driving force for oxygen transfer. C_{lo}^* is the saturated oxygen concentration in the

broth. This is proportional to the relative oxygen concentration in the gas phase and the pressure. C_{lo} is the actual oxygen concentration in the liquid.

The last term gives the changes in concentration due to production or consumption. In case of consumption this term is negative.

The balance equations can be made for both oxygen and carbon dioxide in the liquid phase and in the gas phase. The partial differential equations can be simplified by making a few assumptions. The most important are:

- The linear liquid flow in a bubble column is negligible.
- In steady state all differentials of time are zero. With this assumption all the partial differential equations change to ordinary ones.

The second kind of model that is used is a microbial model. This model describes the growth of the yeast and the uptake rate of substrate. A standard way to describe this is Monod kinetics:

$$\frac{dC_{yeast}}{dt} = \mu_{max} * C_{yeast} * \frac{C_s}{C_s + K_s} \quad (3)$$

$$\frac{dC_s}{dt} = \frac{1}{Y_{yeast,s}} * \frac{dC_{yeast}}{dt} \quad (4)$$

Note that under substrate limited conditions the value of the Monod constant K_s has a large influence on substrate concentration and substrate mixing. High values of Ks lead to high substrate concentrations and therewith an increase in the amount of substrate which is mixed through the fermentor. The Monod constant is strain dependent and is difficult to determine.

By combining the two models it is possible to calculate the substrate concentration at various locations in the fermentor. This can be done by making balance equations for the oxygen, the carbon dioxide and the substrate in more or less the same way as in equation (2).

In order to solve the balance equations the following parameters are measured:

- The oxygen concentration in the gas and the liquid phase over the height of the fermentor.
- The carbon dioxide concentration in the gas and the liquid phase over the height of the fermentor.
- The pressure (related to gas hold-up) over the height of the fermentor.
- The mixing time in the liquid phase.

The following assumptions are made:

- The airflow through the fermentor is assumed to be a plug flow.
- The Carbon Dioxide Production Rate (CPR) is proportional to the Oxygen Uptake Rate (OUR). The ratio between OUR and CPR is called the respiratory quotient (RQ).

All this leads to a set of five differential equations with four unknown parameters. This system is therefore overdetermined. Calculations are performed on two different types of fermentations:

- The first in which the growth is limited by the oxygen transfer rate, e.g. the oxygen concentration in the broth is always zero.
- The second in which growth is limited by substrate. This leads to a gradient in the oxygen concentration.

With poor mixing, high substrate concentrations can occur locally. Due to intermediate alcohol formation this can lead to a reduced yield of biomass on substrate (see Sonnleitner & Käppeli [6]).

Measuring equipment

The experiments were done in a 150 m^2 non-agitated fermentor.

The dissolved oxygen concentration was measured by means of an Ingold DO probe. The dissolved carbon dioxide was measured with a membrane probe, which uses nitrogen as a carrier gas. Gas oxygen and carbon dioxide concentrations were measured with standard off-gas analysis equipment.

Samples of the gas at the different locations in the fermentor were taken by means of a funnel (a technique which is also reported by Oosterhuis [4]). The dissolved oxygen probes, the dissolved carbon dioxide probes and the pH probes were mounted into waterproof cylinders. It was possible to position these cylinders at various locations in the fermentor.

Results.

Oxygen uptake rate.

The oxygen concentration measurements in the gas phase and the pressure measurements show that there is a large gradient in these two properties over the height of the fermentor (figure 5). Since under oxygen limited conditions the driving force for oxygen transfer is proportional to the product of the oxygen concentration and the pressure, there is a large gradient in the driving force for oxygen transfer. In fact, at this scale the driving force varies more than a factor of two over the height of the fermentor (figure 6).

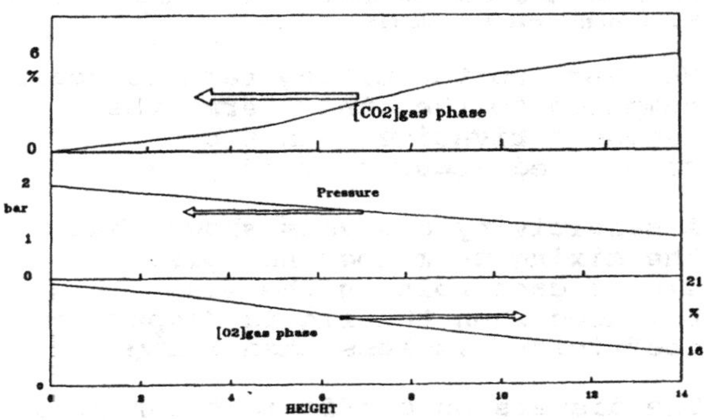

Figure 5:
The oxygen and carbondioxide concentration in the gas phase and the pressure over the height of the fermentor.

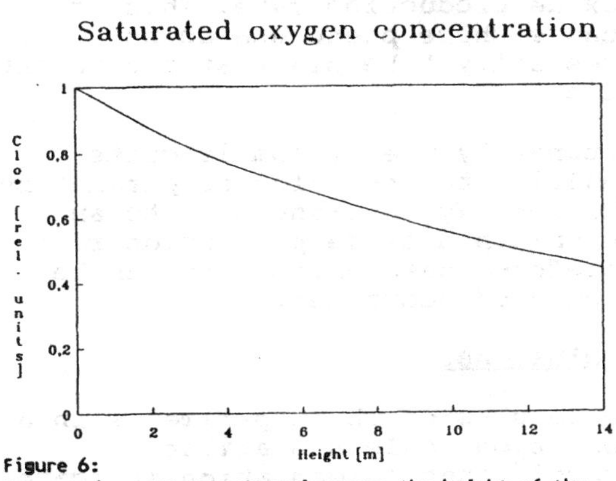

Figure 6:
Saturated oxygen concentration over the height of the fermentor.

From these data the distribution of the oxygen uptake rate can be calculated. This was done for both oxygen and carbon limited conditions (fig. 7).

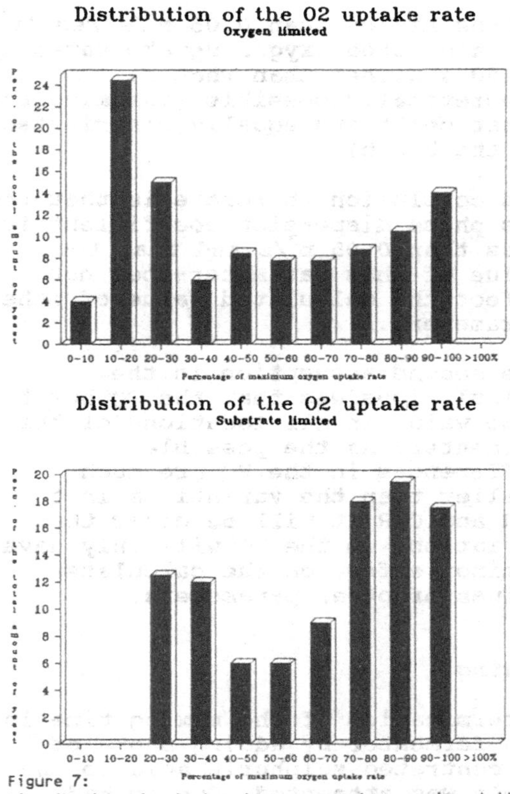

Figure 7:
Distribution of calculated oxygen uptake rate in a bubble column reactor under oxygen and substrate limited conditions.

Fig. 7 shows that the actual oxygen uptake rate of yeast may vary between 30 and 100% of the maximum value. It is very possible that similar variations occur in the growth rate.

The variation in the oxygen transfer can lead to various problems. When a maximum productivity or a maximum yield of biomass on substrate is pursued the substrate uptake rate should be balanced with the oxygen uptake rate. However, in practice this is difficult to achieve. It is likely that the oxygen transfer rate gradients result in oxygen limitations in some locations in the fermenter and substrate limitations in others. The effects of these variations on the productivity of the fermentation and the quality of the yeast are not known.

It has been checked whether the assumption that the gas flow dispersion coefficient is zero has any influence on the calculated results. When the gas phase dispersion coefficient is between 0 and 0.05 m^2/s the mixing term is negligible in the mass balance of the gas phase.

Values higher than 0.05 m^2/s result in calculated oxygen uptake rates up to 50 % higher than what is theoretically possible (assuming the yeast cells are equally distributed in the broth).

The conclusion therefore is that the gas phase dispersion coefficient is less than 0.05 m^2/s and that the value of this parameter does not affect the calculated value of other parameters.

The second assumption in the calculations was that the RQ has the same value in all locations of the fermenter. As the possible differences in the RQ are much smaller than the variations in the OUR and CPR it will be clear that variations in the RQ will only have a minor effect on the calculated values of other parameters.

Mixing.

Determination of the mixing time in the fermentor by adding shots of concentrated sulphuric acid to the broth was attempted. Due to bad micro mixing and because of the large differences in density between the sulphuric acid and the broth this method did not give useful results.

Singh [5] showed that this method can be used in agitated fermentors. The reason for this is that in such fermenters the micro mixing around the impeller is much more intense. This leads to a better pre-mixing of the tracer.
Further developments of methods for measuring mixing times in fermentations are being done in a cooperation between Gist-brocades and the Technical University of Delft.

In theory it should also be possible to calculate the dispersion coefficient from equation 2. Here the mixing term is equal to the product of the second derivative of the concentration to the height and the dispersion coefficient.

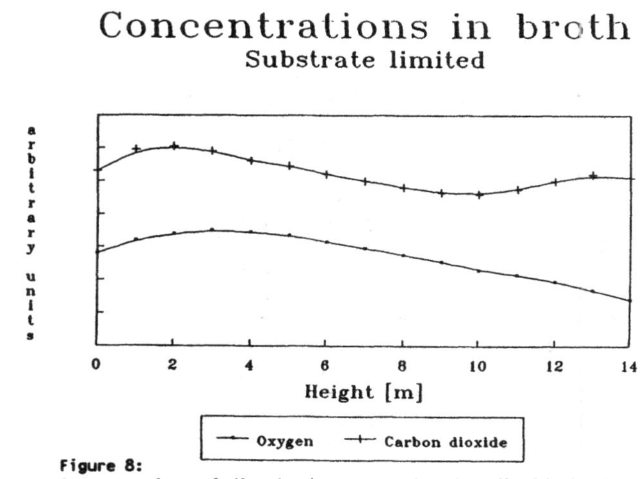

Figure 8:
Concentrations of dissolved oxygen and carbon dioxide in the broth under substrate limited conditions.

However, as the mixing term is small compared to the other terms the second derivative cannot be calculated accurately (fig. 8).

A sensitivity analysis shows that the mixing term does not play a significant role in the mass balances when the liquid dispersion coefficient is less than 5 m^2/s.

The dispersion coefficient can also not be calculated from the knowledge of the oxygen uptake rate, the substrate uptake rate and the carbon dioxide production rate. This is because these phenomena do not necessarily take place at a constant ratio.

Fortunately the system is quite sensitive to the value of parameters like the oxygen transfer rate and the carbon dioxide production rate. Therefore these parameters can be calculated accurately.

Conclusions.

The results of the experiments in a production scale non-agitated bakers' yeast fermentation indicate that regime analysis is a good means of predicting the rate limiting steps of industrial scale fermentation processes.
In a bakers' yeast fermentation process the characteristic time for oxygen consumption is lower than the characteristic time for oxygen transfer.

The experimental results confirm that this leads to possible oxygen limitations and differences in the oxygen uptake rate in the various regions in the fermentor.

Under both oxygen and carbon limited conditions differences in the oxygen uptake rate of a factor 2-3 are present.

Unfortunately in non-agitated fermentors mixing times cannot be measured by using tracers like sulphuric acid. This is because of the bad micro mixing in non-agitated fermentors.

Since in the balance equations for oxygen and carbon the mixing term is only of minor importance, it is impossible to calculate the liquid phase dispersion coefficient from these equations. The order of magnitude of the liquid phase dispersion coefficient is 1-5 m^2/s.

Literature:

[1] Einsele A:
Charakterisierung von Bioreaktoren durch Misch-zeiten.
Chem. Rundschau 25 (1976) p. 53.

[2] Larsson G:
Use of microbial metabolism to predict and control the bioreactor performance.
Ph.D. Thesis, Stockholm, 1990.

[3] Manfredini R, Cavallera V:
Mixing and Oxygen transfer in conventional stirred fermenters. Biotechn. and Bioeng. vol XXV (1983) p 3115-3131.

[4] Oosterhuis N.M.G.:
Scale up of bioreactors: a scale-down approach.
Ph.D. Thesis, Delft, 1984.

[5] Singh V:
On-line determination of mixing parameters in fermenters using pH transients.
Int. Conf. Bioreactor Fluid Dynamics, Cambridge (Eng) 1986.

[6] Sonnleitner,B. Käpelli,O.
Growth of Saccharomyces Cerevisiae is controlled by its limited respiratory capacity: Formulation and verification of a hypothesis.
Biotech. & Bioeng. 28 (1986) p927.

[7] Sweere A.P.J.:
Response of bakers's yeast to transient environmental conditions. Ph.D. Thesis, Delft, 1988.

[8] Vardar F and Lilly M.D.:
Effect of cycling dissolved oxygen concentration on product formation in Penicillin Fermentations.
Eur. J. Appl. Micr. Biotechn. (1982) 14: 203-211.

Table of used symbols.

a specific bubble surface [1/m]
C concentration [mol/l]
C_{so} concentration of substrate addition.
CPR Carbon Dioxide Production Rate
$\mathbb{D}$ dispersion coefficient [m^2/s]
F substrate addition rate
h height in the fermentor [m]
L specific length [m]
OUR Oxygen Uptake Rate
r production rate
t time [s]
v linear velocity [m/s]
Y yield of yeast on substrate

ϵ gas hold-up [-]
μ specific growth rate [1/s]

Superscripts.

* means saturated concentration.

Subscripts.

o oxygen
l liquid
s substrate

Addendum:

- In table 1 the formula for the heat transfer is not given. This because of its size.
- In table 1 D_l en D_g should be $\mathbb{D}_l$ and $\mathbb{D}_g$.
- The coordinate in figure 7 is given in %.

Referees' comments:

1. Page 2, Simulation: You say that simulation is not always possible at a smaller scale. On the contrary, I consider that by modifying the fermentor geometry and/or working conditions, scale down is always possible in principle.

   Answer:
   What is meant under 2 is that it is not possible to design an experiment on a laboratory scale where all phenomena are equal to those on the production scale. When for example the shear stress in an agitated production scale fermentor is scaled down to laboratory scale (the shear stress is proportional to the stirrer speed) the flow in the fermenter changes from turbulent to laminar

We agree that it is possible to scale down individual phenomena.

2. Page 3: What do you call the circulation model? (Please, give a reference)

Answer:
The scheme of a circulation model is given in fig. 3A. The flow is identical to that of a loop reactor. About 90% of the literature on air lift loop reactors deals with this type of model.

3. Page 4: You say that the linear liquid flow is negligible. What are the magnitude of superficial gas velocity, of vvm, of solution viscosity and of H_l/T (H_l = (quiescent) liquid height, T = fermentor diameter)?

Answer:
In the model used, the linear liquid flow accounts for the liquid volume (or height) increase during the fermentation. The liquid volume changes due to the feeds and the evaporation. This flow is not related to any actual local or actual average air flow or liquid flow in the fermentor. Therefore the relation between the quotation of the sentence about the linear liquid flow and the other questions under 4 is not clear to us. The magnitude of the linear velocity is a little less than 1 mm/s, which is negligible compared to the other velocities in the fermenter.
The superficial gas velocity used during the fermentation was 5 ncm/s and higher, the viscosity is about 1-2 mPas, the liquid height varies during the fermentation. It is about 10 m. The tank diameter is a little less than 4 m (see the gross volume and the height of the fermentor).

4. Page 6: For the determination of the mixing time, couldn't you use a spray nozzle dispersing the concentrated sulphuric acid over the solution surface? Moreover, couldn't the literature correlations given a good estimation of the mixing time?

Answer:
In theory it is possible to overcome the problems with the sulphuric acid tracer by using a spray nozzle. However, in practice the use of a spray nozzle is not advisable, as the aerosols, which are present in the air which leaves the broth, in combination with the sulphuric acid lead to a very aggressive sulphuric acid solution. This will damage the metal in the exhaust of the fermentor.
To our knowledge in literature there are no correlations which describe the mixing time in this type of reactor, in this medium and on this scale.

5. Supposing that the unit of the ordinate of your Fig.8 is 10%, could you give the value of saturated concentrations of O_2 and CO_2 at the bottom and at the top of the fermentor, or at least the evolution of the corresponding driving forces?

Answer:
The aim of the paper is to show that there are oxygen uptake rate gradients in a baker's yeast fermentation and that these gradients are different under substrate and oxygen limitation. As the measurements were done in a commercial bakers' yeast fermentation in this paper only that information is given which is necessary to explain the cause of this phenomena.

SOLID DISTRIBUTION IN SLURRY REACTORS STIRRED WITH MULTIPLE IMPELLERS : CONTINUOUS FLOW SYSTEMS

M. NOCENTINI, F. MAGELLI, Dip. di Ingegneria Chimica e di Processo, University of Bologna, I

ABSTRACT

Solid distribution was investigated in slurries stirred by multiple axial impellers in continuous flow equipment. The solid concentration profiles are described by means of the one-dimensional sedimentation-dispersion model. The influence of various parameters is studied theoretically, namely: the Péclet number, the ratio of solid concentration upstream and downstream the exit nozzle, the ratio between the particle settling velocity in the stirred and in a still liquid. The possibility of prevision of the model is then tested by comparing the theoretical concentration profiles – calculated on the basis of independent measurements of the parameters – with selected experiments carried out with slurries of mono-sized spherical particles of different materials and size in water.

INTRODUCTION

In recent years, attention has been focussed on a better understanding and a detailed description of the solid distribution in stirred, solid-liquid contactors. Penaz et al. [1] and Sysova et al. [2] attempted the prevision of solid distribution from first principles. Starting from the continuity equations in a turbulent field, Barresi and Baldi [3] showed that the mean turbulent Péclet number is proportional to the ratio U_t/ND. Brucato and Rizzuti proposed the use of a network-of-zones model [4]: though relatively simple to use, this model can provide reasonable predictions of the solid distribution from basic fluid-dynamic parameters. This approach has successfully been used to describe the behaviour of equipment stirred with multiple Rushton turbines [5].

Tojo and Miyanami [6] adopted the even simpler one-dimensional sedimentation-dispersion model for the description of the solid profiles in vessels stirred with either a propeller or a vibratory disk. The same phenomenological model was applied to vessels stirred with a pitch-blade turbine [7] or multiple radial [8] or axial turbines [9]. Owing to the significant deviation from a monotonic increase in solid concentration from the top to the bottom of the vessel exhibited by any type of impeller, the use of this last model seems justified only for multiple impellers, in which case it fits the overall trend of the profile reasonably well. For these last systems, correlations of the model parameter with the slurry properties and operating conditions have been put forward [8,9] that are also suitable for scale-up. In this respect, it is worth mentioning that the resulting scale-up rule is intermediate between constant tip speed and constant power per unit mass.

The previous studies dealt with batch systems, the only exceptions being the modelling of solids washout [6] and of semibatch systems [8]. In this paper the analysis developed in the past by these authors [8,9] is extended to continuous flow system. The one-dimensional sedimentation-dispersion model is used to model the solid concentration distribution in slurries stirred with multiple impellers. Firstly, the influence of several dimensionless parameters on the solid distribution is studied. The more relevant ones are the Péclet number (as a means of describing the mixing state of the system), the ratio of solid concentration upstream and downstream the exit nozzle (this parameter accounting for non-isokinetic withdrawal), the ratio between particle settling velocity in the stirred and in a still liquid. Then, model previsions made with values of the above-mentioned parameters determined by means of independent measurements are compared with selected experiments performed in a high-aspect-ratio vessel stirred with multiple axial impellers.

THEORY

Liquid phase model

The axial dispersion model is adopted to interpret the liquid phase behaviour. The continuity equation for a non-reacting species is:

$$\frac{\partial C}{\partial t} = D_{eL}\frac{\partial^2 C}{\partial z^2} - U_L\frac{\partial C}{\partial z} \quad . \qquad (1)$$

The evolution in time of the concentration C, as provided by Eqn. (1) subject to a pulse disturbance at t=0 and z=0, has been evaluated in the past and only the final relationships are given here. For batch conditions (U_L=0), the dimensionless solution at z=H is [10]:

$$C^* = 1 + 2\sum_{k=1}^{\infty}\cos\{k\pi\}\cdot\exp\{-k^2\pi^2\theta\} \qquad (2)$$

where $\theta = tD_{eL}/H^2$. For continuous-flow systems, the solution $C^*(\tau)$ evaluated at z=H by adopting the Danckwerts' boundary conditions is [11]:

$$C^* = \sum_{k=1}^{\infty}R_k\cdot\exp\{S_k\tau\} \qquad (3)$$

where:

$$R_k = \frac{2(Pe_L^2+4\alpha_k^2)\cdot\alpha_k\cdot\sin\{\alpha_k\}}{Pe_L(Pe_L^2+4Pe_L+4\alpha_k^2)}\exp\{Pe_L/2\}$$

$$S_k = -\frac{4\alpha_k^2+Pe_L^2}{4Pe_L}$$

and the α_k's are the roots (k=1,2,..., ∞) of the equation:

$$tg\,\alpha = \frac{4\alpha Pe_L}{4\alpha^2-Pe_L^2} \quad .$$

Solid phase behaviour

To interpret the solids distribution profiles in the vessel, the one-dimensional sedimentation-dispersion model is adopted, which was successfully used for describing either bubble columns [12,13] or stirred vessels [6-9]. With the z axis directed upwards (so that z=0 and z=H correspond to the bottom and the top of the vessel, respectively), the steady-state mass balance for the solid phase in the stirred suspension is:

$$D_{es}\frac{d^2\phi}{dz^2} - (U_L+U_s)\frac{d\phi}{dz} = 0 \qquad (4)$$

According to the system of coordinates adopted, U_L is positive for the liquid flowing upwards and U_s is negative for $\rho_P > \rho_L$.

The following boundary conditions are adopted:

$$U_L\phi_i = (U_L+U_s)\,\phi(0^+) - D_{es}\left.\frac{d\phi}{dz}\right|_{z=0+} \qquad (5)$$

$$\phi(H^-) = J\,\phi_o \qquad (6)$$

where the separation coefficient J accounts for solid separation in the exit nozzle due to non-isokinetic withdrawal [12-15]. The additional condition:

$$\phi_i = \phi_o \qquad (7)$$

applies because of steady state conditions.

By defining the following dimensionless parameters:

$$A = 1 + U_s/U_L = 1 + (U_t/U_L)(U_s/U_t) \qquad (8)$$

$$Pe_{Ls} = U_LH/D_{es} \qquad (9)$$

$$\phi^* = \phi/\phi_o \ , \qquad (10)$$

Eqn. (4) can be rewritten in dimensionless form as:

$$\frac{1}{A\cdot Pe_{Ls}}\frac{d^2\phi^*}{d\zeta^2} - \frac{d\phi^*}{d\zeta} = 0 \qquad (11)$$

whose solution is:

$$\phi^*(\zeta) = 1/A + (J-1/A)\exp\{-A\cdot Pe_{Ls}(1-\zeta)\}. \qquad (12)$$

It is worth noting that for a batch system (i.e., 1/A=0, $A\cdot Pe_{Ls}=U_sH/D_{es}$) and a semibatch system (i.e., ϕ_i=0, $A\cdot Pe_{Ls}=F\cdot Pe_s$) Eqn. (12) reduces to the simpler forms reported elsewhere [8].

Theoretical profiles

A detailed, theoretical analysis of the influence of the model parameters on the solid concentration profiles is beyond the scope of this paper. Only a short discussion of the influence of the parameters in Eqn. (12) is given here. For the discussion, reference is made to Figs. 1 to 3. Owing to the definitions adopted for ϕ^* and J, Eqns. (10) and (6), all the dimensionless profiles are characterized by ϕ^*=J at

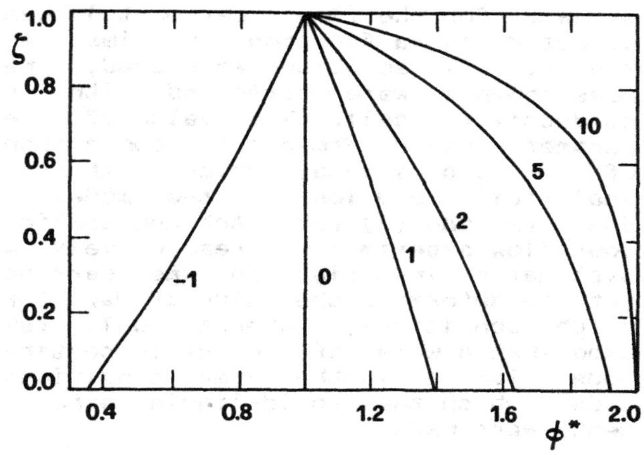

Fig. 1. Influence of Pe_{Ls} on the theoretical solid concentration profiles calculated with Eqn. (12). A=0.5, J=1.

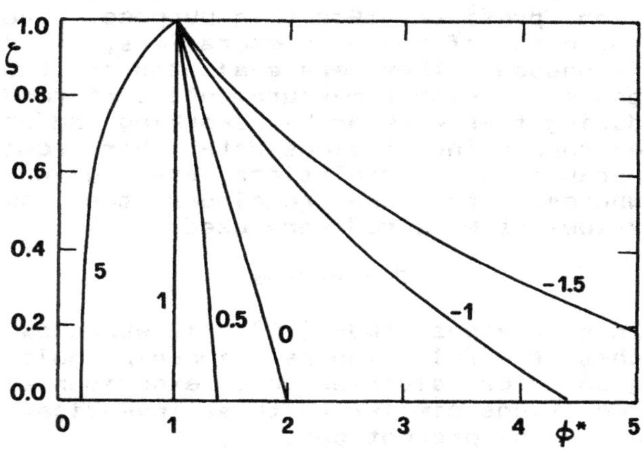

Fig. 2. Influence of A on the solid concentration profiles. Pe_{Ls}=1, J=1.

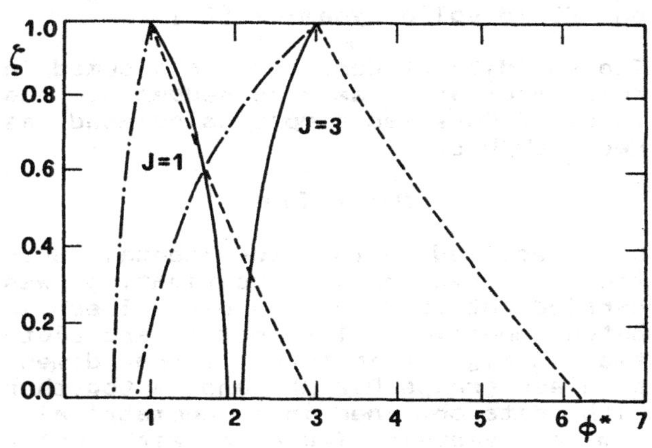

Fig. 3. Influence of J on the solid concentration profiles.
———, A=0.5, Pe_{Ls}=5;
-----, A=0.5, Pe_{Ls}=1;
-·-·-, A=2.0, Pe_{Ls}=1;

z=H.

A reduction of macromixing degree in the system, as given by an increase in Pe_{Ls} value, results in both more inclined profiles and higher solid hold-up (Fig. 1) when the liquid flows upwards in the vessel (Pe_{Ls}>0) and settling solids are considered (A<1).

The condition for a flat profile is A=1/J. An example of the influence of the parameter A is given in Fig. 2 for J=1: for settling solids and the slurry fed at the bottom of the vessel (A<1), the profiles always increase from the top to the bottom, whilst buoyant particles (A>1) exhibit the opposite behaviour; for neutrally buoyant particles (U_s=0, i.e. A=1), the profile is flat. When $U_L=-U_s$ (i.e., A=0), the profile is linear; profiles parallel to that plotted in Fig. 2 are obtained for J≠1, regardless of the value of Pe_{Ls}.

As one would expect, for settling solids and the liquid flowing upwards an increase in J results in higher solid hold-up. More interestingly, the higher the value of J the greater the effect of the other parameters (Fig. 3).

A comparison between semibatch [8] and continuous-flow systems is also worthy of note. As mentioned above, for $U_L=-U_s$ (A=0) the profile is linear in continuous flow, whereas a flat profile is obtained in semibatch conditions. When $U_L>-U_s$ the profile does increase from the bottom to the top of the vessel for semibatch conditions, whereas no profile reversal is apparent for continuous-flow conditions, a downward concavity being simply noticed in this case. This is the obvious consequence of both continuous slurry supply at the base and continuous withdrawal at the top of the vessel.

EXPERIMENTAL

The investigation was carried out in a vertical, cylindrical, flat-bottomed vessel characterized by H/T=4. The vessel was of fully closed design, had pyrex walls and was equipped with four standard baffles of 0.1T width. Agitation was provided with an agitator consisting of a shaft with four identical, evenly spaced Lightnin A310 impellers pumping downwards (see geometry No. 3, Ref. 9). The stirrer diameter was D=9.6 cm. In the turbulent regime the power number for the whole multiple-impeller agitator was N_p=1.2.

The suspensions were made of water and several fractions of spherical solid particles of narrow size distribution, namely: glass beads (d_p=0.33 mm, ρ_p=2.45 kg/l) and PMMA particles (d_p=0.14 mm and d_p=0.23 mm, ρ_p=1.15 kg/l). The terminal velocity of the particles was in the range U_t=0.14-4.4 cm/s. The inlet concentration was in the range 0.5-1.0 grams per litre.

The experiments were carried out at room temperature in a continuous-flow mode, with the slurry inlet at the bottom of the vessel (a distribution plate was placed in such a way as to distribute the feed over the whole plane) and the exit at the top. Slurry withdrawal from the top nozzle was typically non-isokinetic; no attempt was made to investigate the influence of location [17] and inner shape of the outlet pipe. For the sake of operating simplicity, the stream from the exit was recycled directly to the inlet. The feed flow-rate was such as to give a superficial velocity in the vessel in the range U_L=0.15-0.45 cm/s. The rotational speed was always higher than the "just suspended" condition.

The solids profiles inside the vessel were measured by means of the non-intrusive optical technique and the pertinent equipment described by Fajner et al. [18]. A light-emitting diode and a silicon photo-diode were used as light source and receiver, respectively. The light beam (about 4 mm in diameter) passed through the vessel horizontally along a chord about one centimeter off the axis, approximately midway between the vertical baffles. Thus, each measurement could be considered as the mean of solid concentration $\phi(z)$ in the whole horizontal section. For each experimental condition, the solids concentration was measured at 32 positions along the vertical axis of the vessel. The mean solid concentration inside the vessel was in the range 0.5 to 2.6 grams per litre.

The solid concentration at $z=H^-$ (that is, 25 mm above the uppermost measuring point) was estimated by simply extrapolating the measured profile. The solid concentration in the exit stream was measured gravimetrically – i.e. by taking a sample that was then sieved, dried and weighed.

The value of the parameter D_{eL} that characterizes the mixing state of the liquid phase according to the model adopted was determined by means of the classic pulse-response technique. In both flow and non-flow experiments, a rapid injection of a non-reacting liquid tracer was made into the bottom of the vessel and the resulting concentration at the top (or in the outlet stream, for the flow experiments) was detected as a function of time. As tracer, KCl solutions were used; the measurements were performed with a conductivity cell. The value of the parameter was determined by comparison of the experimental curves with the analytical solution of the model – Eqns. (2) and (3) for batch and continuous-flow experiments, respectively. A systematic investigation was carried out to determine the value of D_{eL} for batch conditions, whereas only few experiments were carried out in continuous flow – in the same conditions under which the solid-liquid experiments were made.

THE MODEL PARAMETERS

As will be discussed in the next Section, the model, Eqn. (12), was used to make previsions of the solid concentration profiles. For this purpose, the whole set of the parameters Pe_{Ls}, A, J is needed. They were evaluated on the basis of either measurements performed during this work or by reworking and/or extrapolating previous data. For some parameters correlations are given, whereas for the remainder the raw values as measured were used.

D_{es} values

In a previous study [19], it was shown that the following relationship holds good for slurries and experimental conditions similar to those investigated in the present paper:

$$D_{es} = D_{eL} \quad . \tag{13}$$

The same result had also been obtained in a previous fundamental investigation on fluid-solid dynamics [20].

The validity of Eqn. (13) is assumed in this paper and, as a consequence, the value of Pe_{Ls} was simply calculated as $Pe_{Ls}=U_LH/D_{eL}$.

D_{eL} values

As specified in the Experimental Section, a systematic investigation was carried out to determine D_{eL} values in batch conditions. The results are plotted in Fig. 4 in terms of the dimensionless groups D_{eL}/ND^2 and Re together with data obtained in a geometrically similar vessel (T=0.46 m) with water and viscous solutions [21]. Apparently, the general trend is very similar to that exhibited by multiple Rushton turbines in a vessel with the same aspect ratio [8]. In spite of the scattering of the data, the constant value D_{eL}/ND^2=0.14 can be retained for Re>10^4

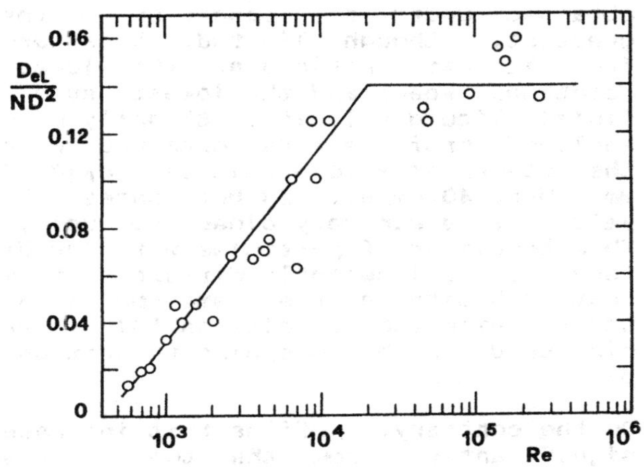

Fig. 4. Dispersion coefficient (liquid phase) for A310 impellers (4 turbines in a H/T=4 vessel).

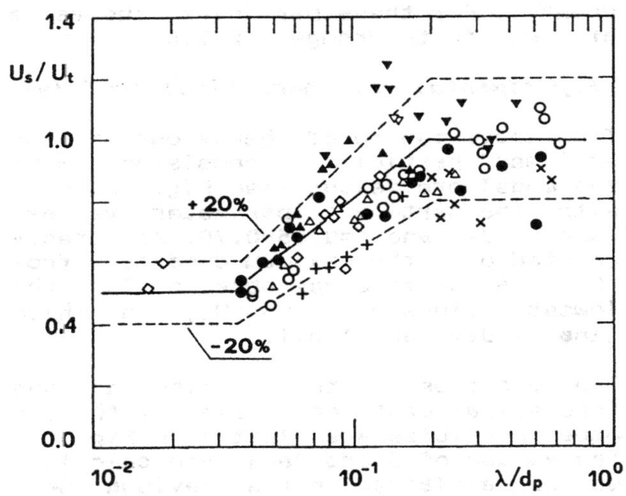

Fig. 5. U_s/U_t data for Rushton turbines [8] and the empirical correlations, Eqns. (14a-c).

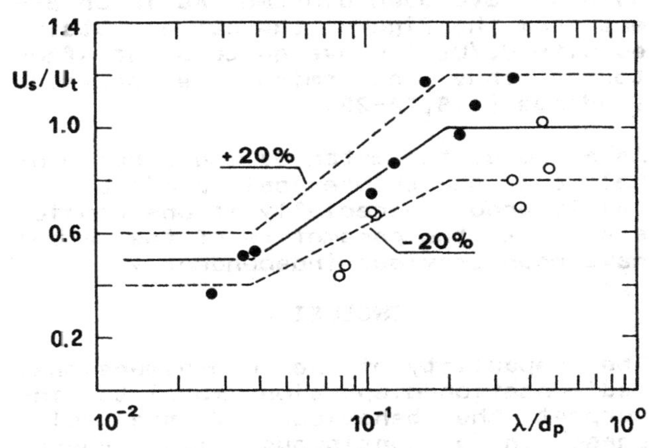

Fig. 6. U_s/U_t data for A310 impellers and the empirical correlations, Eqns. (14a-c). o , d_p=0.33 mm; ●, d_p=0.98 mm.

– at least as a first attempt. Quite surprisingly, this value is very close to that obtained with the Rushton turbines: this result is under investigation, but no explanation is possible at present. For continuous flow conditions, D_{eL} values are slightly affected by the increase in U_L with respect to the value given in Fig. 4 for $U_L=0$. Due to the low number of these last data, no correlation was attempted.

A values

According to Eqn. (8), A is defined in terms of U_L and U_t (that is, operating conditions and slurry properties) as well as the ratio U_s/U_t. It is usually accepted that this last parameter is in the range 0.4-1.0 [1,8,18,22-25]. However, no direct systematic investigation has been made regarding the influence of the relevant parameters on U_s/U_t. Only for Rushton turbines do enough data exist to establish tentatively a correlation. For $Re_p<0.3$, $U_s/U_t=1$ [18]. For $Re_p>0.3$, by interpolating previous data [8], the following relationships can be obtained:

$$U_s/U_t = 1 \tag{14a}$$

for $\lambda/d_p > 0.2$,

$$U_s/U_t = 1.5 + 0.7\,log\{\lambda/d_p\} \tag{14b}$$

for $4\times10^{-2} < \lambda/d_p < 0.2$,

$$U_s/U_t = 0.5 \tag{14c}$$

for $\lambda/d_p < 4\times10^{-2}$.

As shown in Fig. 5, the above empirical equations fit the data fairly well (with an error of ±20%).

Similar resùlts have also been obtained in the present work for multiple A310 impellers. For this purpose, previous data [9] on solids distribution in batch systems – given in terms of $Pe_s=U_sH/D_{es}$ – have been re-elaborated by means of the D_{eL} values of Fig. 4 on the hypothesis that $D_{es}=D_{eL}$. Once U_s was obtained, the U_s/U_t values could be calculated immediately. Though relatively few in number, these data are matched by Eqns. (14a-c) satisfactorily well as shown in Fig. 6.

J values

In spite of interesting investigations on the matter [15,26,27], no general predictive means is available for describing the solid separation at the exit due to non-isokinetic withdrawal. Thus, J was simply calculated as the ratio of the experimental values $\phi(H^-)$

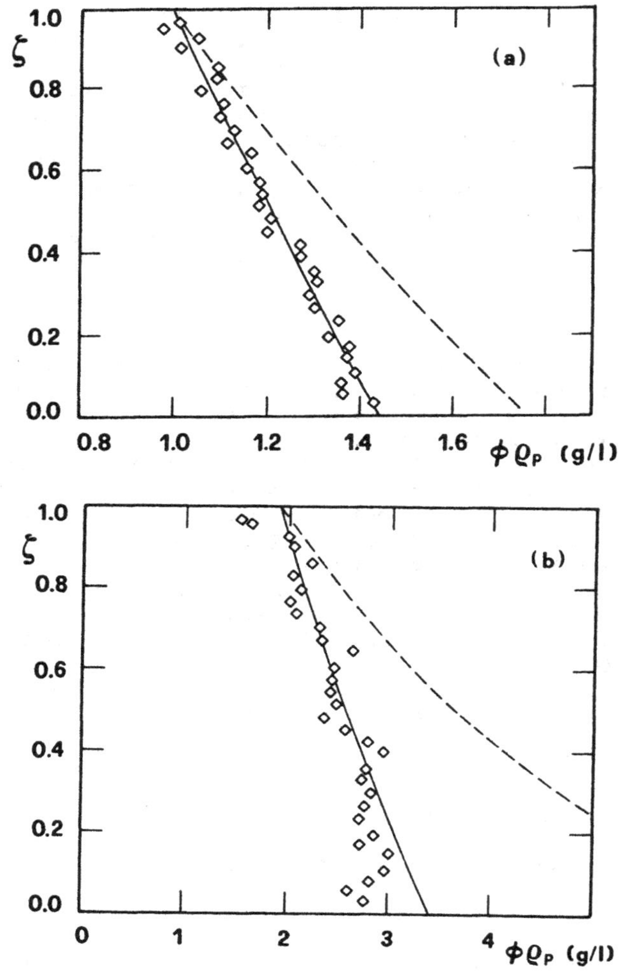

Fig. 7. Comparison of the experimental profiles with the theoretical ones. ———, U_s/U_t given by Eqns. (14a-c); -----, $U_s/U_t=1$.
(a), $d_p=0.23$ mm, $\rho_p=1.15$ kg/l, $\eta=1.1$ mPas, $\phi_i\rho_p=0.20$ g/l, $N=5$ s^{-1}, $U_L=0.15$ cm/s, $U_s/U_t=0.65$, $Pe_{Ls}=0.24$, $A=-0.69$, $J=1$.
(b), $d_p=0.33$ mm, $\rho_p=2.45$ kg/l, $\eta=1.1$ mPas, $\phi_i\rho_p=1.9$ g/l, $N=18.3$ s^{-1}, $U_L=0.46$ cm/s, $U_s/U_t=0.5$, $Pe_{Ls}=0.14$, $A=-3.8$, $J=1.9$.

and ϕ_o.

RESULTS

Qualitative description

The experiments were carried out in the conditions given in the Experimental Section. For the smallest particles ($\rho_p=1.15$ kg/l, $d_p=0.14$ mm, whence $U_t=$ $=0.14$ cm/s) nearly flat profiles were obtained under any condition – the greatest, though limited, departure from it being noticed at the lowest rotational speed and the lowest superficial liquid velocity. Slightly more inclined profiles were obtained with the other plastic particles ($d_p=0.23$ mm, $U_t=0.40$ cm/s). In both cases, the value of J was very close to unity. This behaviour of particles with low U_t (and U_s) is inherently different from that exhibited by the same particles under semibatch conditions [8] – as discussed in the foregoing theoretical part.

On the contrary, profiles that increase significantly from the top to the bottom of the vessel were obtained with glass beads ($U_t=4.4$ cm/s). In the experimental conditions investigated, the influence of U_L on the profiles was moderate, while that of N was much greater. For these particles, the value of J was in the range 1.2-7.5.

Experimental vs. theoretical profiles

The above-mentioned behaviour of the smallest particles is consistent with the model previsions (see Fig. 2) made with the pertinent parameter values, namely $J\approx1$ and $A=0.55$-0.70. As already pointed out, the maximum departure from the flat profile was obtained for the lowest values of N and U_L, in which case $U_s/U_t=1$ and $A=0.1$.

Two examples of the experimental and theoretical profiles obtained with the other particles are plotted in Fig. 7. The values of J and Pe_{Ls} were calculated as explained in the previous Section. For the sake of comparison, the theoretical profiles based on values of A calculated with either Eqns. (8) and (14a-c) or Eqn. (8) and the assumption $U_s/U_t=1$ have been plotted. As is apparent from the figure, the curves obtained with $U_s/U_t=1$ prove quite unsatisfactory, thus confirming the previous findings [1,8,22-25].

As a whole, the match of the experimental and the theoretical profiles is fairly good – especially if one considers that the parameters of the model have been provided independently.

CONCLUSIONS

The capability of the one-dimensional sedimentation-dispersion model to interpret the behaviour of the solid phase in a continuous flow vessel stirred with multiple axial impellers was tested. The model parameters were assessed independently by either resorting to specific measurements or extrapolating previous data. In spite of this, the match of the experimental solid concentration profiles to the theoretical previsions is quite satisfactory.

Certainly, a wider range of experimental conditions should be checked before the outlined procedure is used extensively. Moreover, reliable correlations of all the parameters intervening in the model are required for its use in a fully predictive way. This calls for rational equations to provide the separation coefficient J as well as more U_s/U_t data (especially at low λ/d_p values) and more extensive information on D_{eL} (for various geometrical equipment configurations, under flow and batch conditions).

ACKNOWLEDGEMENTS

This work was financially supported by the Italian Ministry of University and Research (funds "MURST 40%").

The collaboration of Dr. D. Fajner and Mr. D. Cornelli in carrying out the experimental programme is gratefully acknowledged.

LIST OF SYMBOLS

- A parameter, defined in Eqn. (8)
- C concentration of a liquid tracer
- C* dimensionless concentration of a liquid, non-reacting tracer
- D impeller diameter
- D_{eL} dispersion coefficient for the liquid
- D_{es} dispersion coefficient for the solid
- d_p solid particle diameter
- F =1+U/Us, flow factor used in Ref. 8
- H vessel height
- J separation coefficient, Eqn. (6)
- N rotational speed
- N_p $=P/\rho N^3D^5$, power number
- P (average) power consumption
- Pe_L $= U_LH/D_{eL}$, Péclet number (liquid phase)
- Pe_{Ls} $=U_LH/D_{es}$, Péclet number (slurry)
- Pe_s $=U_sH/D_{es}$, Péclet number (solid phase)
- R,S constants
- Re $=ND^2\rho_L/\eta$, rotational Reynolds number
- Re_p $=U_s\rho_L d_p/\eta$, particle Reynolds number
- t time
- t_R $=H/U_L$, holding time
- T vessel diameter
- U_L superficial liquid velocity
- U_s settling velocity of solid particles in a stirred medium
- U_t settling velocity in a still liquid
- V vessel volume
- z vertical coordinate
- α constant
- ϵ $=P/\rho V$, average power consumption per unit mass
- ζ dimensionless vertical coordinate
- θ $=tD_{eL}/H^2$, dimensionless time in Eqn. (2)
- λ $=(\nu^3/\epsilon)^{0.25}$, Kolmogoroff microscale
- η dynamic liquid viscosity
- ν kinematic liquid viscosity
- ρ_L liquid density
- ρ_p solid density
- τ $=t/t_R$, dimensionless time in Eqn. (3)
- ϕ local volumetric solids concentration
- ϕ_i vol. solid concentration at the inlet
- ϕ_o vol. solid concentration at the exit
- ϕ^* $=\phi/\phi_o$, dimensionless solids concentration

REFERENCES

1. Penaz, F., V. Rod and M. Rehakova, 1978, "Concentration distribution of suspension in a mixed vessel". *Coll. Czech. Chem. Commun.*, 43, 848-861.
2. Sysova, M., I. Fort, T. Vanek and V. Kudrna, 1984, "Flow of solid-liquid suspension in agitated system". *8th Congress CHISA* (Prague, Sept. 3-7), paper V3.49.
3. Barresi, A. and G. Baldi, 1987, "Solid dispersion in an agitated vessel". *Chem. Eng. Sci.*, 42, 2949-2956.
4. Brucato, A. and L. Rizzuti, 1988, "The application of the network-of-zones model to solid-liquid suspensions". *Proc. Sixth Eur. Conf. Mixing* (Pavia, May 24-26), 273-280, BHRA, Cranfield.
5. Brucato, A., F. Magelli, M. Nocentini and L. Rizzuti, 1991, "An application of the network-of-zones model to solids suspension in multiple impeller mixers". *Trans. IChemE Part A*, 69, 43-52.
6. Tojo, K. and Miyanami, 1982, "Solids suspension in mixing tanks". *Ind. Eng. Chem. Fund.*, 21, 214-220.
7. Ayazi Shamlou, P. and E. Koutsakos, 1989, "Solids suspension and distribution in liquids under turbulent agitation". *Chem. Eng. Sci.*, 44, 529-542.
8. Magelli, F., D. Fajner, M. Nocentini and G. Pasquali, 1990, "Solid distribution in vessels stirred with multiple impellers", *Chem. Eng. Sci.*, 45, 615-625.
9. Magelli, F., D. Fajner, M. Nocentini, G. Pasquali, P. Ditl and V. Marisko, 1991, "Solids concentration distribution in slurry reactors stirred with multiple axial impellers", *Chem. Eng. and Processing*, 29, 27-32.
10. Siemes, W. and W. Weiss, 1957, "Flüssigkeitsdurchmischung in engen Blasensäulen". *Chemie-Ing.-Techn.*, 29, 727-732.
11. Brenner, H., 1962, "The diffusion model of longitudinal mixing in beds of finite length. Numerical values". *Chem. Eng. Sci.*,

17, 229-243.

12. Kato, Y., A. Nishiwaki, T. Fukuda and S. Tanaka, 1972, "The behaviour of suspended solid particles and liquid in bubble columns". *J. Chem. Eng. Japan*, **5**, 112-118.

13. Rushton, J.H., 1965, "The continuous removal of mixed phases from a mixing tank". *A.I.Ch.E.-IChemE Symp. Ser.*, **No. 10**, 3-7.

14. Rehakova, M. and Z. Novosad, 1971, "The separation effect at the outlet of a vessel with a mechanically agitated suspension". *Coll. Czech. Chem. Commun.*, **36**, 3004-3008.

15. Sharma, R.N. and H.C.L. Das, 1980, "Effect of withdrawal flow velocity on the composition of a two phase system in a mixing tank". *Coll. Czech Chem. Commun.*, **45**, 3293-3301.

16. Joshi, J.B., P.V. Shertukde and S.P. Godbole, 1988, "Modelling of three phase sparged catalytic reactors". *Reviews in Chem. Eng.*, **5**, 71-155.

17. Aeschbach, S. and J.R. Bourne, 1972, "The attainment of homogeneous suspension in a continuous stirred tank". *Chem. Eng. J.*, **4**, 234-242.

18. Fajner, D., F. Magelli, M. Nocentini and G. Pasquali, 1985, "Solids concentration profiles in a mechanically stirred and staged column slurry reactor". *Chem. Eng. Res. Des.*, **63**, 235-240; [supplement in *Chem. Eng. Res. Des.*, **64**, 473 (1986)].

19. Magelli, F., Fajner, D., Nocentini, M. and Pasquali, G., 1986, "Solids concentration profiles in vessels stirred with multiple impellers". *Proc. World Conf. III of Chem. Eng.* (Tokyo, September 21-25), paper 8k-204, vol. 3, pp. 362-365, Soc. Chem. Eng. Japan, Tokyo.

20. Soo, S.L., 1978, "Diffusivity of spherical particles in dilute suspensions". *AIChE Symp. Ser.*, **74** (No. 174), 184-185.

21. Bacchelli, F., 1987, *M Thesis* in Industrial Chemistry. University of Bologna.

22. Schwartzberg, H.G. and R.E. Treybal, 1968, "Fluid and particle motion in turbulent stirred tanks". *Ind. Eng. Chem. Fundls*, **7**, 6-12.

23. Levins, D.M. and J.R. Glastonbury, 1972, "Particle-liquid hydrodynamics and mass transfer in a stirred vessel. Part I: Particle-liquid motion". *Trans. IChemE*, **50**, T32-T41.

24. Kuboi, R., I. Komasawa and T. Otake, 1974, "Fluid and particle motion in turbulent dispersion. II. Influence of turbulence of liquid on the motion of suspended particles". *Chem. Eng. Sci.*, **29**, 651-657.

25. Nienow, A.W. and R. Bartlett, 1974, "The measurement and prediction of particle-fluid slip velocities in agitated vessels". *Proc. First Eur. Conf. Mixing and Centrif. Separ.* (Cambridge, September 9-11), p. B1-1 to B1-15, BHRA, Cranfield.

26. Baldi, G. and R. Conti, 1978, "Continuous removal of a solid suspension from stirred tanks". *Proc. Int. Symp. on Mixing* (Mons, February 21-24), paper B2, Faculté Polytechnique de Mons.

27. Machon, V., V. Kudrna and V. Hudcova, 1980, "Wash-out of solid phase from the mechanically stirred charge". *Coll. Czech. Chem. Commun.*, **45**, 2152-2163.

SECTION 2. MIXING OPERATIONS

THE MIXING OF MISCIBLE LIQUIDS WITH LARGE DIFFERENCES IN DENSITY AND VISCOSITY

S.S.H. BURMESTER, C.D. RIELLY, Dept. of Chemical Engineering, University of Cambridge
M.F. EDWARDS, Unilever Research, Port Sunlight Laboratory, GB

Summary

The mixing of miscible fluids with significant density differences, but with similar viscosities, may lead to stratification, in which case buoyancy effects must be overcome to achieve a homogeneous mixture. Differences in viscosity between the added and the bulk fluid cause a reduction in the strain rate experienced by the more viscous fluid; furthermore, large viscosities damp the inertial effects which would cause entrainment and mixing. Stirred tanks have commonly been used to perform such batch mixing processes but they are not always the most effective devices.

Experimental studies were carried out to determine the conditions under which a viscous and dense fluid would be easily mixed into a low viscosity bulk, in fully turbulent flow (in all cases, the length scale of the viscous addition was larger than the integral velocity length scale for the flow, so that the bulk fluid appeared fully turbulent to the added material). The mixing time was determined experimentally from the concentration time-history of tracer fluid using conductivity micro-probes. Preliminary studies of the mixing of non-Newtonian fluids showed the difficulty in determining an apparent viscosity at the strain rate actually experienced by the added fluid.

Dimensional analysis has been used to interpret these mixing time results in terms of the magnitude of the bulk inertia forces relative to the initial buoyancy forces and to viscous forces within the added fluid. A critical Reynolds number, which depends on the viscosity ratio of the two fluids, has been proposed as the criterion for the operating conditions under which the mixing rate is extremely slow and the mixing time is much longer than that required for mixing fluids with similar properties.

1. Introduction

The mixing of miscible liquids with very different physical properties is a common operation in many chemical and food process industries. The applications range from the mixing of various gasoline fractions in large storage tanks, to the homogenisation of ingredient fluids in soft drinks production, or the blending of domestic detergent products. The mixing operation considered here is the addition of a small quantity of a high viscosity liquid into a low viscosity bulk, which is agitated by a mechanical stirrer and is in turbulent flow. The two fluids are fully miscible, but the effects of viscous and buoyancy forces must be overcome by the inertial forces generated by the impeller, for homogenisation to take place. When the two fluids have large differences in density and viscosity, very long mixing times can result.

Mixing rates in batch stirred tanks are however impracticable to predict theoretically and experimentally derived mixing times have been used to determine the efficiency of the operation. A fundamental experimental and theoretical understanding of the forces and mixing mechanisms involved in overcoming the retarding forces (viscous and buoyancy) present is relevant to the improvement of the stirred tank mixing equipment traditionally used in these processes.

A study of the mixing of liquids stratified by density, and with similar volume fractions and viscosities, was carried out by van de Vusse (1955), using a variety of impeller types and geometries in unbaffled tanks. van de Vusse proposed a correlation for the dimensionless mixing time in terms of the ratio of inertia to buoyancy forces, the Froude number; the dimensionless mixing time was found to decrease with decreasing Froude number, *i.e.* with increasing density difference. Similar behaviour was observed by Ahmad *et al* (1985) for thermally stratified layers with similar viscosities in baffled tanks agitated by a downward pumping marine impeller. Rielly & Pandit (1988) investigated the additional effect of viscosity differences, often present in stratified liquid systems, on the mixing rate with a Rushton turbine in a standard geometry baffled tank. Their experimental results for the individual effect of density differences on mixing times were comparable to those obtained by van de Vusse (1955) and Ahmad *et al* (1985); they also concluded that viscosity differences between the fluids restrained deformation of the 'interface' between the two stratified layers and increased mixing times. Rielly & Pandit (1988) proposed that there was a critical value of the ratio of inertia forces in the turbulent bulk flow to the viscous forces in the heavier fluid which had to be exceeded for rapid mixing to occur. All of these studies referred to above were carried out in a fully turbulent bulk flow, with Reynolds number, $Re_1 = \rho_1 N D^2/\mu_1 > 10^4$ — the notation here uses subscripts 1 and 2 to represent the bulk fluid and the added fluid, respectively, N is the impeller speed and D is the impeller diameter.

The effect of mixing small quantities of a high viscosity material into a low viscosity, fully turbulent, bulk flow in a stirred tank was discussed by Smith and Schoenmakers (1988) who injected small pulses (typically 0.1 % of the tank volume) of viscous solution (thick sugar syrups / 1-2 % w/w aqueous CMC solutions) into a 0.29 m diameter baffled tank containing water, stirred by a 6-bladed Rushton turbine. They concluded that for injection of a pulse of viscous material, the mixing rate was independent of the viscosity of the added fluid over a range of viscosity ratio, μ_2/μ_1 from 1 to 10^3. These workers also added the material as drops or 'clumps' and noted that often these 'clumps' would adhere to the tank bottom resulting in very long mixing times. The adhesion of the viscous material to the wall was found to be influenced by the location of the addition at the free surface.

Bouwmans and van den Akker (1990) examined the effects of viscosity ratio, in the range μ_2/μ_1 from 10^{-2} to 10, on mixing time for two distinct cases:

i) the addition of a viscous liquid into a low viscosity bulk, in a similar way to Smith and Schoenmakers (1988); and,

ii) the addition of a low viscosity liquid into a high viscosity bulk.

Bouwmans and van den Akker (1990) added Poly-Vinyl-Pyrrolidone (PVP) solutions to a bulk mixture of water, ethanol and glycerol; the PVP was added as drops into a 0.29 m diameter baffled tank agitated by a downward pumping pitched blade impeller. They also studied the effect of the density of the added liquid on the rate of mixing for positively and negatively buoyant additions. At low impeller speeds, materials less dense than the bulk tended to remain near the free surface and became only slowly dispersed, but addition at the impeller plane caused rapid dispersion and gave much shorter mixing times. They concluded that the mixing rate depended on the addition point and the Richardson number, but was also affected by the viscosity of the added fluid. (The Richardson number is related to the reciprocal of the Froude number and is defined in eqns. (7) and (8).) This suggests that the mixing rate not only depends on the Richardson number, but also on the Reynolds number of the flow (even for fully turbulent flow) and the viscosity ratio, μ_2/μ_1, of the two fluids.

The previous work has shown that, for small length scale additions and for small values of the viscosity ratio, there is little effect of a difference in physical properties on the dimensionless mixing time. The purpose of the present work is to determine the range of density differences and viscosity ratios over which this is true, and to explain the mechanism by which turbulent flow generated in the bulk becomes ineffective at dispersing a small addition of viscous material.

2. Theory

The turbulent mixing of miscible liquids in stirred tanks depends largely on convection, due to the pumping action of a rotating impeller, and local turbulent dispersion created by transfer of the impeller energy to large scale turbulent eddies. The largest eddies cascade their energy down to the smallest eddies, resulting in a distribution of the turbulent kinetic energy over a range of length scales. When a clump of viscous fluid is added into the turbulent bulk fluid the strain applied by the smaller eddies may not be sufficient to disrupt the high/low viscosity 'interface'; the retarding effects of the viscous forces cause an additional resistance to the straining effects exerted by the bulk flow. In these circumstances only large stresses generated by large scale energy containing eddies are able to overcome these forces and achieve good mixing.

A criterion for the conditions at which viscosity effects limit the mixing rate (in the absence of buoyancy effects) can be obtained using estimates of the magnitudes of the forces causing and retarding deformation of the viscous fluid, based on measurable quantities (*i.e.* impeller speed and diameter). The additions considered here have length scales which are larger or comparable to the integral velocity length scale of the flow and are very much larger than the Kolmogorov scale. The turbulent shear stresses or Reynolds stresses in the bulk fluid, which cause deformation and dispersion of the added fluid, are

$$\tau_1 = \rho_1 \overline{u_1 v_1} \tag{1}$$

In eq. (1) u_1 and v_1 are fluctuating velocity components in orthogonal directions within the turbulent bulk fluid. The velocity correlation $\overline{u_1 v_1}$ is related to the rms fluctuating velocities, u′ and v′, via the correlation coefficient, R

$$R = \frac{\overline{u_1 v_1}}{u_1' v_1'} \tag{2}$$

For isotropic turbulence $u_1' = v_1'$, but even for the anisotropic flow inside a stirred tank we can write that $u_1' \sim v_1'$, so that

$$\tau_1 \approx c\, \rho_1 u_1'^2 \tag{3}$$

where the constant c is related to the velocity correlation coefficient, R, and is of order 1 (Batchelor, 1951). The rms velocity fluctuation u_1' is spatially distributed throughout the tank, the largest values being found around the impeller; however, Benayad *et al* (1985) showed that at geometrically similar positions within the tank, u_1' can be scaled on the impeller tip speed (πND).

The viscous shear stresses, τ_2, in the added, high viscosity fluid, which oppose deformation by inertial effects, are given by

$$\tau_2 \approx \mu_2 \frac{u_1'}{l_1} \tag{4}$$

where u_1'/l_1 is the strain rate exerted by turbulent eddies with characteristic velocity u_1' and length scale l_1 acting on the viscous fluid. The size of the large eddies, l_1, is again varies spatially, but at geometrically similar positions within the flow it may be taken to scale on impeller diameter D. (Benayad *et al.* (1985) showed that in fully turbulent flow the integral velocity length scale was about 0.1D close to the impeller and increased with distance away from the impeller). Scaling u_1'/l_1 on ND/D gives that the rate of strain is proportional to N. This was experimentally justified by Weetman *et al.* (1988), who showed that the average rate of strain calculated from the mean velocity field (obtained using laser Doppler anemometry) at any point in the tank could be correlated as a linear function of the impeller speed.

For the deformation of the interface to occur the inertial (turbulent) stresses, τ_1, must be greater than the viscous shear stresses, τ_2, so that

$$c\, \rho_1 (ND)^2 > K \mu_2 \frac{ND}{D} \tag{5}$$

where the constant K accounts for the proportionality in the estimates of the u′ and l_1 scaled on the impeller speed and diameter. Since u_1' and l_1 are not uniformly distributed over the flow domain, the constants K and c would depend on the position within the flow that mixing takes place. An overall criterion for mixing could be based on the locations at which u_1' is a maximum and l_1 is a minimum, *e.g.* close to the impeller. Inequality (5) may then be written in terms of the Reynolds number of the turbulent fluid:

$$Re_1 = \frac{\rho_1 N D^2}{\mu_1} > \frac{K}{c}\frac{\mu_2}{\mu_1}, \qquad (6)$$

which shows that there exists a critical region for $Re_1 << \frac{K}{c}\frac{\mu_2}{\mu_1}$ below which the mixing would be very slow. It should be noted that this analysis ignores the buoyancy effects, which would further retard the distortion of the interface generated by inertial forces.

The total resistance to deformation of the viscous, inertialess addition also depends on the surface area exposed per unit volume (*i.e.* a characteristic length scale), which is related to the initial scale of segregation defined by Danckwerts (1951). A larger surface area of addition exposed to the bulk fluid would have more viscous fluid elements which would be simultaneously strained, improving the rate of deformation of the high/low viscosity 'interface'. Hence, the initial condition of the viscous addition, in terms of the interfacial area between fluids per unit volume of addition, is important in determining the mixing time. In this work the characteristic length scale of the added fluid is taken to be its volume divided by its initial surface area.

For Newtonian fluids, large differences in viscosity may also be associated with large density differences. Very viscous additions tend to maintain their identity as a single element and remain undeformed and little new surface area is generated as they circulate within the flow. In this case an initially large surface area per unit volume is required for the bulk inertial forces to reduce the length scale of the added fluid (by deformation of the high / low viscosity 'interface'), before the negatively buoyant forces cause the more dense fluid to settle to the base of the tank. The large strains experienced in the vicinity of the impeller are extremely important in causing a reduction in the length scale of the added, viscous fluid elements. If the viscous fluid is added in such a way as to avoid passing through the impeller region as it falls the base of the tank, then very poor mixing should result. Also, since the initial ratio of surface area per unit volume ($1/l_2$) depends on the method and volume of addition, then these can easily be manipulated to improve the mixing rate.

For geometrically similar tanks (constant impeller and tank geometric ratios) and location of addition, the dimensionless mixing time, $N\theta$, depends on:

$$N\theta = f\left(\frac{\rho_1 N D^2}{\mu_1}, \frac{\Delta\rho g l_2}{\rho_1 N^2 D^2}, \frac{\mu_2}{\mu_1}, \frac{l_2}{D}\right) \qquad (7)$$

or

$$N\theta = f\left(Re_1, Ri, \frac{\mu_2}{\mu_1}, \frac{l_2}{D}\right) \qquad (8)$$

where Ri is the Richardson number, and l_2 is the initial length scale of the added fluid. (This analysis ignore the effects of molecular diffusion, which would only lead to very slow mixing rates).

When the bulk fluid is in fully turbulent flow, the viscous forces in the bulk are negligible and eqn. (8) can be simplified to:

$$N\theta = f\left(Re_1 \times \frac{\mu_1}{\mu_2}, Ri, \frac{l_2}{D}\right), \qquad (9)$$

using the result of eqn.(6).

For $Re_1 \times \mu_1/\mu_2$ well above the critical value of K/c in eqn. (6) the interface is easily disrupted and the viscous material is rapidly dispersed. The mixing rate should not depend on the viscosity of the added fluid and only on the Richardson number and initial length scale l_2; l_2 would be affected by the method of addition, *e.g.* addition as a narrow jet or as a single mass. There is some difficulty in estimating an exact value for this length scale, however for any one method and volume of addition it should be constant, so that the dimensionless mixing time should be correlated as a function of Re_1, μ_1/μ_2 and Ri only.

3. Mixing Experiments with Differences in Density and Viscosity

Experiments on the effects of viscosity and density differences on the mixing time were carried out in a 0.29 m diameter standard geometry tank (Fig. 1) agitated by a Rushton turbine at rotational speeds, N, between 100 and 400 rpm. A few experiments were also performed with a 6-bladed 45° pitched turbine for comparison with the Rushton turbine.

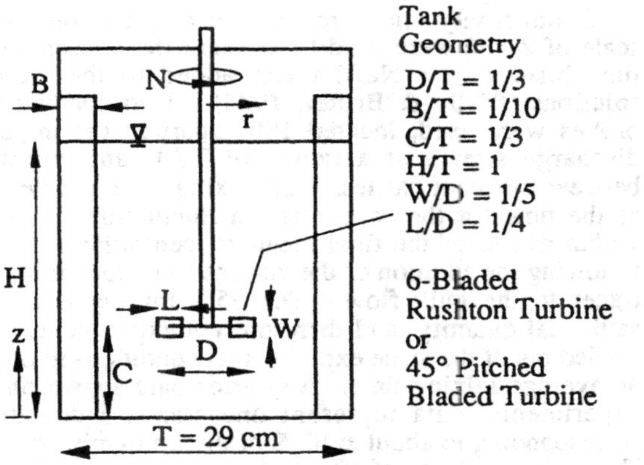

Figure 1 The standard tank configuration

Water was the bulk fluid for all the experiments and for all impeller speeds the bulk Reynolds number, Re_1 was greater than 10^4 (*i.e.* fully turbulent flow). Dense salt solutions — NaCl, $CaCl_2$ and $ZnCl_2$ — having ρ_2 from 1100 to 1800 kg/m^3, but with similar viscosities to water, were used for the study of the influence of buoyancy forces (Richardson number) on the dimensionless mixing time. The viscous fluids added were aqueous glucose syrup solutions (Newtonian) with μ_2 in the range 1 to 8 Pa s and $\rho_2 \approx 1400$ kg/m^3. Lower density aqueous PVP solutions with apparent viscosities from 1 to 5 Pa s (shear thinning) were used to decrease the density to 1100 kg/m^3, whilst maintaining

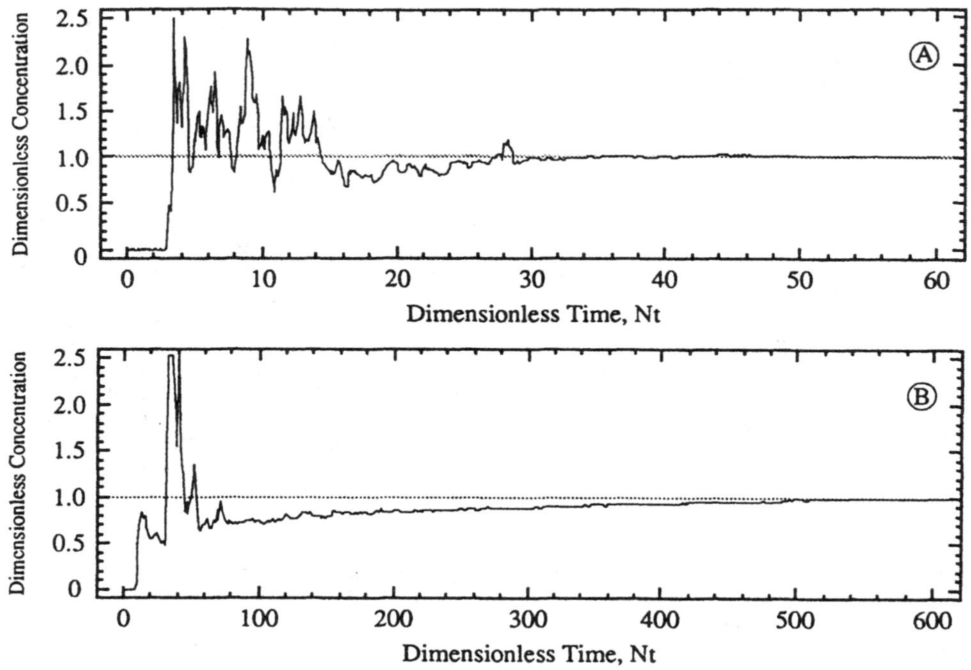

Figure 2 Comparisons of concentration time histories for (A) a low viscosity addition, and (B) a high viscosity addition

approximately the same range of apparent viscosities as the glucose syrup solutions.

Conductivity microprobes with a resolution length scale of 2 mm were used to measure the concentration time history of a NaCl tracer added to the viscous solutions (Rielly & Britter, 1985). Two conductivity probes were used, located 180° apart in the impeller discharge stream at a radius of T/4, and midway between adjacent baffles. The mixing time was defined as the time for the concentration fluctuations to decay within ± 5 % of the final mean concentration (Fig. 2), following the addition of the viscous fluid containing the tracer to the bulk flow. The 95% mixing time is a statistical quantity, and therefore ten experiments were carried out at the same experimental conditions to obtain an average mixing time. Any error bars shown on the experimental data represent one standard deviation, corresponding to about ± 10 % of the mean mixing time. Typically a volume of 10 ml of viscous tracer fluid was added as a single mass at the free surface, at a point T/4 from the wall. In a few cases the viscous liquid was added as a narrow stream from a large syringe, allowing the length scale of the addition to be varied, but maintaining the same volume of addition.

4. Analysis of the Experimental Results

4.1 *The Influence of Density Differences*

Negatively buoyant additions at the free surface, with similar viscosities to the bulk, follow the motions of the turbulent bulk fluid; the density difference between the added fluid and the bulk gradually becomes less intense as the interface of the initial clump is eroded and entrained into the bulk fluid. Mixing takes place due to distortion and stretching of the tracer fluid by large and small velocity fluctuations within the bulk fluid, in a similar way to the turbulent mixing of similar property fluids. Figure 3 shows that there is no effect of Richardson number on the dimensionless mixing time, $N\theta$, over the range of initial length scales and density differences investigated. These results are restricted to additions with a small length scale, l_2, and it might be expected that for larger volume fraction additions, there would be an effect of Ri on the dimensionless mixing time.

Rielly and Pandit (1988) and van de Vusse (1955) found that for initially stratified liquids with large volume fractions, the dimensionless mixing time increased with increasing Richardson number. The release of larger volumes from the free surface would allow us to increase the range of the Richardson number to give similar values to those for initially stratified layers (a worst case); this is the subject of further work.

4.2 *The Influence of Viscosity Differences*

In practice it is difficult to obtain Newtonian fluids with large differences in viscosity but the same density as the bulk fluid, water. The study using glucose syrup solutions allowed the viscosity ratio to be varied whilst maintaining an *approximately* constant density difference, so that for a constant impeller speed, the Reynolds and Richardson numbers could be maintained

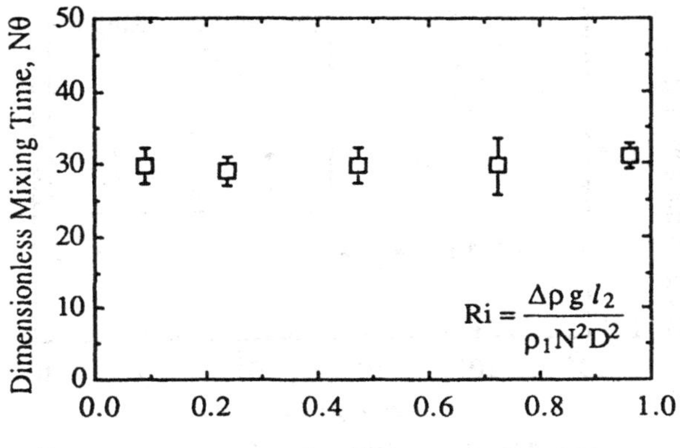

Figure 3 The effect of Richardson number on Nθ for a viscosity ratio of $\mu_2/\mu_1 \approx 1$ and $Re_1 = 1.8 \times 10^4$. 10 ml of salt solutions was added at the free surface (Rushton turbine).

constant, whilst varying the viscosity ratio, μ_2/μ_1. The rates of mixing obtained with these glucose syrup solutions could also be compared to those of $CaCl_2$ solutions with similar Richardson numbers, but in the latter case, with a viscosity ratio of about one.

Visual observations and photographs show that when $Re_1 \times \mu_1/\mu_2$ was greater than about 10^2 small scale eddies could not distort the clump and only large scale deformations were evident. The initial shape of the addition was approximately maintained until the added fluid reached the impeller plane where large scale turbulence disrupted and split the single mass into a number of elements. If these elements still had a sufficiently large length scale, negative buoyancy would cause a considerable fraction to fall to the tank bottom, where they were only slowly dispersed by low shear within the circulating flow.

Figure 2 gives a comparison of the concentration time histories for (A) a low viscosity addition, and (B) a high viscosity addition. The former shows the rapid decay of small scale concentration fluctuations of the added tracer, and fairly rapid mixing, whereas, (B) shows very few concentration fluctuations and a slow approach to homogeneity. In the latter case the added fluid was not well dispersed as it passed through the impeller and it settled to the base of the vessel. For the large viscosity ratio in (B) the turbulent shear stresses exerted by the bulk on the viscous fluid on the base of the tank were not sufficiently strong to rapidly distort the viscous interface, and mixing was slow.

Figure 4 shows that at low values of the viscosity ratio the dimensionless mixing time was short ($N\theta \approx 28$ for the disc turbine) and independent of μ_2/μ_1. For the glucose syrup solutions with $Re_1 \times \mu_1/\mu_2 < O(10^2)$, the dimensionless mixing time was strongly affected by the viscosity ratio. The slow dispersion process increased the mixing time by up to one order of magnitude, compared to situation where the added fluid was completely dispersed in the bulk flow and did not adhere to the base of the tank. These results are in qualitative agreement with the findings of Smith & Schoenmakers (1988), although these workers presented no indication of the range of conditions under which mixing was poor.

Non-Newtonian aqueous PVP solutions were used (as the viscous addition) in some experiments to study the effect of smaller density differences on the mixing time, for the same range of viscosities as the Newtonian fluids previously studied. This gave $\Delta\rho \approx 100\ kg/m^3$ for PVP, compared with $\Delta\rho \approx 400\ kg/m^3$ for glucose syrup solutions. The apparent viscosity of the PVP solutions were calculated at the bulk fluid average shear rate calculated by the Metzner & Otto (1957) method, *i.e.*

$$\dot{\gamma} = \beta N, \tag{10}$$

where the shear rate constant, β, depends on impeller type. (Strictly this method is only valid in the laminar regime, but has been used here to obtain an estimate of the average shear rate experienced by the added fluid.) Typical average shear rates in these experiments were 20 to 100 s^{-1}. Photographs and observations of the mixing process showed that at low impeller speeds the clump of added fluid was advected by the mean flow to the impeller region. Close to the impeller the clump became finally divided and eventually a fraction of the viscous material fell to the bottom, where it was slowly dispersed. In this case the mixing times were much longer than for fluids, with the same Newtonian viscosity as the apparent viscosity of the PVP solutions. However, figure 4 shows that at a slightly larger Reynolds number, the PVP solutions were easily dispersed, giving dimensionless mixing times only slightly larger than those for fluids with similar properties.

These anomalous results can be explained by closer examination of the shear thinning behaviour of aqueous PVP solutions, particularly at the low strain rates which would be typical found near the base of the tank. Figure 5 shows that although this PVP solution has an approximately constant viscosity over the tank average shear rates, its viscosity is much larger at low shear rates. At low impeller speeds this PVP solution was not fully dispersed as it passed through the impeller region, and a large fraction settled to the base of the tank, where it had a high apparent viscosity because of the low rates of

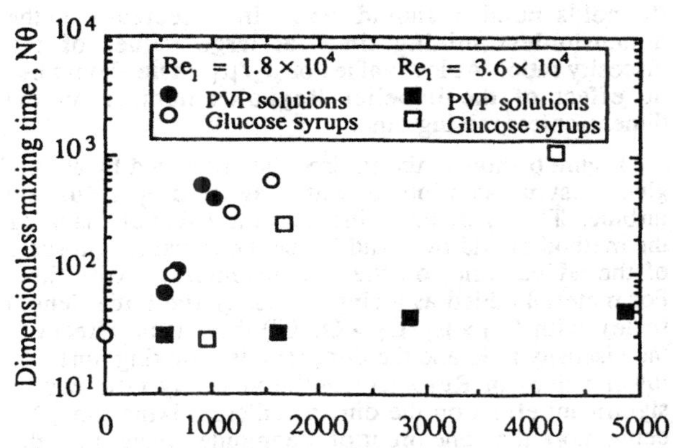

Figure 4 The effect of density differences on the dimensionless mixing time, for different viscosity ratios at approximately constant Re and Ri (Rushton turbine). For the glucose syrup solutions $Ri \approx 0.47$ and for the PVP solutions $Ri \approx 0.12$.

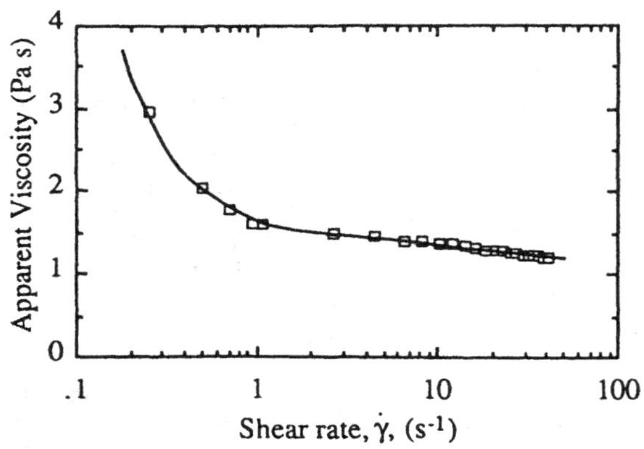

Figure 5 The apparent viscosity of a 13 % w/w aqueous PVP solutions at various strain rates.

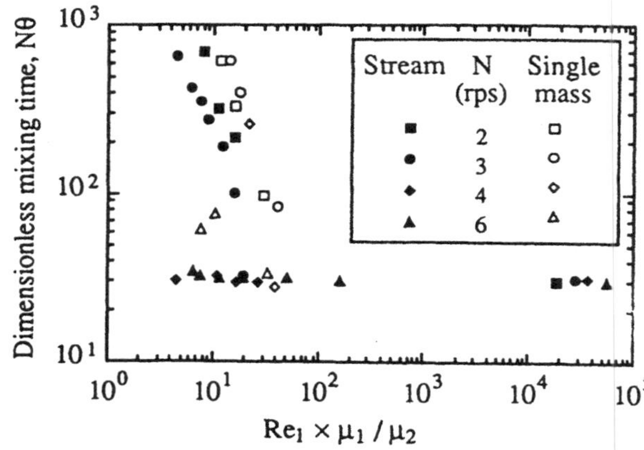

Figure 6 The effect of the group $Re_I \times \mu_I / \mu_2$ on the dimensionless mixing time, using aq. glucose syrup solutions added at the free surface, either added as a stream or as a single mass (Rushton turbine).

shear within the flow. At higher impeller speeds, the same PVP solution was better dispersed in the impeller region, reducing the local length scale of the added fluid. In this case buoyancy did not much affect the added fluid; it did not settle to the base and circulated with the bulk flow. There it had a low apparent viscosity, because it was advected in high shear regions, and it had a short dimensionless mixing time.

A thorough comparison between the mixing rate of the Newtonian fluids and the PVP solutions was not possible because the effect of a change in the Richardson number was confounded by the varying apparent viscosity of the non-Newtonian fluid, depending on whether it circulated with the bulk flow or settled to the base of the tank.

4.3 *The Effect of Impeller Speed*

Eqn.(5) is a statement that the inertial stresses generated in the bulk fluid must exceed the viscous restraining stresses in the viscous fluid, for rapid mixing to occur. Since the Reynolds stresses scale on N^2 and the viscous stresses on N, then an increase in impeller Reynolds number should result in a decrease in the dimensionless mixing time, at large values of the viscosity ratio. At low values of μ_2/μ_1, there should be no effect of the impeller Reynolds number on the dimensionless mixing time.

Figure 6 shows mixing time data obtained by adding glucose syrup solutions to water, agitated by a Rushton turbine. The figure also illustrates the effect of changing the method of addition, and hence the initial length scale of the added fluid, on the dimensionless mixing time. For material added as a single mass (large initial length scale), with $Re_1 \times \mu_1/\mu_2 > O(10^2)$ there is no effect of the viscosity ratio and the dimensionless mixing time. At lower values of $Re_1 \times \mu_1/\mu_2$ the viscosity ratio has a significant effect on the dimensionless mixing time, Nθ being more than one order of magnitude greater than the value for mixing of similar property fluids. The data does not collapse onto a single line in this region, illustrating the additional effect of the Richardson number on the dimensionless mixing time, *when viscosity differences are also present.* Experiments performed at low values of the Richardson number, *i.e.* at a high impeller speed, with a small density difference, or with a small initial length scale for the added fluid have a small value of Nθ. This demonstrates that if buoyancy effects are negligible, then the viscous material is convected with the mean bulk flow and is easily dispersed; however, if buoyancy effects are significant, they will cause the viscous addition to adhere to the base of the tank, in regions where the intensity of turbulence and shear rates are weak, and very much longer mixing times result. The overall conclusion from these results is that density differences between the fluids exacerbate the increase in mixing time, caused by having a viscosity ratio between the added and bulk fluids. In contrast to the situation described in §4.1, the Richardson number has a significant effect on the dispersion of a material which has a *large* viscosity ratio with the bulk fluid.

4.4 *The Effect of the Method of Addition*

Figures 6 and 7 demonstrate that addition of the same volume of viscous, dense fluid as a stream rather than as a single clump leads to better mixing performance.

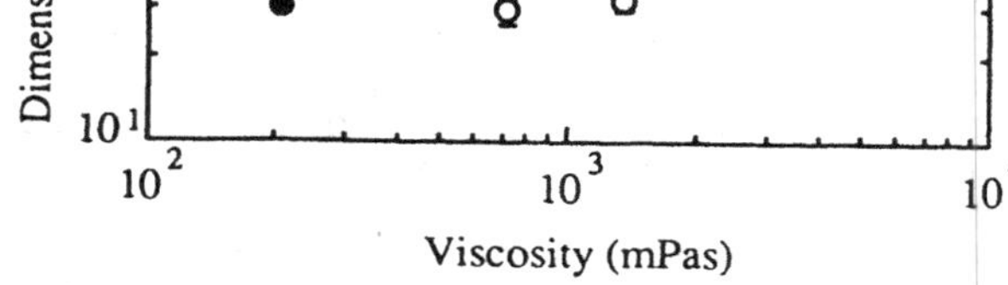

Figure 7 The effect of the method of addition on the dimensionless mixing time for different viscosity solutions at $Re_I = 3.6 \times 10^4$ (Rushton turbine).

When added as a stream, the large surface area of high viscosity material exposed per unit volume increases the mixing rate as more fluid elements are simultaneously strained, *i.e.* the added fluid has a smaller characteristic length scale (and hence also a smaller Richardson number) than when added as a single clump. The buoyancy effect on the fluid added as a stream is reduced and smaller fractions of the viscous fluid fall to the tank bottom, giving shorter dimensionless mixing times. Moreover, addition as a stream causes some degree of initial dispersion, before the added fluid is acted upon by the bulk flow. By comparison, when added as a clump, the large volume per unit surface area gives a large initial length scale; buoyancy forces cause the fluid to settle to the bottom of the tank, and longer mixing times result.

4.5 *The Effect of Location of Addition*

The effect of the location of the addition was studied when the fluid was added as a stream at the free surface. The positions investigated were: (i) at the wall; (ii) at a radius of T/4; and, (iii) close to the shaft. Figure 8 shows that a strong effect on the mixing rate was observed at low impeller speeds (N = 120 rpm), the longest time resulting from the addition at the mid-point between the wall and the shaft. Visual observations show that at position (ii) the stream is not much deformed until it reaches the impeller plane, where a fraction gets dispersed and the remaining volume elements enter the lower circulation loop and are convected to the tank bottom. By comparison, for an addition close to the wall, the upward circulation flow along the wall causes additional dispersion of the fluid. The resulting decrease in the length scale of the added fluid reduces the negative buoyancy effects. For location iii), close to the shaft, the stream is convected to the impeller where it is easily mixed due to the high shear rates and turbulence intensity in this region. At high impeller speeds it was found that the point of addition had a reduced influence on the dimensionless mixing time, as might be expected due to the more intense agitation throughout the tank. The best addition point was always found to be close to the shaft.

4.6 *The Influence of the Impeller Type*

Experimental measurements and numerical predictions of the flow field generated by a Rushton turbine in a stirred tank show the presence of two circulation loops. In the mixing experiments described here, any fluid elements of the viscous material which are not sufficiently dispersed during their first passage through the impeller, enter the second circulation loop below the impeller, where they adhere to the tank bottom. This results from the additions being negatively buoyant, and retaining a large enough local Richardson number after passage through the impeller, for buoyancy effects to be significant.

For the pitched-blade impeller an axial flow is induced, with only a single circulation loop. When pumping downwards an addition from the free surface is drawn into the impeller region where disruption of the viscous material occurs.

The measurements of the dimensionless mixing times

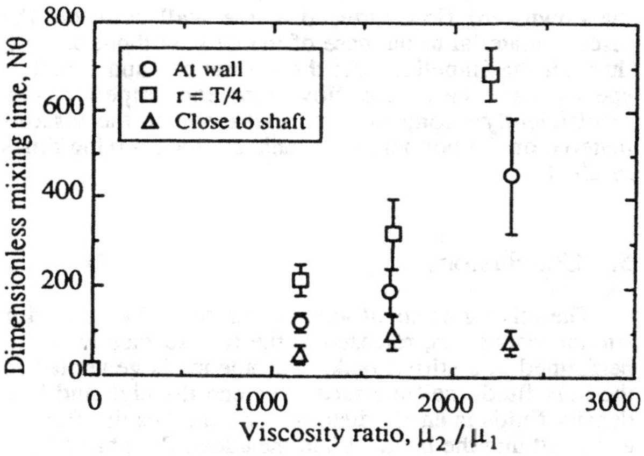

Figure 8 The effect of the addition point on the dimensionless mixing time when the viscous fluid was added as a stream ($Re_1 = 1.8 \times 10^4$, $Ri = 0.05$) (Rushton turbine).

for these two impellers are shown in figure 9, for different viscosity ratios. At low values of $Re_1 \times \mu_1/\mu_2$ the pitched-blade turbine (downward flow) has slightly shorter dimensionless mixing times than the Rushton turbine. This is because the added viscous fluid is drawn into the suction side of the impeller flow and only a small fraction settles to the bottom. Furthermore, the downward axial flow from the pitched-blades impinges directly on any viscous material which has settled on the base of the tank. However, at higher values of $Re_1 \times \mu_1/\mu$ the dimensionless mixing time is about 50 % longer for the pitched-blade turbine than for the Rushton turbine; this is in good agreement with the correlations by Prochazka & Landau (1961) for mixing of similar property fluids.

For the pitched-blade impeller pumping *upwards*, a large fraction of fluid fell to the bottom and the dimensionless mixing times were increased significantly, compared to the downward pumping impeller. The upward discharge flow carries the negatively buoyant addition away from the impeller region, and moreover,

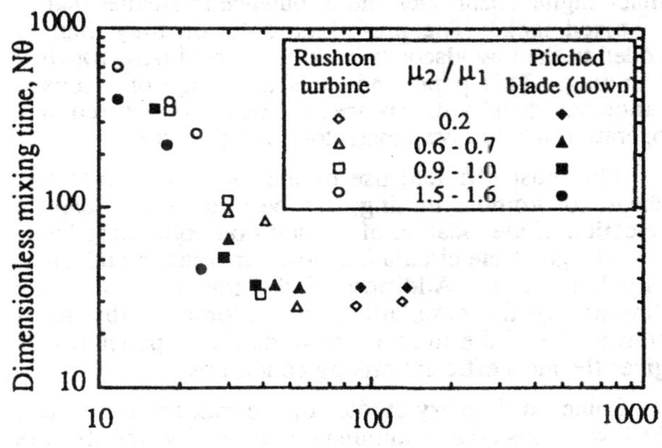

Figure 9 A comparison of the dimensionless mixing time for a Rushton turbine and a downward pumping pitched-blade turbine, over a range of viscosity ratios and impeller speeds.

the downward flow induced at the wall convects the viscous material to the base of the tank, without passing through the impeller. At the clearances and impeller speeds used here, the flow into the impeller was insufficiently strong to shear and disperse the viscous material on the bottom of the tank and long mixing times resulted.

5. Conclusions

The mixing of small volumes of dense liquids, with similar viscosities, released at the free surface is easily performed in a stirred tank. Turbulence is generated in the bulk fluid, the 'interface' between the high and low density fluids is easily distorted and the heavier fluid is entrained into the bulk. When the added fluid has a high viscosity compared to the bulk, viscous stresses within the added fluid restrain its deformation and turbulence cannot be generated in the added fluid. In this situation only small length scales additions (large surface area exposed per unit volume), with a low Richardson number, remain in the bulk fluid and get easily mixed. In the early stages of mixing, shear and elongation are predominant, but as new surface area is generated, local turbulent dispersion and convection mechanisms become important . For large initial length scales of added fluid and for high viscosity ratios, a fraction of viscous material falls to the tank bottom and the mixing time can increase by more than one order of magnitude. This would occur when the group $Re_1 \times \mu_1/\mu_2$ is less than a critical value which depends on the value of the Richardson number, and for our experiments is $O(10^2)$. In this case the inertial effects experienced by the added fluid close to the impeller are not sufficient to disperse the interface with the bulk fluid, and hence stirred tanks are not adequate mixing devices.

The requirement of large shear and elongational stresses to decrease the initial length scale of the added fluid (or the initial striation thickness) of the added viscous fluid can be achieved by pre-mixing in an appropriate device, *e.g.* a static or dynamic in-line mixer. These mixers use high specific rates of energy dissipation, but within a small overall volume, to achieve much higher shear rates and turbulence intensities than in a stirred tank. The pre-mixed solution may then be added to the low viscosity bulk in a stirred tank, for final blending. This paper indicates the range of viscosity ratios and density differences for which the stirred tank operation would give satisfactory performance.

The most efficient use of the bulk turbulence and shear, to promote mixing, can be obtained by careful selection of the location of the addition point, based on a knowledge of the circulation flow generated by different impeller types. Addition of the tracer fluid in the direction of the circulation flow to obtain subsequent convection of the material towards the impeller region gives the most efficient mixing conditions.

Some preliminary studies of the mixing of aqueous PVP solutions (shear-thinning) solutions showed that the wide range of shear rates within the stirred tank flow can cause a non-Newtonian material to exhibit a much higher apparent viscosity than might be calculated from the Metzner & Otto (1957) approach. If the non-Newtonian fluid is trapped in low shear regions of the vessel it behaves as a very viscous fluid and has a slow mixing rate.

6. References

Ahmad, S. W., Latto, B. & Baird, M. H. I., 1985, *Mixing of stratified liquids*, Chem. Eng. Res. Des., **63**, 157-167.

Batchelor, G. K., 1951 , Proc. Cambridge Phil Soc Maths Phys. Sci. **47**, 359.

Benayad, S., David, R. & Cognet, G., 1985, *Measurement of Coupled Velocity and Concentration Fluctuations in the Discharge Flow of a Rushton Turbine in a Stirred Tank*, Chem. Eng. Process, **19**, 157-165.

Bouwmans, I. & van den Akker, H. E. A., 1990, *The Influence of Viscosity and Density Differences on Mixing Times in Stirred Vessels.*, Fluid Mixing IV.

Danckwerts, P.V., 1951, *The Definition and Measurement of some Characteristics of Mixing*, Appl. Sc. Research, A3, 279-296.

Harnby, N., Edwards, M.F. & Nienow, A.W., 1985, *Mixing in the Process Industries* , Butterworths.

Metzner, A. B. & Otto, R. E., 1957, *Agitation of non-Newtonian Fluids* , A.I.Ch.E.J., **3** (1), 3.

Prochazka, J. & Landau, J., (1961), Coll. Czech. Commun., **26**, 2961.

Rielly, C.D. & Pandit, A.B., 1988, *The Mixing of Newtonian Liquids with Large Density & Viscosity Differences* , 6th Eur. Conf. Mixing, Pavia, Italy, pp. 69-77.

Rielly, C.D. & Britter, R.E., 1985, *Mixing Times for passive Tracers in Stirred Vessels*, 5th Eur. Conf. on Mixing, Wurzburg, FRG, pp. 365-376.

Smith, J.M. & Schoenmakers, A. W., 1988, *Blending of Liquids of Different Viscosity* , Chem. Eng. Res. Des., **66**, 16-21.

van de Vusse, J.G., 1955, *Mixing by Agitation of Miscible Liquids* , Chem. Eng. Sci., **4**, 178-200.

Weetman, R. J. & Oldshue, J. Y., 1988, *Power, Flow and Shear Characteristics of Mixing Impellers* , 6th Eur. Conf. Mixing, Pavia, Italy, pp. 43-50.

7. Notation

c	Constant in eq. (3)	
D	Impeller diameter	m
g	Gravity acceleration	m/s^2
K	Constant in eq. (5)	
l_1	Length scale of eddies	m
l_2	Initial length scale of the added fluid	m
N	Impeller speed	rps
R	Velocity correlation coefficient	
Re	Reynolds number	$(\rho ND^2/\mu)$
Ri	Richardson number	$(\Delta\rho g l_2/\rho_1 N^2 D^2)$
t	Time	s
u′, v′	Rms fluctuating velocity component	m/s

Greek

β	Shear rate constant in eq.(10)	
$\Delta\rho$	Density difference	kg/m^3
μ	Dynamic viscosity	Ns/m^2
θ	Mixing time	s
ρ	Density	kg/m^3
τ	Shear stress	N/m^2

Subscripts

1	Bulk fluid
2	Secondary fluid to be mixed

POWER INPUT AND GAS HOLD UP IN GAS LIQUID DISPERSIONS AGITATED BY AXIAL FLOW IMPELLERS

V. MACHON, Prague Institute of Chemical Technology, CS
C.M. MCFARLANE*, A.W. NIENOW, University of Birmingham, GB

ABSTRACT

The power and gas hold up characteristics of a five blade Maxflo T hydrofoil impeller, (Prochem Ltd., D=0.45T), and of a six blade, 45° mixed flow impeller, (D=0.5T) have been studied in a cylindrical baffled vessel (T=0.56m) of aspect ratio 1:1. The influence of the interfacial properties of the liquid on power input and gas hold up were investigated in de-ionized water (coalescing system); and in solutions of sodium sulphate and poly-propylene glycol (non-coalescing systems).

Studies were performed on the Maxflo T impeller in a downward pumping mode and on the mixed flow impeller in both a downward and upward pumping mode. Comparisons were also made between these impellers and the Rushton disc turbine in an air-water system. The hold ups in each case were approximately the same except for the Maxflo T which gave higher ones, this effect increasing as the system became more resistant to coalescence. This finding may go some way to explaining the improved fermentation performance of the Maxflo T which has been reported.

INTRODUCTION

Recently, research was started on gas-liquid dispersion in stirred vessels using new types of mechanical stirrers. In the past, the Rushton turbine has usually been recommended for the dispersion of gas in liquids and power input, gas-hold-up, the conditions of flooding and mass transfer coefficients were studied in detail[1,2,3,4]. Later, it was discovered that good gas dispersion could be achieved by stirrers with combined radial-axial flow of liquid. Such stirrers could be constructed as a turbine with inclined blades[5] or with shaped blades of more complicated design[6].

Such inclined blade impellers could operate in two ways: the liquid could be pumped down (MFD) or up (MFU). Using these stirrers, a good gas dispersion could be achieved as well as good top-to-bottom circulation of the liquid[7].

* present address, University of Surrey, U.K.

These stirrers usually have lower power numbers than Rushton turbines. In addition, the decrease of power during gas dispersion is often lower[7]. On the other hand at the same power input and superficial gas velocity, equal k_La's[6,8] and hold-ups[5,9] have been found during experiments conducted in water. However, work in fluids containing coalescence inhibitors has not been conducted comparing different agitators.

This paper reports the results of experimental studies of gas hold-up and power input in gas-liquid dispersion using pitched blade turbines with inclined blades pumping water up or down and Prochem stirrers with shaped blades pumping down. The influence of the presence of a dissolved inorganic salt as well as an antifoam agent was studied too.

EXPERIMENTAL

The power input and gas hold-up were measured in a flat-bottomed cylindrical vessel with liquid depth equal to the vessel diameter equipped with four vertical wall baffles. (Fig.1). Two types of stirrer were used: a five bladed Prochem Maxflo T with shaped blades (Fig. 2a) and a six bladed, 45°-pitch, turbine (Fig. 2b). The Prochem stirrer always pumped the liquid downwards; the pitch blade turbine up or down. The power input was measured by a strain-gauge mounted on the shaft and the torque was transmitted by telemetry. The full details of the measuring method was described elsewhere[10]. The gas hold-up was calculated from the measurements of average height of the gas-liquid dispersion level using an ultrasonic echo unit, Nivosonic FMU 2780. The stirrer speed was in the range 100-450 rpm and the gas flow rate used was 0-2.0 vvm.

The liquids were used deionised water, Na_2SO_4 solution at concentrations of 0.05, 0.1 and 0.2 moles per litre and a water solution of polypropylene glycol (PPG) at 0.1 vol %. The density of all solutions were approximately equal to the density of the water. The experiments were performed at 20^oC. The values of the surface tension of Na_2SO_4 solutions were very close to those of water (water = 72.8×10^{-3} kg s^{-2}; 0.2 m Na_2SO_4, 69.0×10^{-3} kg s^{-2}) whilst the presence of 0.1%

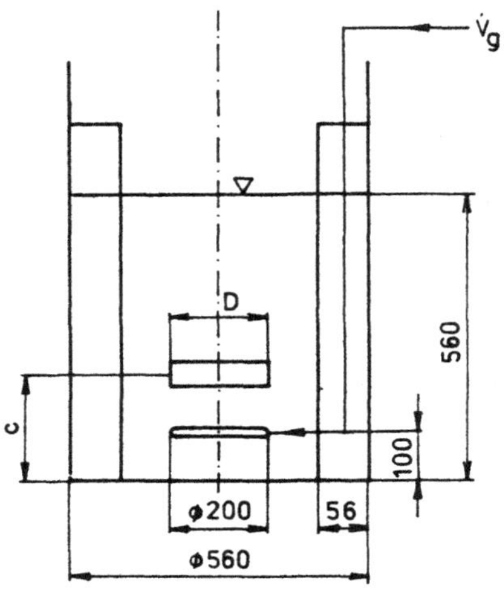

	PROCHEM	PBT
D	251.5	280
c	252	140

Fig.1 **The Overall Rig**

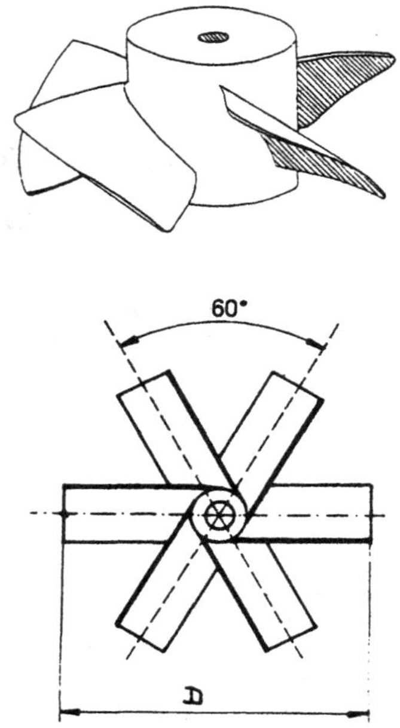

Fig.2 **The Prochem Maxflo T and Pitched Blade Turbine**

polypropylene glycol strongly influenced the surface tension (38.7×10^{-3} kg s^{-2}).

RESULTS AND DISCUSSION

The ungassed power numbers were found to be independent of the liquids for all impellers. The measured values of power numbers (defined as $Po = P/\rho N^3D^5$) was for the Prochem stirrer Po = 1.35; for pitch blade turbine pumping up (MFU), Po = 1.72; and for pitch blade turbine pumping down (MFD), Po = 1.85. At higher speeds, the suction of air through the surface was observed which caused a decrease of power number.

The results of power measurement in gas-liquid dispersions are presented in the form of relative power, P_g/P versus gas flow number ($N_F = \dot{V}_g/(ND^3)$) where P_g is the power input into the aerated liquid.

In Figs. 3,4,5 are plotted P_g/P vs N_F at constant gas flow rate 1.5 of vvm. There is a distinct minimum in the curves for the Prochem and for the MFD in contrast with the MFU showing a smooth decrease in P_g/P with decreasing values of N_F, e.g. with increasing stirrer speed as found previously(5,6,7). With the Prochem and MFD, there is a sharp

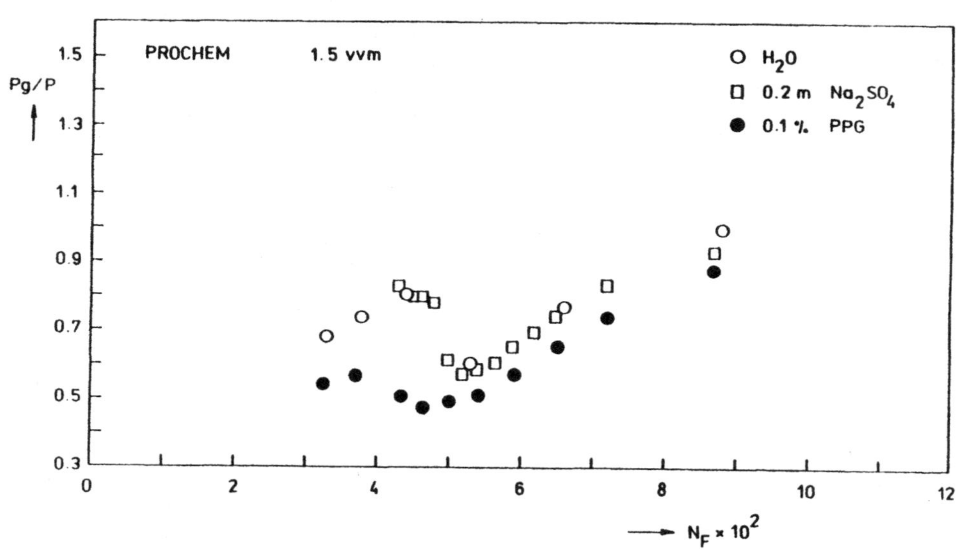

Fig.3 P_g/P versus Flow Number at Constant Air Flow for the Prochem

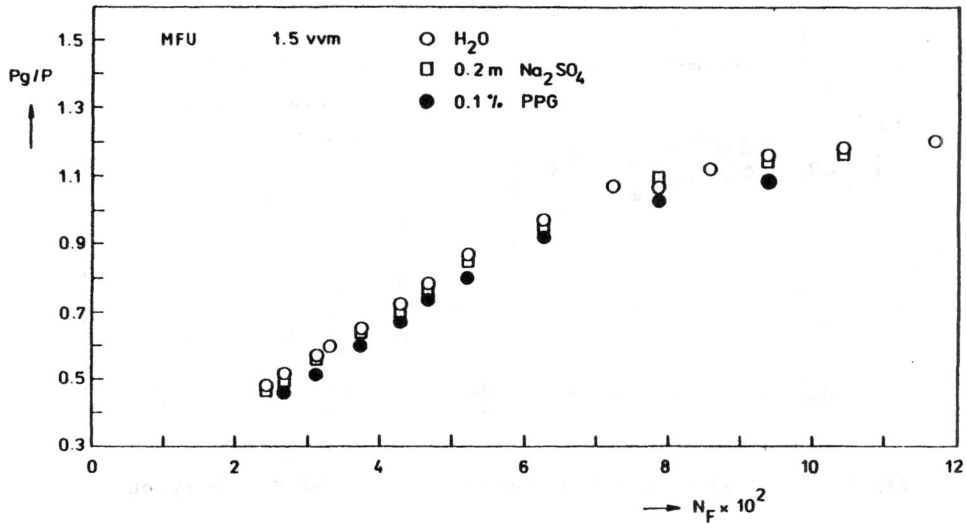

Fig.4 P_g/P versus Flow Number; MFU

increase in P_g/P at the lowest value of N_F i.e. at high speeds, with a minimum at $N_F = 4.5 \times 10^{-2}$ for the Prochem and at $N_F = 3.5 \times 10^{-2}$ for the MFD. These changes relate to direct -indirect loading and torque instabilities[5,8,11].

The influence of the physical properties of liquids on gassed power input is not very strong for Rushton turbines[2]. The same is true here as can be seen from Figs 3,4 and 5. However, as seen in Figs 7-12, the physical property has a profound effect on hold-up. This is because in pure water (a so-called coalescent system), the bubbles tend to coalesce as they touch each other, giving large bubbles in such dispersions. Sodium sulphate solutions are typical of non-coalescent systems, in which the bubbles do not coalesce easily and the dispersion consists of smaller bubbles (typically 2-3mm diameter). Though polypropylene glycol is an antifoam agent, it generates a very low surface tension, very tiny bubbles and a very high hold-up in water.

Clearly the power input is dependent mainly on the sparged gas flowrate, the influence of the liquid properties on power input being significant only in the region of the minima of the power curves (Figs. 3 and 5) i.e., under conditions of indirect loading where gas enters the impeller after recirculation. On the other hand, Fig. 6 is a plot of P_g/P vs N_F for a constant stirrer speed for Prochem stirrers for these liquids. The curves all have the usual shape, i.e., fall in power with increasing Q_G with a rapid drop marked by a dashed line at the indirect-direct loading transition. In PPG solutions, the small bubbles recirculate so that at low gas flow rates there is a much sharper decrease of P_g/P ratio. It can be concluded that the physical properties of the liquid have an effect on power input mainly during high levels of gas recirculation.

The gas hold-up, Z, (volumetric ratio of the gas present in the gas dispersion) data are

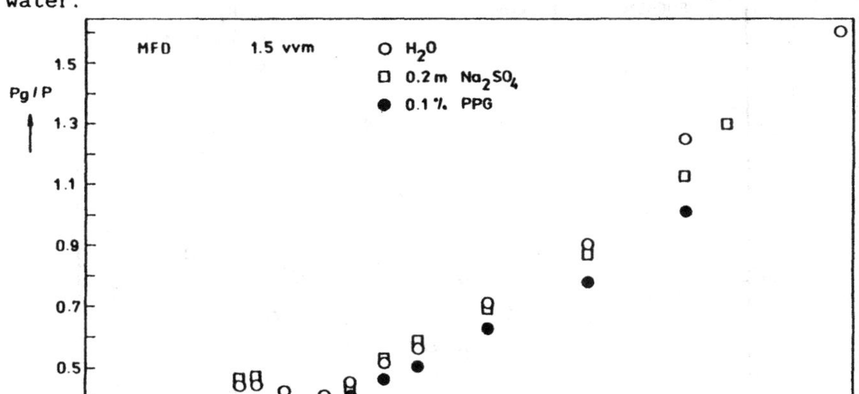

Fig.5 P_g/P versus Flow Number; MFD

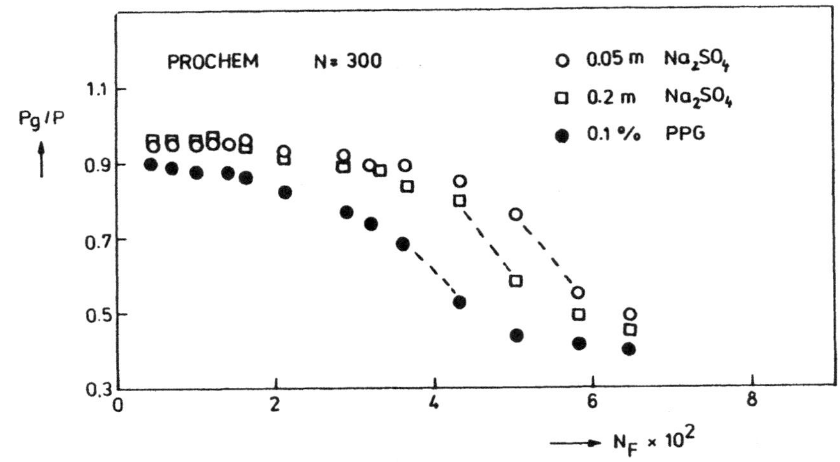

Fig.6 P_g/P versus Flow Number at Constant Speed; Prochem

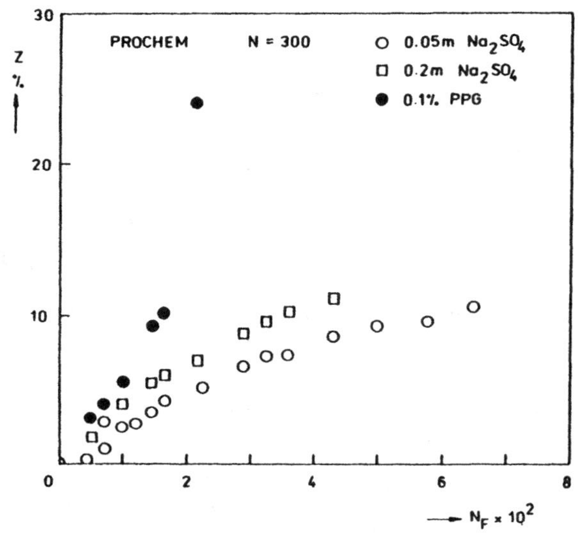

Fig.7 Hold-up versus Flow Number; Prochem

given in Figs. 7-13. In Figs 7-11, the data are shown for each type of impeller for each fluid. Z data are plotted for variable speed and variable flow rate. In all cases, the sodium sulphate solution produces a slightly higher hold-up than in water; and the PPG dramatically increases it. The effect is shown in more detail in Fig. 12. This finding fits in well with earlier work on the effect of coalescence inhibitors[12]. On the other hand, Fig. 13 compares the hold-up at 1.5 vvm from the different impellers in the different fluids at equivalent power inputs. It includes data from a Rushton turbine in water[11]. This figure shows that in water all the impellers produce broadly the same hold-up. Such a result fits in well with earlier results on water[5,8,9]. However, as the fluids become increasingly non-coalescent, the greater is the hold-up with the Prochem impeller. Indeed, there is progressively a significant difference between the different agitators, with hold-up at the same conditions greatest in the order Prochem - MFD-MFU,

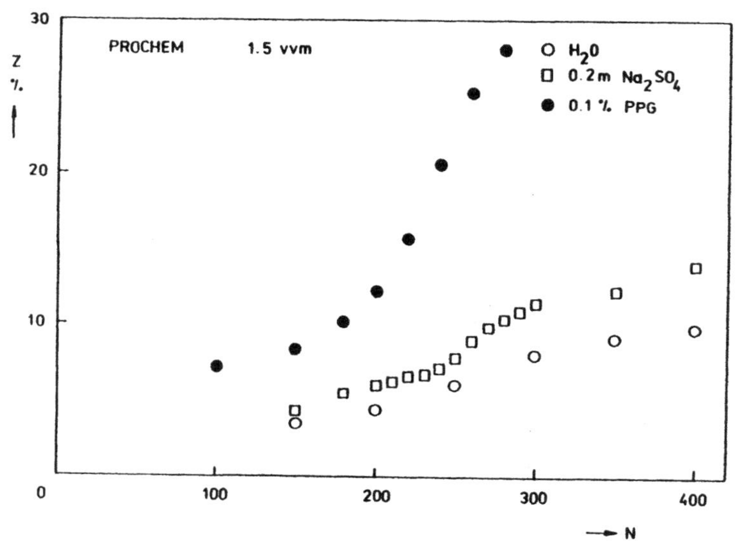

Fig.8 Hold-up versus Speed; Prochem

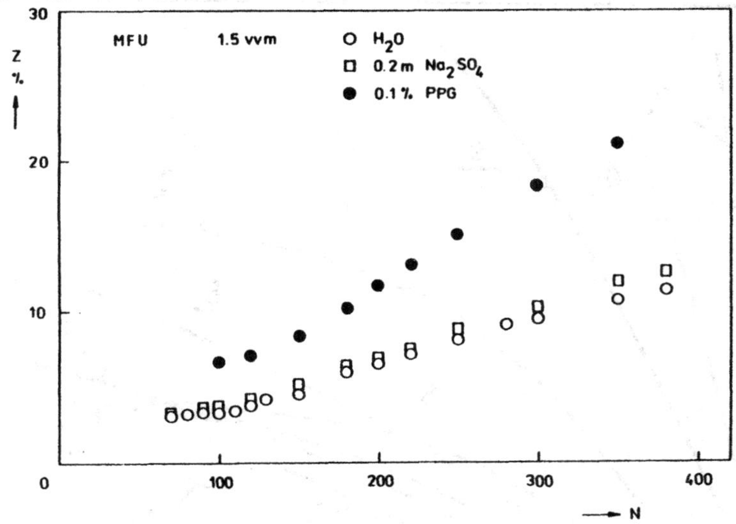

Fig. 9 Hold-up versus Speed; MFU

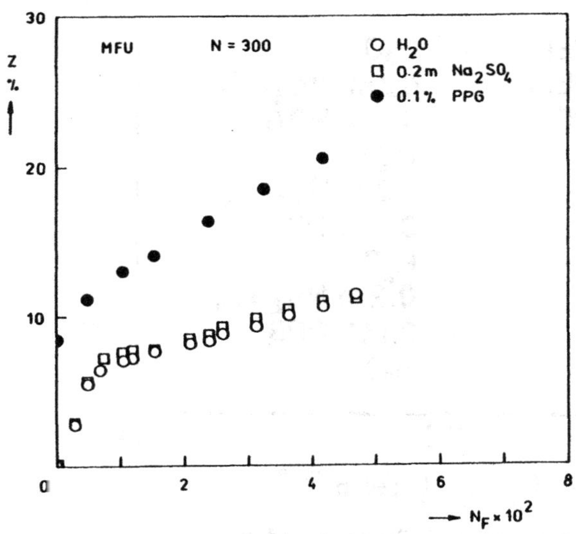

Fig. 10 Hold-up versus Flow Number; MFU

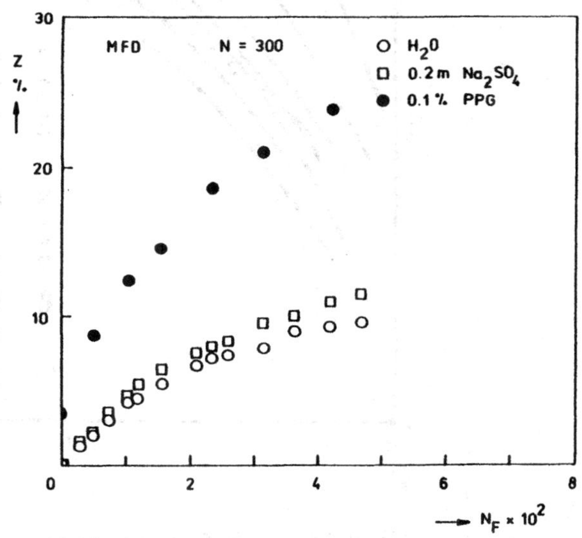

Fig. 11 Hold-up versus Flow Number; MFD

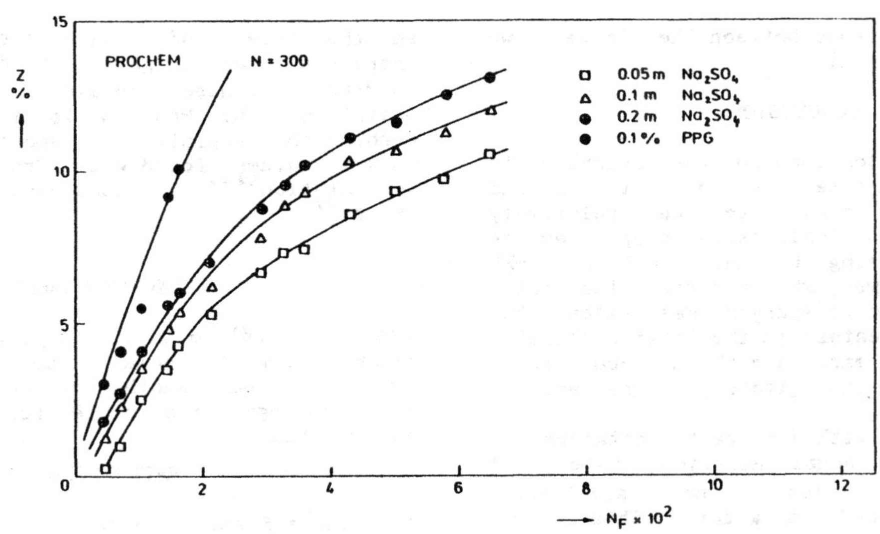

Fig. 12 Hold-up versus Flow Number for All Solutions; Prochem

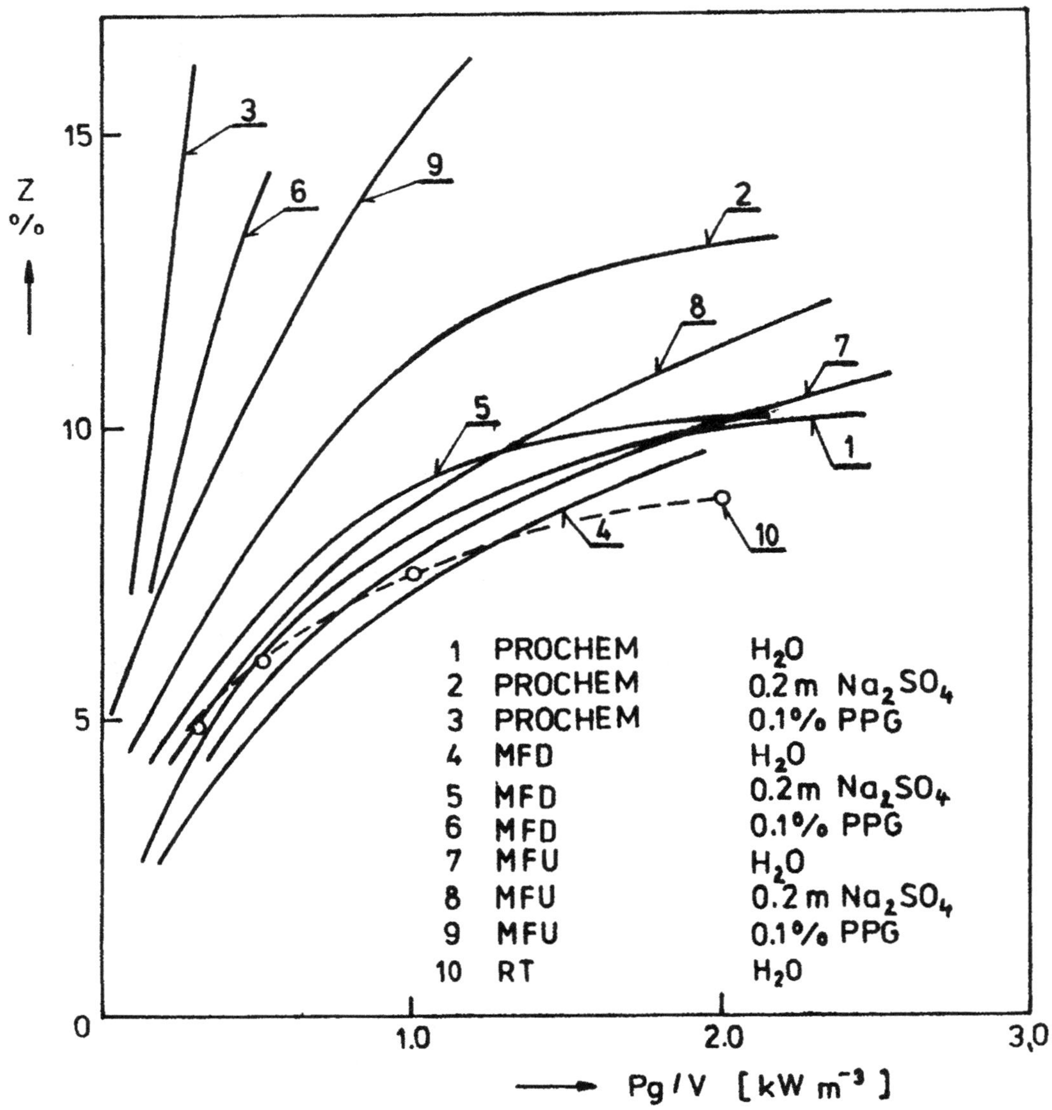

Fig. 13 **A Comparison of All Impellers and All Fluids:Hold-up versus Power/Unit Volume at 1.5 vvm**

though the difference between the latter two is still rather small.

CONCLUSIONS

As with the Rushton turbine, the hydrodynamic parameters which determine flow patterns and power draw are shown to be relatively insensitive to the coalescence properties of the fluid when using Prochem, MFU and MFD agitators. However, when recirculation rates are high compared to sparged gas rates, the power fall is greatest in the least coalescing fluid because in each case the hold-up, i.e., gas flow through the agitator, is greatest.

When the hold-up with different agitators is compared at an equal aeration rate and specific power input, no significant difference is found in water. This result also accords with earlier findings. However, as the level of coalescence inhibition increases, very significant differences in hold-up produced develop between the agitators. The Prochem is most effective. Perhaps this explains the reported improvement in k_La values[13] found with Prochems in real fermentation, an improvement not found for water[6].

ACKNOWLEDGEMENT

One of us (VM) would like to thank the British Council for financial support for this project, and another (CMM) the SERC Biotechnology Directorate for a research studentship.

REFERENCES

1. van't Riet, K. and Smith, J.M., Chem. Eng.Sci., 28, 1031, (1973).

2. Nienow, A.W., Wisdom, D.J., and Middleton J.C., Proc.2nd European Mixing Conf., BHRA Cranfield, 1977; Paper F1.
3. Machon, V. and Vlcek, J., Collection of Czech.Chem.Commun., 50, 2863, (1985).
4. Machon, V., Vlcek, J. and Skrivanek, J., Proc. 5th Euro. Mixing Conf., BHRA Cranfield, 1985, pp 155-170.
5. Bujalski, W. Konno, M. and Nienow, A.W., Proc.6th Euro.Mixing.Conf., BHRA, Cranfield, 1988; pp 389-398.
6. Balmer, G.J. Moore, I.P.T. and Nienow, A.W., "Biotechnology Processes; Scale-up and Mixing", (Ed. C.S. Ho and J.Y. Oldshue), A.I.Ch.E. New York, pp 117-127.
7. Hass, V.C., and Nienow, A.W., Chem.Ing. Technik, 61, 152, (1988).
8. Bujalski, W., Nienow, A.W. and Liu Huoxing, Chem.Eng.Sci., 45, 415, (1990).
9. Chapman, C.M., Nienow, A.W., Cooke, M. and Middleton, J.C., Chem.Eng.Res.Des., 61, 182, (1983).
10. Kuboi, R., Nienow, A.W. and Allsford, K.V., Chem.Eng.Comm., 22, 29, (1983).
11. McFarlane, C.M., Ph.D. Thesis, University of Birmingham, 1991 (to be published).
12. Lee, J.C. and Meyrick, D.L., Trans.I. Chem.E., 48, T37, (1970).
13. Buckland, B.C., Nienow, A.W. et al, Biotech.Bioeng., 31, 737 (1988).

NOMENCLATURE

c	impeller clearance from base of vessel, m
D	impeller diameter, m
H	liquid height, m
N	stirrer speed, s^{-1}
N_F	gas flow number, $\dot{V}_g/(ND^3)$
P	power input into ungassed liquid, W
P_g	power input into gassed liquid, W
Po	power number $P/(\rho N^3 D^5)$
Re	Reynolds number, ND^2/η
T	tank diameter, m
$\dot{V}_g$	gas flow rate, $m^3 s^{-1}$
vvm	(volume of gas/min)/volume of liquid
Z	gas hold-up in %: (volume of gas/volume of dispersion) x 100
η	liquid kinematic viscosity, $m^2 s^{-1}$
ρ	liquid density, kg m^{-3}

REFEREES' QUESTIONS AND AUTHORS' ANSWERS

Question 1

Could you give the molecular weight of the polypropylene glycol used? This would be useful for future comparison.

Answer 1

The molecular weight of the polypropylene glucol was 2000.

Question 2

On your p.1, not only is the density of all the solutions approximately equal to that of the water, but also the viscosity.

Answer 1

Yes

IMPROVING XANTHAN FERMENTATION PERFORMANCE BY CHANGING AGITATORS

Z. XUEMING, Tianjin University, China
A.W. NIENOW, C.A. KENT, S. CHATWIN, University of Birmingham, GB
E. GALINDO, Universidad Nacional de Mexico

ABSTRACT

In order to improve fermentation performance, much effort in the laboratory is put into strain and media development including feeding strategies. However, such work may not give good results unless associated with the correct fluid dynamic environment, e.g., mixing conditions. Previous mixing studies have suggested that for Xanthan fermentations, better mixing conditions are achieved by large impeller-to-fermenter diameter ratios. Utilising a unique accurate in-fermenter strain gauge/telemetry technique for power measurement developed at the University of Birmingham, this work shows that by a combination of fed-batch strategy and good mixing (using a D/T ratio of 0.63), Xanthan concentration was increased by about 50%, the total energy requirement reduced by 40% and the fermentation time reduced by 20%. Overall heat transfer coefficients during sterilisation are twice as high as compared to a D/T ratio of 0.42 at the same specific power input. The results agree well with the cavern theory for the mixing of yield stress fluids

INTRODUCTION AND LITERATURE

Mixing effects due to inhomogeneities are of great significance in large scale fermentations and often lead to a performance which is inferior to that obtainable on the bench scale. However the impact of such mixing is difficult to show on the pilot scale except when high viscosity, shear thinning, yield stress broths are produced. Xanthan fermentations are an excellent, commercially significant, example of such broths. Studies on reconstituted Xanthan solutions which have similar rheological properties to those of the broth have shown that improved mixing is obtained by replacing the traditional small diameter, D, (one-third vessel diameter, T) Rushton turbines by larger ones[(1)]. However, because of the very different torque-power characteristics of these two agitators, simple retrofitting is expensive especially on the large scale. On the other hand, retrofitting with low power number agitators of large diameter also gives improved mixing in all broths of similar rheological properties[(2)]. Large scale mycelial fermentation studies have shown such an improvement on replacing ~1/3T Rushton turbines by ~0.45T Prochem Maxflo T agitators[(3)].

The motors of small scale fermenters are often sufficiently overpowered to allow different size Rushton turbines to be used with the same drive system. In that case, though, power measurement is rarely accurate enough to properly assess the relative performance of the two. In this work, a 150 litre LSL fermenter has been fitted with an accurate steam-sterilisable strain gauge telemetry system,[(4,5)] thereby allowing such a comparison to be made.

Of course, often improved mixing can only be shown to give an improved fermentation performance if the fermentation is sufficiently demanding; for example, if an improved medium or strain leads to an enhanced oxygen demand. Xanthan fermentations are not particularly demanding in the latter respect. However, De Vuyst et al[(6)] pointed out that during a batch fermentation of <u>Xanthomonas campestris</u>, two phases can be distinguished: a) the trophophase in which there is cell propagation and during which there is a relatively high oxygen demand and the medium should have a low C/N ratio; b) the idiophase during which gum is produced, the oxygen demand is low and a high C/N is needed. On this basis, they developed a two step batch process but were left with a high (~16%) residual sugar level. It can be postulated that this unsatisfactory performance was due to inadequate mixing. Funahashi et al[(7)] have reached Xanthan concentrations up to 40g/L by intermitent addition of glucose using a Rushton turbine at D/T=0.5 in a 6L tank; however, they have not reported the power consumption nor considered mixing aspects of the process.

This paper shows that a sugar feeding strategy is successful <u>provided</u> indeed good mixing is maintained throughout the run; and that such mixing is achieved with less energy by using a large impeller.

EXPERIMENTAL

MICROORGANISM

A variant of <u>Xanthomonas campestris</u> NRRL-B-1459 from Centro de Investigación Sobre Ingeniería Genética y Biotecnología, Universidad Nacional Autónoma de Mexico was used. The strain recovered from a lyophilized tube, was maintained on YM (see below) slants at 4°C.

MEDIA

The seed culture medium (YM) contained (g/litre): glucose, 10; peptone, 5: yeast extract 3; malt extract, 3. The production medium formulation was that developed by Galindo et al[8]: glucose, 24; $(NH_4)_2SO_4$, 1; $MgSO_4$, 0.23; KH_2PO_4, 3.9; citric acid, 1.3; $CaCO_3$, 0.0024; H_3BO_4, 0.0048; ZnO, 0.0072; $FeCl_3$, .0014.

INOCULUM PREPARATION

There were three steps (all at 29°C) for 24h, agitated at 200 rpm: 200 ml flask; 2 litre flask; and a 5 litre fermenter. The final culture was used as the innoculum for the 150 litre fermenter.

FERMENTATION

A 150 litre LSL fermenter (T=0.49m) agitated with a Rushton turbine was linked through a 3D-THINKLAB interface to an IBM-AT computer for process management via LSL MENTOR. Temperature was controlled at 29°C and pH at 7 (via 2N KOH and 2N H_2SO_4). The air flow rate was held at 0.5 vvm and the agitator speed was increased step wise based on two criteria. In the main, the speed was increased so that the dO_2 >30%. Towards the end of the run when the broth became very viscous and had a yield stress, visual observation of the broth surface indicated the speed necessary to keep the broth in motion, if possible. With all these changes, power was measured by strain gauge/telemetry installed in the fermenter shaft[5], giving as in associated non-fermentative studies, an accuracy of about ± 5%[4]. Oxygen and CO_2 gas analysers were linked to the computer so that derived variables could be determined using LSL MENTOR calculation facilities. These data are not reported here.

Three fermentations were undertaken: a) A batch fermentation with the medium nominally as given above with an agitator 0.42T; b) A fed-batch with the above medium and agitator but with an additional volume of glucose being added to increase its level by approximately 5g/litre when the residual glucose fell to about 5g/litre; and c) a similar fed batch using a 0.63T agitator.

OFF-LINE ANALYSIS

a) Cell growth by optical density with a 565 nm U.V. scanning spectrophotometer; b) Xanthan gum by KCl/ ethanol precipitation; c) residual glucose by DNS colorimetric assay; and d) rheological properties by a Rheomat 30 rheometer from 0.09 to 622 s^{-1} shear rate.

RESULTS

BATCH FERMENTATION, D/T = 0.42

This fermentation lasted some 90 hours, with the agitator speed increased from 200 rpm to 550 rpm as described above. A final Xanthan concentration of 18g/litre was produced with the broth rheological properties expressed as a power law fluid, $\tau = K\dot{\gamma}^n$, changing with time as shown in Fig.1(a). During the run, the gassed power number, $Po_g = P_g/\rho N^3 D^5$ remained almost constant at about 2.6. The Reynolds number, on the other hand, defined[9] as $Re = \rho N^{2-n} D^2/K k_s^{n-1}$ where $k_s \cong 11.5$, fell progressively to about 10^3.

FED BATCH, D/T = 0.42

The progress of the fermentation is shown in Fig. 2a with a shot of glucose being added after about 48 hours. All the glucose was utilised after about 96 hours and a Xanthan concentration of about 27 g/litre was obtained (Fig. 2a). The power and speed variation with time is shown in Fig. 3 with the aerated power number again remaining approximately constant at 2.6. The change of K and n with time is shown in Fig. 1(b).

FED BATCH, D/T = 0.63

The progress of this fermentation is shown in Fig. 2b. This time the additional shot of glucose was added after only 40 hours by which time the glucose was down to 3.5 g/litre. The total fermentation time was reduced by about 24 hours and the overall yield was about the same. The dO_2, power and speed variation with time are shown in Fig. 4. The aerated power number was again about 2.6 with Re falling to about 350. The change of rheological properties with time is shown in Fig. 1(c)

OVERALL HEAT TRANSFER COEFFICIENT, U

These were measured by an unsteady state technique during steam heating and cooling of water and of the final broth during the sterilisation cycle at the end of the run. U values are given along with specific power draw at the time of the run in Table 1.

DISCUSSION

YIELD, BATCH TIME AND POWER CONSUMPTION

The result for the initial batch fermentation

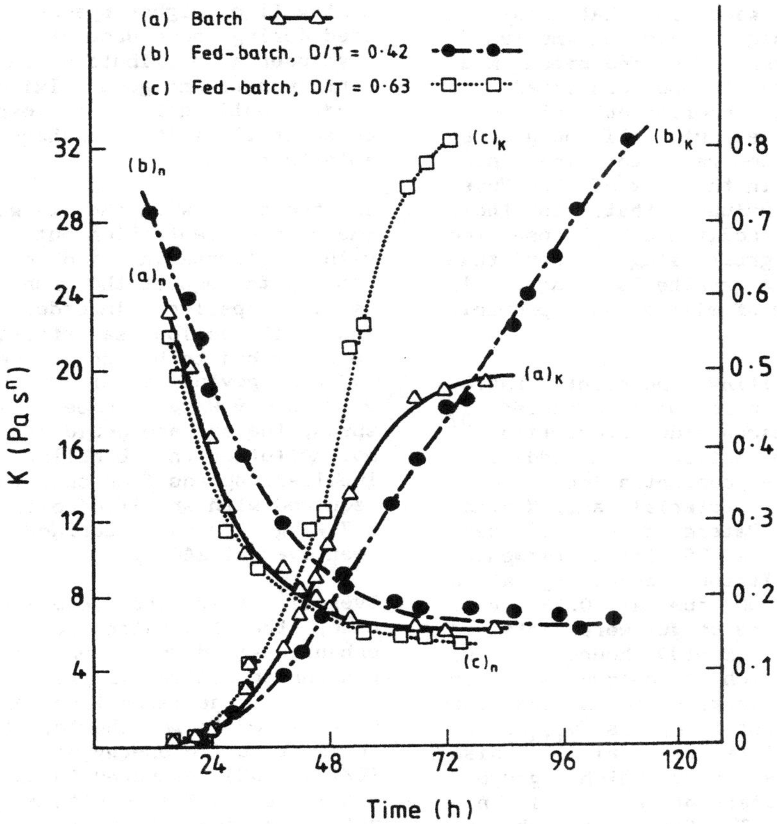

Fig.1 Rheological properties (K and n) versus time for the three fermentations

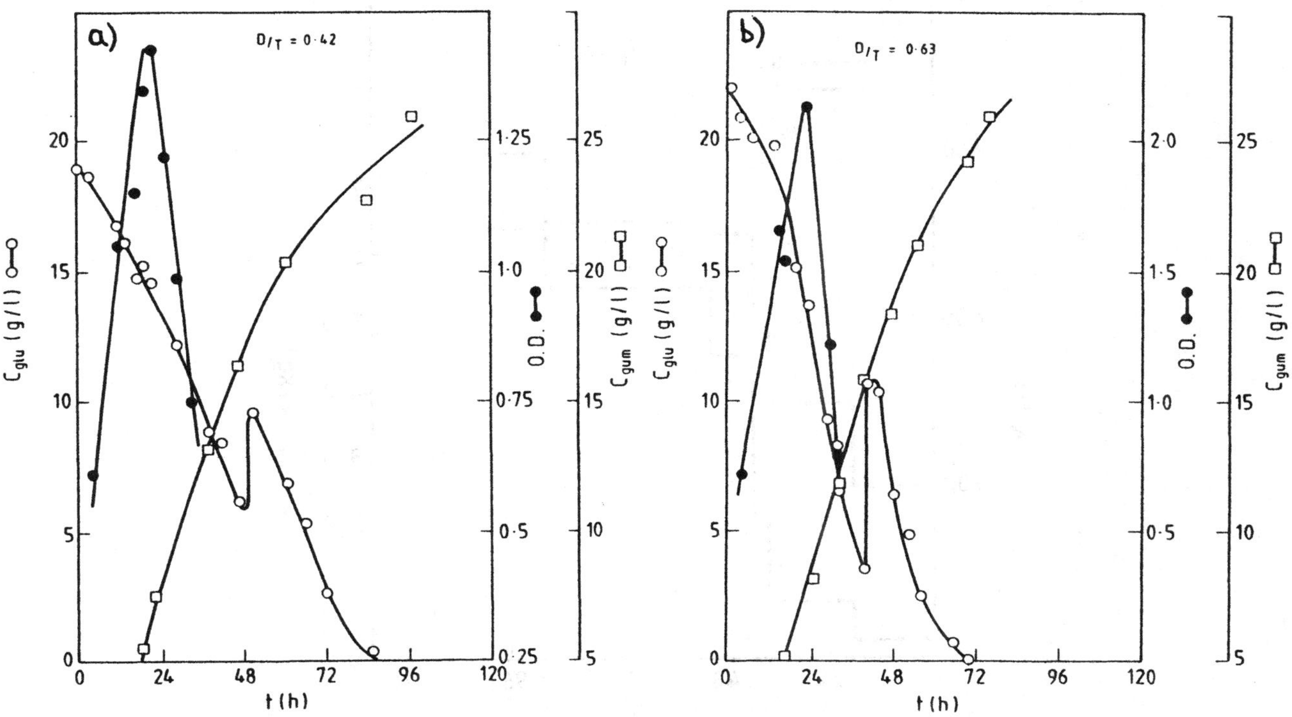

Fig.2 Fed-batch fermentation: O.D. and concentrations of glucose and gum

was essentially the same as that achieved previously with the same strain and medium[8] in a 25 litre fermenter. The fed-batch mode enabled more glucose to be converted to Xanthan with a yield coefficient of about 0.8(g Xanthan/g glucose), with all the glucose being consumed in contrast to the poor performance reported in the work of De Vuyst et al.[6] It is concluded that, in their case, they added the additional glucose too soon during the cell growth stage rather than waiting for the gum synthesis stage. In addition, the quality of mixing was probably inadequate.

A high carbon to limiting nutrient ratio, often nitrogen, has also been proposed, in order to favour polysaccharide production[10] during a batch fermentation. In addition, high initial glucose concentration can be inhibitory, both for bacterial and Xanthan production[7]. Thus, Peters et al.[11] used 55 g/litre glucose in a 25 litre fermenter with two Intermig agitators operating at a constant speed of 300 rpm at 0.25 vvm. However, only 19 g/litre of gum were obtained and the fermentation took ~100 hours. Thus, it seems that a fed-batch technique such as that reported in this work, based on the two phases of the fermentation, is a better strategy. The data of Peters et al.[11] also show that using a very high glucose concentration at the start of a run is not good for cell growth. The fact that enhanced mixing (i.e. higher speeds and powers) was not used during the course of the run[11] may also have been a contributory factor to the bad performance; and good mixing, on the other hand, could help to explain the high concentrations that Funahashi et al[7] were able to reach.

The fed-batch with the larger impeller gave the same concentration but in a shorter time, with K increasing and n decreasing more rapidly throughout the run compared to the smaller impeller. In order to keep the dO_2 and bulk motion satisfactory, the 0.42T impeller had to be run at speeds from 200 to 600 rpm, giving specific power inputs from 0.22 to 6.9 W/kg. Indeed, even at the highest speed, the surface could not be seen to be moving (other than bubbling). With the 0.63T impeller, speeds from only 100 to 275 rpm were required with specific power inputs of 0.23 to 5.7 W/kg. In fact, surface motion could be seen even at 240 rpm.

Overall, therefore, compared to the batch case, the fed-batch fermentation gave an enhanced yield on consumed sugar and a higher gum concentration (27g/litre) was achieved. The use of the large impeller enabled this to be achieved in a shorter time. Thus, the total specific energy used was 225 kW h/m^3 (D/T = 0.63) compared to 353 kW h/m^3 (D/T = 0.42) i.e. ~40% less with a productivity per unit time some 20% greater with the larger D/T ratios.

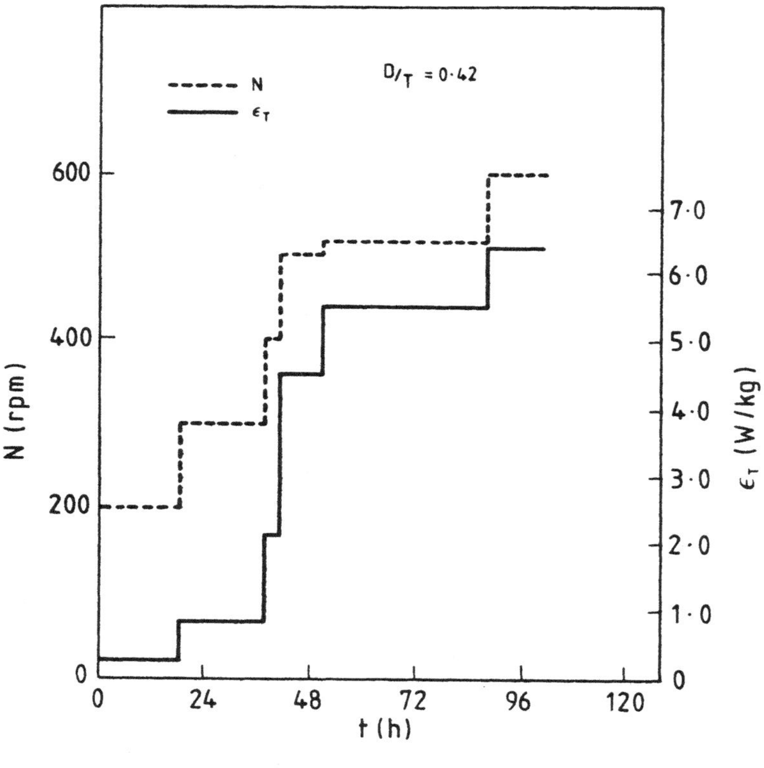

Fig.3 Fed-batch fermentation: impeller speed and specific power

MIXING CONSIDERATIONS

The size of the well-mixed cylindrical cavern with a fluid with a yield stress can be estimated[12] from

$$(D_c/D)^3 = \left(\frac{1.36}{\pi^2}\right)\left(\frac{Po_g\ \rho N^2 D^2}{\tau_y}\right)$$

where τ_y is the yield stress. τ_y was estimated as 30 Pa using the Herschel- Bulkley model to curve-fit the rheogram for the broth containing 27 g/litre Xanthan. Taking the gassed power number as 2.6, this gives a cavern reaching the top surface (H = 0.65m) at about 600 rpm for D/T = 0.42 and 220 rpm for D/T = 0.63. These values fit in well with observations of broth movement and earlier work on reconstituted Xanthan solutions[13].

HEAT TRANSFER COEFFICIENTS

The U values with both impellers were very similar in water especially when compared relative to the power input. Those in the sterilised and heat treated Xanthan were much lower which can be entirely attributed to the lowering of the vessel side film heat transfer coefficient. However, the 0.42 T impeller when compared at equal power gave U values 50% or less than those with the 0.63T one[9] even though, based on an average shear rate of 11.5 N(s^{-1}), the viscosity is about 1000 mPas with the 0.63T and only 500 mPas with the 0.42T. This improvement is associated with the much greater fluid movement at the wall due to the larger cavern in spite of the higher average viscosity with the 0.63T impeller.

CONCLUSIONS

A fed-batch Xanthan fermentation has been shown to give a higher yield of Xanthan with respect to consumed glucose and also allows it to reach a higher concentration (27 g/litre) of Xanthan than a batch one (18 g/litre). The use of a 0.63T Rushton turbine impeller has enabled this fed-batch performance to be achieved with 40% less power in 20% less time compared to a 0.42T one. The larger impeller also enables overall heat transfer coefficients about twice as great to be obtained at the same power though they are only some 15% of those in water. The use of accurate in-fermenter strain gauging has enabled these improvements to be quantified and the performance to be related to the cavern theory of mixing with yield stress fluids.

ACKNOWLEDGEMENTS

Thanks are due to the SERC Biotechnology Directorate for support for S.C. for the development of in-fermenter strain gauging; and to British Council for support for Z.X.; and to the E.E.C. for support for E.G. under the scheme "International Scientific Co-operation-Mexico".

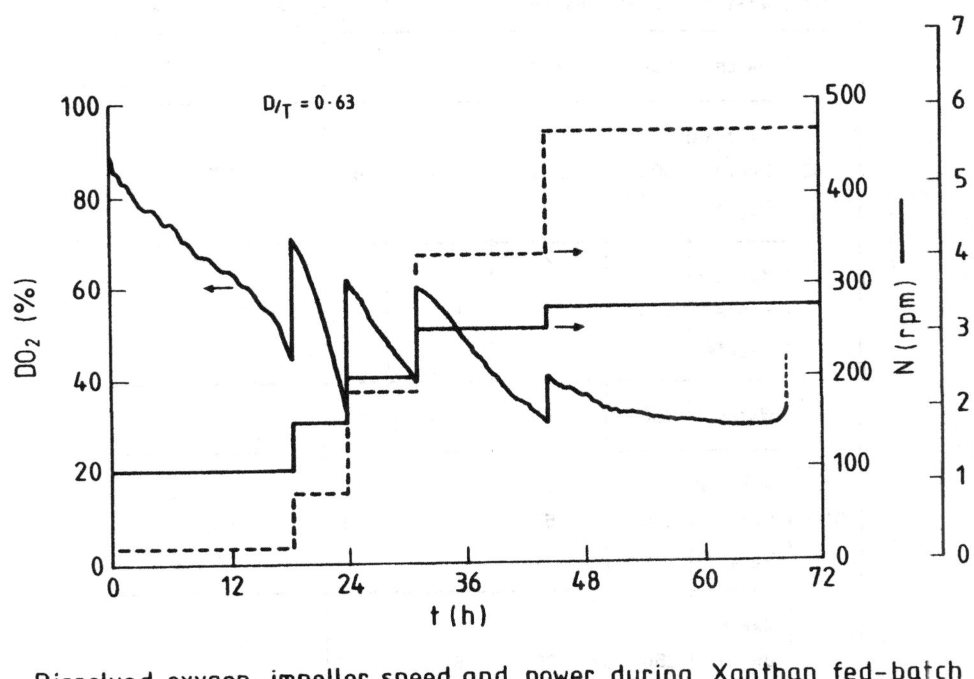

Fig.4 Dissolved oxygen, impeller speed and power during Xanthan fed-batch fermentation

REFERENCES

1) Nienow, A.W., "The World Biotechnology Report, Vol.I, Europe, Online, U.K., 1984, pp.293-304.
2) Nienow, A.W., Trends in Biotech. 8, (1990), 224.
3) Buckland, B.C., Gbewonyo, K., DiMasi, D., Hunt, G., Westerfield, G. and Nienow, A.W., Biotechnol.Bioeng., 31, (1988), 737.
4) Kuboi, R., Nienow, A.W. and Allsford, K.V., Chem.Eng.Comm., 22, (1983), 29.
5) Nienow, A.W., Chatwin, S., Zhao Xueming, Kent, C.A. and Dawson, M., Proc.Int. Workshop on Bioreactor Performance. OGBT, Graz, Austria, P.164-165.
6) De Vuyst, L., Vann Loo, J. and Vandamme, E., J.Chem.Tech. Biotechnol., 39, (1987), 263.
7) Funahashi, H., Yoshimida, T. and Taguchi, H., J.Ferment.Technol., 65, (1987), 603.
8) Galindo, E., Torrestiana, B. and Garcia-Rejòn, A., Bioprocess Eng., 4, (1989), 113.
9) Metzner, A.B. and Otto, R.E., AIChEJ, 3, (1957), 3.
10) Margaritis, A. and Pace, G.W. in Comprehensive Biotechnology (Ed. M. Moo-Young), Pergamon Press, 1985, Vol.3, pp.1005.
11) Peters, H.U., Herbst, H., Hesselink, P.E.M., Lünsdorf, H., Schumpe, A. and Deckwer, W.D., Biotechnol.Bioeng., 34, (1989), 1393.
12) Nienow, A.W. and Elson, T.P., Chem.Eng. Res.Des., 66, (1988), 15.
13) Galindo, E., Badham, R.S. and Nienow, A.W., Proc. 2nd Int.Conf. on Bioreactor Fluid Dynamics, BHRA Cranfield, (1988), pp.65-78.

NOMENCLATURE

C_{glu}	concentration of glucose, g/litre
C_{gum}	concentration of Xanthan gum, g/litre
D	impeller diameter, m
D_c	cavern diameter in a yield stress fluid, m
H	broth height, m
K	consistency index, Pa s^n
n	flow behaviour index,
N	agitator speed, rev/s
OD	optical density
P_g	power input under aerated conditions, W
Po_g	power number under aerated condition
Re_g	Reynolds number
T	fermenter diameter, m
U	overall heat transfer coefficient, W/m^2 K
ε_T	specific power input, W/kg
$\dot{\gamma}$	shear rate, s^{-1}
ρ	density, kg/m^3
τ	shear stress, N/m^2
τ_y	yield stress, N/m^2

TABLE 1 OVERALL HEAT TRANSFER COEFFICIENTS AT DIFFERENT CONDITIONS

D/T	SYSTEM	N(rpm)	U (W/m^2K)	ε_T (W/kg)	$\frac{U}{\varepsilon_T}$	$\left(\frac{U}{\varepsilon_T}\right)$Xanthan/$\left(\frac{U}{\varepsilon_T}\right)$water
0.42	WATER	500	420	8.1	52	-
	XANTHAN BROTH C_{gum}=18	500	61	7.4	8.2	16%
	XANTHAN BROTH C_{gum}=27	500	22	6.7	3.3	6.3%
0.63	WATER	250	380	7.4	51	-
	XANTHAN BROTH C_{gum}=18	200	57	2.8	20	39%
	XANTHAN BROTH C_{gum}=27	250	38	6.0	6.3	12%

REFEREES' QUESTIONS AND AUTHORS' ANSWERS

Question 1

In your Table 1, you give values of the overall heat transfer coefficient U. Your measurements performed during steam heating and cooling water allow to determine the process side heat transfer coefficient H_j. Only H_j depends on the agitation, aeration and rheological properties of the broth! The characteristic variable for h_j is not C_T!

Answer 1

a) The overall heat transfer coefficient is a function of all the resistances between the bulk of the broth ($1/h_j$) and the heat transfer media. In this work, insufficient experiments were done to obtain h_j. However, especially with the Xanthan solution, $U \simeq h_j$ because $1/h_j$ is much greater than all other resistances. These present data however are only meant to show trends.

b) I believe that the referee has mis-read the text and C_T is ε_T, the specific power input (W/kg) measured by the strain gauge/telemetry technique. Thus, Table 1 compares U under different conditions by normalising against specific power input.

Question 2

Could you give some information about mixing times in Xanthan solutions?

Answer 2

Until the cavern extends to the extremities of the vessel contents, the mixing time is infinity, i.e., the stagnant zone outside the cavern is never blended with the rest. Once the whole vessel is in motion, the mixing time could, in principle, be estimated by a suitable correlation for high viscosity, shear thinning-fluids.

DETERMINATION OF THE MEAN SPECIFIC POWER INPUT IN STIRRED VESSELS WITH SUSPENSIONS

R.K. GEISLER, home address : München, D
C. BUURMAN, Koninklijke/Shell-Laboratorium, Amsterdam, NL
A.B. MERSMANN, Technical University Munich, D

1. INTRODUCTION

Mixing of solid/liquid suspensions in stirred vessels is very essential for a variety of chemical or physical processes, such as crystallization and precipitation, dissolution processes, ion-exchange and adsorption, as well as slurry polymerisation, bio-fermentation or heterogeneous catalytic reactions. In most cases, the fluid motion is highly turbulent ($Re > 10^4$). The wide range of applications is covered by suspensions with small solid particles in very low concentrations up to large particles in high concentrations.
For the design of large stirred vessels the tip-speed of the impeller or vice versa the mechanical power input, necessary for complete suspension, is of fundamental interest.
For many decades most of the prevailing investigations for a scale-up rule from small vessels with laboratory scale to large stirred vessels in a production plant have resulted in global empirical correlations based on a large number of experimental data.
Zwietering [1] belongs to the first authors who introduced the empirical scale-up equation

$$n\ D^{\kappa} = \text{const.} \qquad (1)$$

Many other investigations are based on this type of equation, without taking into account the complex fluid dynamics of the solid particles in the suspension flow of small and large stirred vessels. Therefore a great confusion exists up to now for the extrapolation of the mean specific power input in stirred vessels from a small tank diameter to a larger one.
Fig. 1 shows a great variety of different scale-up predictions for stirred suspensions, which can be found in literature.

One of the main objectives of this investigation is to reveal the discrepancies between the different authors for scale-up criteria (e. g. [1], [2], [3], [4], [5], [9], [11], [16], [18], [19], [20], [21], [22]) and to develop a physical model suitable for scale-up of stirred suspension vessels. This paper distinguishes between different suspension regimes, based on the relative size of the stirred vessel. In these regimes different physical phenomena determine the suspension of solid particles.

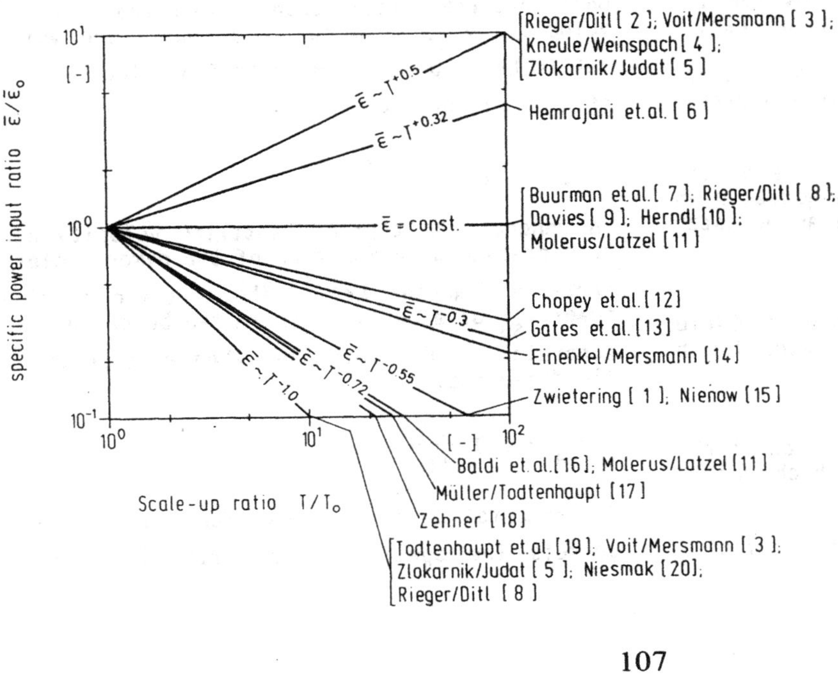

Fig. 1:

Suspension harp according to different scale-up predictions

2. SUSPENSION CRITERIA

The desired state of suspension is very essential for the design of the stirred vessel and the determination of the necessary power input. In some cases a small part of deposited solid particles is permissible, whereas for other applications of solid suspensions, e. g. heterogeneous catalytic reactions, all particles should be distributed within the entire stirred liquid.
Therefore the following suspension criteria can be distinguished:

- o Incomplete suspension;
- o Complete suspension;
- o Homogeneous suspension.

In general a considerably higher power input for homogeneous suspensions, in comparison to complete suspensions, is required. Therefore, the necessary tip-speed or the corresponding power input is practically calculated as a compromise between the desired state of suspension and the required power input.

3. MODELLING OF THE MEAN SPECIFIC POWER INPUT IN STIRRED SUSPENSIONS

In general the power input for suspending solid particles depends on particle properties (d_p, ρ_D, φ), on fluid properties (ν_C, ρ_C), as well as on geometrical data (T, D/T, h/T, type of impeller).
For modelling the mean specific power input in stirred vessels it is necessary to take into account the different modes of particle transportation. As a general rule three different particle movement conditions in stirred vessels with suspensions can be distinguished:

- o lifting of the deposited solid phase from the bottom;
- o generation of an upward fluid flow for the vertical hydraulic transportation of particles;
- o mean specific power input necessary to overcome the specific power of the sinking particle swarm.

With respect to the Reynolds number Re of the stirrer two different cases are illustrated in fig. 2 for suspending particles in an agitated vessel:

- o case a:
  off-bottom lifting and prevention of particle deposition on the bottom, which leads to an inhomogeneous suspension;
- o case b:
  avoidance of settling and compensation of the sinking velocity of particles, which result in a complete suspension.

The Reynolds number, Re, is depicted in fig. 2 as a function of the Ar-number for different ratios T/d_p according to experimental results.

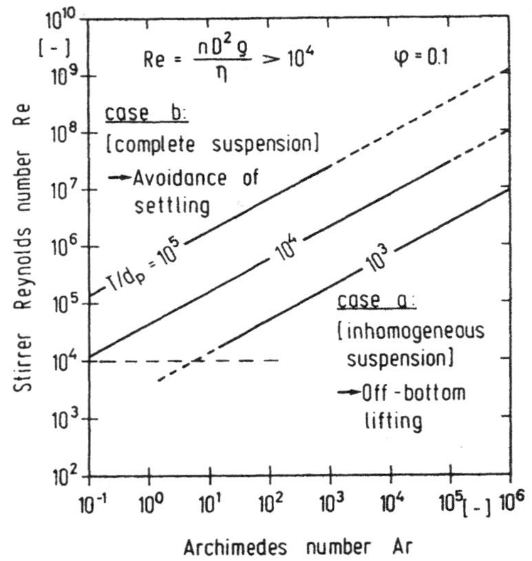

Fig. 2: Arrangement of the different cases for suspensions in stirred vessels depending on the Reynolds number and on the Archimedes number for different ratios T/d_p

For the turbulent agitation of suspensions in cylindrical vessels a typical Re-number of about $Re \approx 10^4$ should be regarded as minimum condition, because for $Re < 10^4$ the turbulent flow, i. e. three-dimensional fluctuating velocity and scale of turbulence, is not sufficient for off-bottom lifting

One of the most important parameters for the determination of the required power input is the calculation of the settling velocity or the specific power of solid particles falling in the gravitational field.
The well-known equation of Richardson/Zaki [23] describes the relationship between the settling velocity, v_{ss}, of a particle swarm, the volumetric hold-up φ and the settling velocity, v_s, of a single particle:

$$v_{ss} = v_s\,(1 - \varphi)^s \tag{2}$$

In general the terminal settling velocity can be expressed as a function of the dimensionless particle diameter $Ar^{1/3}$. The specific sinking power ϵ_{ss} of a particle swarm can be calculated according to Einenkel [25] when considering a fluidized bed:

$$\epsilon_{ss} = \varphi\, g\, \frac{\Delta\rho}{\rho_C}\, v_{ss} \tag{3}$$

Experimental data show that for small stirred vessels the specific power input $\overline{\epsilon}$ is by a

factor of up to 1000 higher than the swarm sinking power ϵ_{ss}, whereas for the same particle size distribution in large vessels $\overline{\epsilon}$ is of the same order of magnitude as ϵ_{ss}.
A new approach will be presented in the following chapter in order to give a reasonable interpretation of the existing discrepancies of the different scale-up rules.

This approach is based on the following simplifications:

- o spherical, monosized particles;
- o stationary flow of fluid and particles in the vessel;
- o constant Power number Ne for all suspensions.

a. Large particles in small vessels

In the case of relatively large particles in small stirred vessels it can be observed that the solid phase is completely deposited on the vessel bottom for a short period and then suddenly lifted up, which results in complete suspension. After a period of 0.1 up to 2 sec the particles settle again on the bottom. Similar operating conditions can be observed for fluidized beds, in which large high density particles are fluidized (the so-called aggregative fluidization). This behaviour leads always to a heterogeneous solid distribution in stirred vessels. For this oscillating, pulsating up- and down movement of the solid phase a certain amount of energy, called lift energy, is necessary.
This behaviour of the suspension flow can be observed mainly in very small stirred vessels ($T < 0.5$ m) with large solid particles ($d_p >$ 1 mm) in a range of $T/d_p \lesssim 500$ and should not be regarded as relevant for scale-up rules.

b. Particles in medium-sized and large vessels

In this case ($T/d_p > 500$) it is possible to create a stable, almost homogeneous suspension flow. This flow behaviour of fluidized beds may be compared with particulate fluidization.
In this case the specific power input is composed of two different contributions

- o specific power ϵ_{circ} for circulation of the suspension;
- o specific sinking power ϵ_{ss} of the particle swarm,

according to

$$\overline{\epsilon} = \epsilon_{circ} + \epsilon_{ss} \quad (4)$$

The volumetric flow rate, Q_s, of suspension through the stirrer region with the axial fluid velocity v_1 (see also fig. 3) amounts to

$$Q_s = N_{qs}\, n\, D^3 = \frac{\pi}{4} D^2 v_1 \quad (5)$$

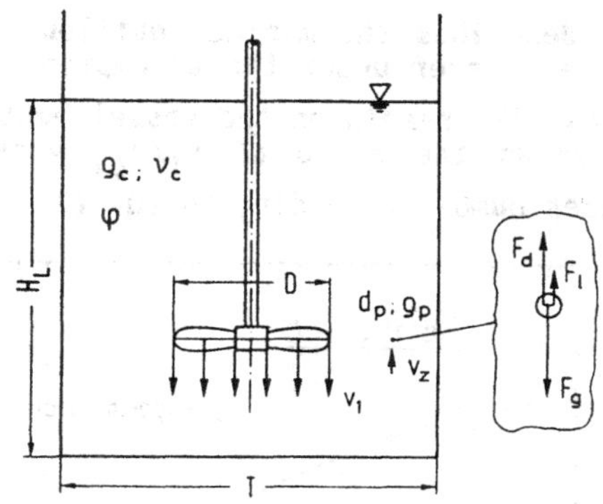

Fig. 3: Force-balance for a particle in complete stirred suspension

Based on empirical data from Einenkel [25] and own experimental data the volumetric Flow number, N_{qs}, of suspensions can be expressed in terms of the Flow number, N_{qo}, of the single phase flow and the volumetric hold-up φ:

$$N_{qs} = \frac{N_{qo}}{1 + 14\left[\frac{\varphi}{\varphi_0}\right]^{3/2}} \quad (6)$$

The parameter φ_0 represents the solid hold-up of a fixed bed and is in most cases in the range of $0.55 \le \varphi_0 \le 0.8$.

In the upward flow the fluid velocity v_z should be at least at the same level as the particle swarm sinking velocity v_{ss} in order to obtain a force balance of the existing drag forces and the potential forces of the particles and the fluid. Especially for stirrers with an axial outflow direction a widening of the outflow cross-section can be observed (divergence zone), which leads to an additive kinetic energy dissipation. If the fluid outflow velocity v_1 is too small, no stable and complete fluid flow without any stagnant region can be achieved.
On this account it becomes clear that also for very small particles, suspended in an agitated vessel, a minimum outflow velocity v_1 of the stirrer is necessary for off-bottom lifting. As a realistic simplification, supported by experimental data of measurements of the local fluid velocity in stirred suspensions [24], the following empirical equation of the ratio of the fluid outflow velocity, v_1, of the stirrer and the sinking velocity, v_{ss}, of a particle swarm was developed ($Re > 10^4$):

$$\frac{v_1}{v_{ss}} = \frac{160 + 5\,Ar^{0.6}}{0.5 + Ar^{0.6}} \qquad (7)$$

Eq. (7) describes the minimum outflow velocity v_1 of the stirrer under the assumption that no particles will remain on the vessel bottom. Fig. 4 shows the ratio of v_1/v_{ss} versus the Archimedes number according to eq. (7).

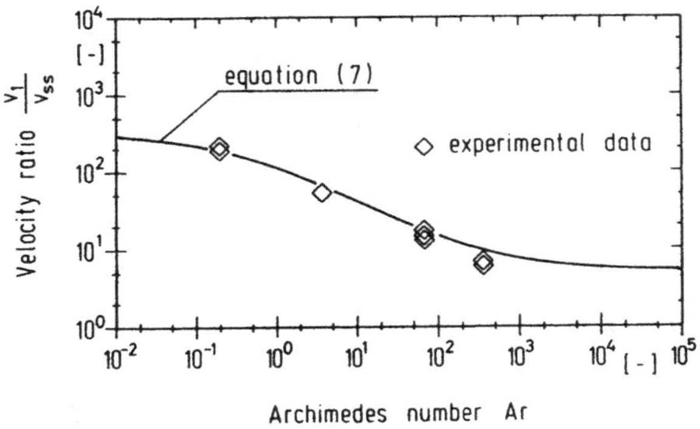

Fig. 4: Dependency of the ratio v_1/v_{ss} versus the Ar-number (exp. data from [24])

This means that a minimum discharge velocity, v_1, is necessary to obtain a stable upward transportation of all particles within stirred vessels in the turbulent flow regime ($Re > 10^4$) and to avoid settled particle layers.
In the case of very large particles or Archimedes numbers the ratio v_1/v_{ss} approaches the constant value of 5. This result is in accordance with data of Brauer [26], who gave a criterion for the vertical hydraulic transportation of solid particles in a pipe.
For complete turbulent fluid flow v_1 is proportional to the stirrer tip-speed u_{tip}, which was scrutinized by different independent investigations in stirred vessels (see e. g. [27], [28], [29]).
Combination of eq. (5) with eq. (7) leads to the following expression:

$$u_{tip} = \frac{\pi^2}{4\,N_{qs}}\,\frac{160 + 5\,Ar^{0.6}}{0.5 + Ar^{0.6}}\,v_{ss} \qquad (8)$$

In contrast to some models published in literature the new approach in this paper takes into consideration the real swarm sinking velocity, v_{ss}, according to [23] and not the settling velocity according to Stokes law, valid only for $Re_p < 1$.
With the relation of eq. (9) for the mean specific power dissipation for cylindrical vessels ($H_L/T = 1$)

$$\epsilon_{circ} = \frac{4\,Ne}{\pi^4}\left[\frac{D}{T}\right]^2 \frac{1}{T}\,u_{tip}^3 \qquad (9)$$

the following general equation for the circulation power dissipation, necessary for the suspension flow, is derived by combination of equations (8) and (9):

$$\epsilon_{circ} = \frac{\pi^2\,Ne}{16\,N_{qs}^3}\left[\frac{D}{T}\right]^2 \frac{1}{T} * \left[\frac{160 + 5\,Ar^{0.6}}{0.5 + Ar^{0.6}}\right]^3 v_{ss}^3 \qquad (10)$$

The power contribution ϵ_{circ} depends on the geometry of the stirred vessel and its impeller and on properties of the fluid and particles. For a given suspension (d_p, $\Delta\rho$, ρ_C, Ar = const.) and given geometrical ratios (D/T = const.) and constant Power and Flow number the specific power input ϵ_{circ} is inversely proportional to the vessel diameter T. This means that ϵ_{circ} approaches zero for $T \longrightarrow \infty$. However, the specific power input, ϵ_{ss}, of a particle swarm, settling in the gravitational field, is the minimum value for a stationary flow of suspensions. Therefore the specific sinking power ϵ_{ss} of a particle swarm according to equation (3) is valid for suspensions, stirred in large agitated vessels, for which the tank diameter T is large in comparison with the particle diameter d_p, i. e. $T/d_p \longrightarrow \infty$.

The total specific power input $\bar{\epsilon}$, according to eq. (4), of the contribution ϵ_{circ}, due to the circulation flow of the suspension, and the contribution ϵ_{ss}, necessary to overcome the specific power of settling particles, can be normalized by the specific energy of a sinking particle with C_D = const.

$$\bar{\epsilon}^* = \epsilon^*_{circ} + \epsilon^*_{ss} = \frac{\bar{\epsilon}}{\left[g^3\,d_p\left[\frac{\Delta\rho}{\rho_C}\right]^3\right]^{1/2}} =$$

$$= \frac{\pi^2\,Ne}{16\,N_{qs}^3}\left[\frac{d_p}{T}\right]\left[\frac{D}{T}\right]^2\left[\frac{160 + 5\,Ar^{0.6}}{0.5 + Ar^{0.6}}\right]^3 * \left[\frac{v_{ss}^2}{d_p\,g\left[\frac{\Delta\rho}{\rho_C}\right]}\right]^{3/2} + \varphi\left[\frac{v_{ss}^2}{d_p\,g\left[\frac{\Delta\rho}{\rho_C}\right]}\right]^{1/2} \qquad (11)$$

As a conclusion of eq. (11) the two different contributions of the specific power input show quite different physical importance, depending on the diameter ratio T/d_p. The contribution, necessary for circulation, prevails in small vessels, whereas in large vessels the second contribution is dominant.

The specific power contribution, ϵ_{circ}, for the circulation of the suspension flow increases with decreasing tank diameter T, because for a given mean upward or axial flow velocity $\overline{v}_z$ the friction pressure drop Δp_{circ} of the circulating suspension is higher with respect to a higher ratio of wall surface per unit volume of the suspension. Furthermore the pressure drop, caused by the diversion of the suspension flow (especially for suspensions with large volumetric hold-ups of solid particles and the increased apparent viscosity of the suspension with Non-Newtonian fluid behaviour) increases when the vessel diameter T is reduced.
On the other hand, ϵ_{circ} becomes very small for $T \longrightarrow \infty$. However, the axial fluid velocity $\overline{v}_z$ must be at least equal to the settling velocity v_{ss} of the particle swarm. As a consequence the contribution ϵ_{ss} for compensation of the settling velocity is equal to the mean specific power input $\overline{\epsilon}$ with the result that $\epsilon_{circ} << \epsilon_{ss}$ or $\overline{\epsilon} \approx \epsilon_{ss}$ for $T \longrightarrow \infty$.

4. COMPARISON WITH EXPERIMENTAL RESULTS

In fig. 5 the normalized specific power input $\overline{\epsilon}^*$ is plotted against the expression $\left[\frac{T}{d_p}\right]\cdot\left[\frac{T}{D}\right]^2$ for two different Ar-numbers (Ar = 10; Ar = 10^4) and different volumetric hold-ups. In both cases a Power number Ne = 0.5 and a Flow number N_{qo} = 0.5 was assumed.
Curves for other parameters can be obtained from equation (11).
This diagram is approximately valid for complete suspensions under optimum geometrical conditions. For homogeneous suspensions the mean specific power input might exceed considerably the amount of $\overline{\epsilon}^*$ according to equation (11). In addition fig. 5 reveals an increasing specific power input with increasing solid hold-up φ up to about 0.3. For larger values of $\varphi > 0.3$ a further increase of the mean specific power input $\overline{\epsilon}$ in stirred suspensions is not necessary.

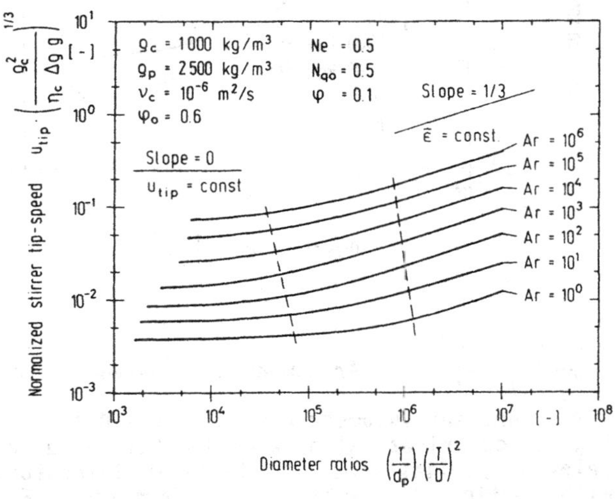

Fig. 5: Calculated normalized specific power input in stirred suspensions for two different Archimedes numbers (Ar = 10; Ar = 10^4) and for different volumetric hold-ups φ (according to eq. (11))

The dependency of the mean specific power input of the Archimedes number is elucidated in fig. 6. With the same geometrical conditions and physical properties of the suspension the normalized mean specific power input was calculated as a function of the diameter ratios

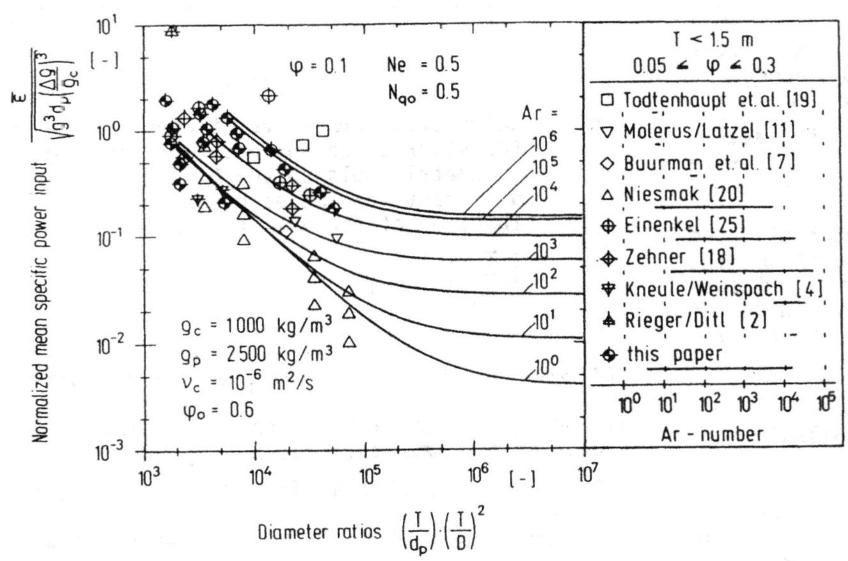

Fig. 6:

Calculated normalized specific power input in stirred suspensions for a constant volumetric hold-up φ = 0.1 and for different Archimedes numbers (experimental data in the range of T < 1.5 m and 0.05 ≤ φ ≤ 0.3)

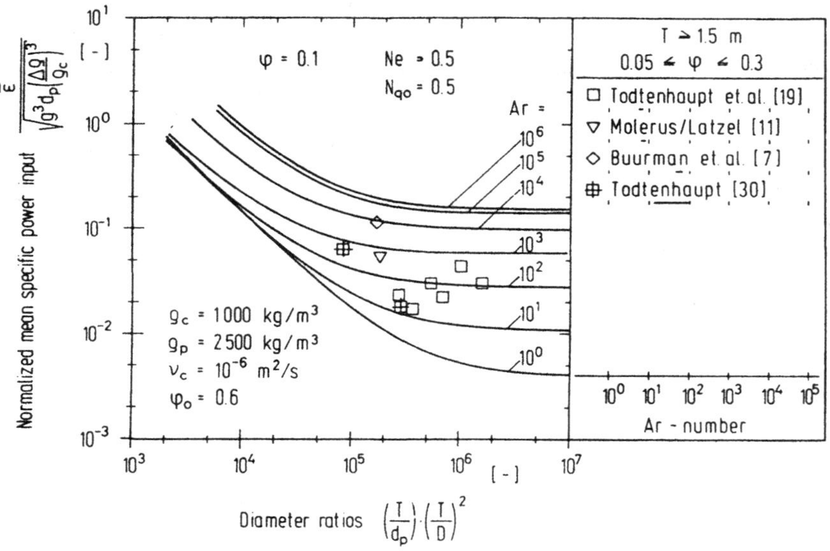

Fig. 7:

Calculated normalized specific power input in stirred suspensions for a constant volumetric hold-up $\varphi = 0.1$ and for different Archimedes numbers (experimental data in the range of $T \geq 1.5$ m and $0.05 \leq \varphi \leq 0.3$)

$\left[\frac{T}{d_p}\right]\cdot\left[\frac{T}{D}\right]^2$ with Ar-number as a parameter and for a constant volumetric hold-up $\varphi = 0.1$.

Fig. 6 comprises also experimental data of relevant publications, available in literature with sufficient quantitative information. For this comparison all data were used without regard of the suspension criteria, the type of the stirrer and with a solid hold-up φ in the range $0.05 \leq \varphi \leq 0.3$.

Whereas fig. 6 represents experimental data for stirred vessels up to a tank diameter $T < 1.5$ m, fig. 7 includes data of medium-size and large tanks ($T \geq 1.5$ m). In addition fig. 6 and 7 show the range of the Ar-numbers in which experiments were carried out by various authors.

The new approach makes very clear that a scale-up criterion according to

$$\bar{\epsilon} \sim T^{\alpha} \tag{12}$$

with a constant exponent α is not reasonable since this exponent with $0 < \alpha < 1$ depends on the tank diameter T or on the ratio T/d_p.

The first papers dealing with a variable scale-up rule, dependent on the diameter ratio, were published by Voit/Mersmann [3] or Molerus/Latzel [11].

This paper now presents a more sophisticated approach for a scale-up rule, based on the different forces of particles in stirred suspensions. This first approach was made with the restriction to axial impellers but it can also be transferred to radial impellers.

5. DETERMINATION OF THE NECESSARY STIRRER TIP-SPEED

If the mean specific power input in stirred suspensions is known, the stirrer tip-speed can be calculated according to eq. (9) and eq. (11). Fig. 8 shows the dependency of the normalized stirrer tip-speed u_{tip} of the diameter ratios $\left[\frac{T}{d_p}\right]\cdot\left[\frac{T}{D}\right]^2$.

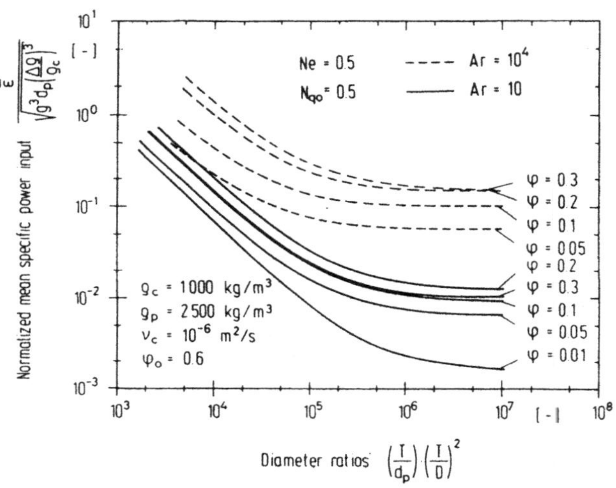

Fig. 8: Calculated normalized stirrer tip-speed for stirred suspensions for a constant volumetric hold-up $\varphi = 0.1$ and for different Archimedes numbers as a function of the diameter ratios $\left[\frac{T}{d_p}\right]\cdot\left[\frac{T}{D}\right]^2$

For small ratios of $\left[\frac{T}{d_p}\right]\cdot\left[\frac{T}{D}\right]^2$ the stirrer tip-speed is constant, which represents a decrease of the mean specific power input according to $\bar{\epsilon} \sim T^{-1}$, whereas for very large ratios of $\left[\frac{T}{d_p}\right]\cdot\left[\frac{T}{D}\right]^2$ the stirrer tip-speed is approximately proportional to $T^{1/3}$ and $\bar{\epsilon}$ remains constant. The curves of the stirrer tip-speed in fig. 8 are valid for the indicated combination of the fluid and particle values and represent optimized geometrical conditions for stirred vessels with suspensions. However for other arrangements of the stirrer, clearance height, bottom shape of the tank or suspension criterion, the stirrer tip-speed can be much higher as the data given in this paper.

6. CONCLUSION

Most existing discrepancies for a scale-up prediction of stirred suspension vessels can be traced back to experimental data, obtained in small vessels, and explained by an unpropitious interpretation of the data with respect to scale-up.
The results of this approach can be summarized in the following way:

- o Calculation of the specific power input as a summation of the specific circulation power and the specific sinking power.
- o Consideration of different Archimedes numbers.
- o Presentation of the dimensionless specific power input as a function of the diameter ratios $\left[\frac{T}{d_p}\right] \cdot \left[\frac{T}{D}\right]^2$ to indicate that both the tank diameter and the particle size influence the power input necessary for suspensions.

The most important conclusion is that there is no simple and constant scale-up rule possible in order to describe the power input for a large range of suspension properties, tank size, particle size, geometrical conditions or comparable suspension criteria.

SYMBOLS USED

Ar	[-]	Archimedes number
C_D	[-]	drag coefficient
D	[m]	agitator diameter
d_p	[m]	particle diameter
F_d	[N]	drag force
F_g	[N]	gravitational force
F_l	[N]	lift force
g	[m/s²]	gravitational acceleration
H_L	[m]	height of total vessel contents
h	[m]	clearance height of impeller
Ne	[-]	Power number
N_{qo}	[-]	Flow number of a homogeneous fluid
N_{qs}	[-]	Flow number of a suspension
n	[rev/s]	agitator speed
P	[W]	mechanical power
Δp_{circ}	[Pa]	pressure drop of the circulating suspension
Q_C	[m³/s]	volumetric flow rate
Q_s	[m³/s]	volumetric flow rate of a suspension
Re	[-]	Reynolds number of the stirrer
Re_p	[-]	particle Reynolds number
s	[-]	exponent
T	[m]	tank diameter
u_{tip}	[m/s]	stirrer tip-speed
V_L	[m³]	vessel contents
v_s	[m/s]	terminal falling velocity of a single particle
v_{ss}	[m/s]	terminal falling velocity of a particle swarm
v_z	[m/s]	fluid velocity in the upflow region
$\bar{v}_z$	[m/s]	mean fluid velocity in the upflow region
v_1	[m/s]	fluid velocity in the stirrer outflow region
$\bar{\epsilon}$	[W/kg]	mean specific power input
ϵ_{circ}	[W/kg]	specific fluid circulation power
ϵ_{ss}	[W/kg]	specific swarm sinking power
$\bar{\epsilon}^*$	[-]	normalized mean specific power input
ϵ^*_{circ}	[-]	normalized specific circulation power
ϵ^*_{ss}	[-]	normalized specific swarm sinking power
η	[Ns/m²]	dynamic viscosity
φ	[-]	volumetric concentration of the dispersed phase
φ_o	[-]	volumetric concentration of a fixed bed
κ	[-]	empirical exponent
ν	[m²/s]	kinematic viscosity
ρ_C	[kg/m³]	fluid density
ρ_D	[kg/m³]	solid density
ρ_s	[kg/m³]	density of a suspension
$\Delta\rho$	[kg/m³]	density difference

Subscripts:

lam	laminar
C	continuous phase
D	dispersed phase
circ	circulation
s	suspension
o	model

Dimensionless Groups and Formula:

$$Ar = \frac{d_p^3 \, g \, \Delta\rho \, \rho_C}{\eta_C^2} \qquad \text{Archimedes number}$$

$N_{qo} = \frac{Q_C}{n\ D^3}$ Flow number

$Ne = \frac{P}{\rho\ n^3\ D^5}$ Power number

$Re = \frac{n\ D^2}{\nu}$ stirrer Reynolds number

$\bar{\epsilon} = \frac{P}{\rho\ V_L}$ mechanical specific power

$\rho_s = \varphi\ \rho_D + (1 - \varphi)\ \rho_C$ density of the suspension

REFERENCES

[1] Zwietering, T. N.: Suspending of solid particles in liquid agitators. Chem. Eng. Sci. 8, (1958), pp. 244-253

[2] Rieger, F.; P. Ditl: Suspension of Solid Particles in Agitated Vessels. 4th European Conference on Mixing; Leeuwenhorst, (1982)

[3] Voit, H.; A. Mersmann: General Statement for the Minimum Stirrer Speed During Suspension. Ger. Chem. Eng. 9, (1986), pp. 101-106

[4] Kneule, F.; P. M. Weinspach: Suspendieren von Feststoffpartikeln im Rührgefäß. Verfahrenstechnik 1, (1967), 12, pp. 531-540

[5] Zlokarnik, M.; H. Judat: Stirring. In: Ullmann's Encyclopedia of Industrial Chemistry. Vol B2, 25, Weinheim, (1988)

[6] Hemrajani, R. R.; D. L. Smith; R. M. Koros; B. L. Tarmy: Suspending Floating Solids in Stirred Tanks - Mixer Design, Scale-up and Optimization. 6th European Conference on Mixing; Pavia, (1988)

[7] Buurman, C.; G. Resoort; A. Plaschkes: Scaling-up Rules for Solids Suspension in Stirred Vessels. 5th European Conference on Mixing; Würzburg, (1985)

[8] Rieger, F.; P. Ditl; O. Havelkova: Suspension of Solid Particles - Concentration Profiles and Particle Layer on the Vessel Bottom. 6th European Conference on Mixing; Pavia, (1988)

[9] Davies, J. T.: Particle Suspension and Mass Transfer Rates in Agitated Vessels. Chem. Eng. and Process. 20, (1986), 4, pp. 175-181

[10] Herndl, G.: Stoffübergang in gerührten Suspensionen. PhD-thesis Technical University Munich, (1982)

[11] Molerus, O.; W. Latzel: Suspension of Solid Particles in Agitated Vessels. Part I+II. Chem. Eng. Sci. 42, (1987), 6, S. 1423-1437

[12] Chopey, N. P.; T. G. Hicks: Handbook of Chemical Engineering Calculations. New York, (1984)

[13] Gates, L. E.; J. R. Morton; P. L. Fondy: Selecting agitator systems to suspend solids in liquids. Chem. Eng. 83, (1976), pp. 144-150

[14] Einenkel, W. D.; A. Mersmann: Erforderliche Drehzahl zum Suspendieren in Rührwerken. vt-Verfahrenstechnik, 11, (1977), 2, pp. 90-94

[15] Nienow, A. W.: Suspension of solid particles in turbine agitated baffled vessels. Chem. Eng. Sci. 23, (1968), pp. 1453-1459

[16] Baldi, G.; R. Conti; E. Alaria: Complete suspension of particles in mechanically agitated vessels. Chem. Eng. Sci. 33, (1978), pp. 21-25

[17] Müller, W.; E. K. Todtenhaupt: Großrührwerke in der Grundstoffindustrie. Aufbereitungs-Technik 1, (1972), pp. 38-42

[18] Zehner, P.: Fluiddynamik beim Suspendieren im Rührbehälter. Chem.-Ing.-Tech. 58, (1986), 10, pp. 830-831

[19] Todtenhaupt, P.; P. Forschner; D. Grimsley; A. B. M. Lases: Design of Agitators for the ERGO Carbon-in Leach Plant. GOLD 100. Proceedings of the International Conference on Gold. SAIMM, (1986), pp. 195-207

[20] Niesmak, G.: Feststoffverteilung und Leistungsbedarf gerührter Suspensionen. PhD-thesis Technical University Braunschweig, (1982)

[21] Liepe, F.; R. Koschek: Untersuchungen zum Stoffvereinigen in flüssiger Phase. Chem. Tech. (Leipzig) 29, (1977), 7, pp. 378-391

[22] Kraume, M.; P. Zehner: Suspendieren im Rührbehälter - Vergleich unterschiedlicher Berechnungsgleichungen. Chem.-Ing.-Tech. 60, (1988), 11, pp. 822-829

[23] Richardson, J. F.; W. N. Zaki: Sedimentation and Fluidisation. Part I. Trans. Instn. Chem. Engrs. 32, (1954), pp. 35-53

[24] Geisler, R.: Fluiddynamik und Leistungseintrag in turbulent gerührten Suspensionen. PhD-thesis Technical University Munich (to be published)

[25] Einenkel, W.-D.: Beschreibung der fluiddynamischen Vorgänge beim Suspendieren im Rührwerk. VDI-Forschungsheft Nr. 595, Düsseldorf, (1979)

[26] Brauer, H.: Turbulenz in mehrphasigen Strömungen. Chem.-Ing.-Tech. 51, (1979), 10, pp. 934-948

[27] Geisler, R. K.; A. B. Mersmann: Local Velocity Distribution and Power Dissipation Rate of Suspensions in Stirred Vessels. 6th European Conference on Mixing; Pavia, (1988)

[28] Wu, H.; G. K. Patterson: Laser-Doppler Measurements of Turbulent-Flow Parameters in a Stirred Mixer. Chem. Eng. Sci. 44, (1989), 10, pp. 2207-2221

[29] Plion, P.; J. Costes; J. P. Couderc: Study by Laser Doppler Anemometry of the Flow Induced by a Propeller in a Stirred Tank - Influence of Baffles. 5th European Conference on Mixing; Würzburg, (1985)

[30] Todtenhaupt, P.: EKATO - internal communication

<u>ADDENDUM</u> : Summary of the paper

One of the most essential problems, dealing with solid-liquid suspensions in stirred vessels, is the determination of a reliable scale-up rule from small stirred tanks to large vessels in an industrial production scale.
According to a new approach, based on physical modelling of the input complex fluid dynamics, the specific power input in stirred suspensions can be calculated as a summation of the specific calculation power and the specific sinking power of a particle swarm. The submitted results, which are compared with a great number of experimental data in literature, reveal that there is no simple and constant scale-up rule possible in order to describe the power input for a large range of suspension properties, tank sizes, geometrical conditions or comparable suspension criteria.

A COMPARISON OF THE EFFICIENCY OF FIVE AGITATOR DESIGNS FOR SOLIDS SUSPENSION DUTIES AT HIGH SOLIDS CONCENTRATIONS

N.I. HEYWOOD, S. REHMAN, R.G. WHITTEMORE, Warren Spring Laboratory, Stevenage, GB

ABSTRACT

The performance of five agitator designs for solid suspension duty has been assessed using sand slurry in a 900 mm diameter tank. One-dimensional vertical solids concentration profiles were measured using a γ-ray backscatter technique for sand of two different size fractions (300 to 500 μm and 50 to 200 μm) at four bulk solids concentrations from 50% to 65% by weight.

Two different measures of suspension efficiency were calculated from each profile: a suspension index expressed as a percent and the standard deviation of local solids concentration about the bulk value. The power efficiency of three proprietary agitator designs (Denver 'MIL' propeller, Ekato 'Intermig', Scaba 'SHP') were compared with that of a marine propeller and a 4-blade, 45° pitched turbine (PBT). In general all three commercial designs tended to be the most power efficient, although under some conditions the marine propeller performed best. The PBT generally performed relatively poorly.

INTRODUCTION

Much published research in the suspension of solids in agitated tanks has focused on the condition when agitation is just sufficient for all particles to be suspended. This condition is generally referred to as 'just-off-bottom-suspension' or JOBS. However, in the case of rapidly-settling suspensions, JOBS does not represent very good mixing and typically only around 20% of the solids present are lifted into the top half of the tank at this condition. The agitator speed required to achieve JOBS does not, in itself, convey any information about the solids distribution in the tank. This information is important to design feed and withdrawal methods for either batch or continuous slurry tank operation.

Traditional agitators such as the marine propeller and the four-bladed, 45° pitched turbine (PBT) are frequently used for solid suspension duties when pumping in a downward direction. In recent years, several other commercial agitator designs have become available. Manufacturers of these agitators sometimes claim that their agitator will require less power to achieve a certain level of solid suspension compared with the power requirements of their competitors' novel designs or of the traditional designs.

Although some published studies have looked at one or two-dimensional solids concentration profiles in agitated tanks, these have normally involved at the most two or three agitator types and little attempt has been made to compare the performance of these agitators through reference to these concentration profiles and the power levels necessary to achieve them. Thus there is almost no information to assist the design engineer in selecting an appropriate agitator to meet a specified solid suspension duty.

The purpose of the present study is to begin to fill this information gap by providing comparative data on the performance of different agitator designs when suspending the same concentration of sand in water. This was done using a novel technique for measuring the local solid concentration of relatively high concentrations of sand-water slurry. One-dimensional vertical solids concentration profiles have been measured at various elevations in a tank using a non-invasive γ-ray backscatter technique. These profiles then formed the basis for two suspending indices which were related to the power drawn by the agitators. The experimental apparatus and some of the data interpretation techniques used in the present study were developed by Christie (1), whose main aim was the development of scale-up procedures for specified degrees of solid suspension. The Christie work is still largely unpublished.

PREVIOUS STUDIES

Significant research effort has been focused on the measurement of the spacial distribution of solids suspended in an agitated tank in the form of either one- or two-dimensional concentration profiles. These studies have generally involved the use of comparatively small tanks containing low solids concentrations. These restrictions are partly the result of the technique used to measure local solids concentration. The result is that little information exists to allow the prediction of the spacial solids distribution in tank sizes of practical importance. In addition, little attempt has been made in the past to compare the power efficiencies of alternative agitator designs.

Previous studies aimed at one-dimensional concentration distributions can be categorised into two classes. The first involves the measurement of the complete profile through the full height of the slurry by measuring the local solids concentration at a number of discrete, usually equi-spaced positions in a vertical line at some specified distance from the agitator shaft. The second focuses on the detection of the height of suspended solids below which it is assumed that the solids are relatively uniformly mixed and above which the liquid is relatively free of solids.

A variety of techniques has been used to measure the local solids concentration. These may be classified into three main categories: local sample withdrawal at the vessel wall, radiation transmittance or reflection, and conductivity.

Einenkel (2) reports vertical profiles for 200 μm glass beads in a 790 mm diameter vessel at concentrations up to 25% by volume. It is not clear what measurement technique was used. Pavlushenko et al (3) took local samples at four vertical positions 35 mm from the wall of a 300 mm diameter vessel. Barresi and Baldi (4,5) used 10 equi-spaced tank wall withdrawal points. Four agitators types were compared, including a Lightnin A310, using a 390 mm vessel, and concentrations of glass ballotini up to 5.1% by wt were used. Baldi et al (6) also used a sample withdrawal technique at five vertical locations on a 175 mm vessel wall. Buurman et al (7,8) sampled at the wall at three positions using vessels of three different sizes (including a 4.5 m^3 tank). A single agitator type (45° pitched, 4-bladed turbine) suspended four different solids types: sand at concentrations up to 60% by volume, polystyrene beads up to 65%, coal up to 75%, and spent catalyst at up to 70%. The data were used to define an agitator speed for "complete suspension".

Yamazaki et al (9) used glass ballotini and silica sand concentrations up to 45% by wt; solids concentration was measured by a light reflection technique using a photoelectric cell. A light backscattering technique has been used by Bos and Heerens (10) to measure the local concentration of 3 mm PTFE beads in a 236 mm diameter vessel at concentrations up to 20% by wt. Both Bohnet and Niesmak (11) and Shamlou and Koutsakos (12) were able to use a He-Ne laser transmittance technique because their experiments employed glass ballotini at low concentrations of 1.5% and from 1% to 6% by wt respectively. Fajner et al (13) and Magelli et al (14) used a light transmittance technique for several solids types in 132 and 227 mm diameter vessels with their heights four times their diameter and using 4 Rushton turbines mounted equally-spaced on a single shaft. The use of a transmittance rather than reflectance technique facilitates the measurement of the mean solids concentration in a horizontal plane rather than a local or point value.

Mak and Ruszowski (15) and Rieger et al (16) have employed conductivity probes to measure local solids concentration. Mak and Ruszowski (15) used 610 and 1830 mm diameter vessels to determine scale-up criteria. Sand-water mixtures up to 30% by wt were agitated using a 45° pitched, 4-bladed turbine. Rieger et al (16) used four flat-bottomed vessels with diameters from 100 to 400 mm. Glass beads of diameter 0.185 and 1.36 mm were agitated separately at concentrations from 2.5 to 10% volume using a single agitator type (six blade turbine).

This survey of previous work demonstrates that the measurement of local solids concentration at the vessel wall by a γ-ray backscatter technique is quite novel. In addition, little attention has been given in past work to the comparison of several alternative agitator designs for suspending highly concentrated slurries in a relatively large tank size.

EXPERIMENTAL APPARATUS

Agitators

Five agitators were selected for this study: three proprietary designs, the Denver 'MIL' propeller, the Ekato Intermig and the Scaba 'SHP', and two 'conventional' designs, the marine propeller and the 4-bladed, 45° pitched turbine (PBT). The dimensions of the agitators are given in Table 1. Fig. 1 shows their designs.

Table 1. Details of the Agitators and their Installation

Agitator	Diameter (mm)	Number of Blades	Blade Shape	Clearance (mm) *	Number of Units on Shaft
Pitched-blade Turbine (PBT)	341	4	Flat	300	1
Ekato 'Intermig'	600	2	Flat	200/650	2
Marine Propeller	300/400	3	Curved	300	1
Denver 'MIL' Propeller	381	3	Curved	248/568 **	2
Scaba 'SHP'	400	3	Curved	150	1

* Measured from mid-point of each agitator to tank bottom (see Fig. 2)
** Two Impeller Units

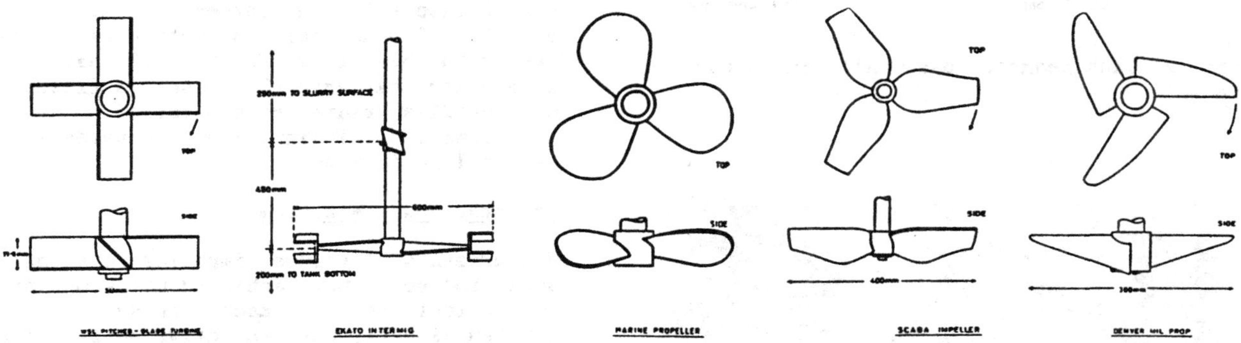

Fig. 1. The design of the five agitators

Slurry Tank

The vessel used was cylindrical with a flat bottom, 900 mm in diameter by 1200 mm deep, and was fitted with four baffles, arranged at right angles. Each baffle was 75 mm wide, extended from 75 mm above the tank base to 75 mm below the slurry surface and was inset 12.5 mm from the vessel wall as shown in Fig. 2. For all tests, the vessel was filled to 900 mm depth, which equates to a volume of 572 litres, and was always mounted centrally under the agitator drive. A polycarbonate window was fitted in the vessel wall for both visual observation and measurement of the solids concentration using a γ-ray backscatter technique.

Measurement of Local Solids Concentration

The γ-ray backscatter technique measured the solids concentration in the agitated slurry at one elevation at any one time.

A γ-radiation source, shielded to produce a collimated beam, was directed through the polycarbonate window in the vessel wall. A scintillation detector mounted alongside the source was used to measure the radiation scattered back from the slurry in the tank. The source holder and detector were mounted on a carriage that was moved vertically up the tank window at 21 equi-distant steps by a ball screw driven by a stepper motor (see Fig. 3). The stepper motor was controlled from a drive unit connected via an interface to a computer.

The relationship between solids present and intensity of backscattered radiation is complex because of two competing effects. γ-radiation is most effectively scattered by the lighter elements. The net result is that the radiation received back at the detector falls with increased solids in the path of the beam. Thus the predominant effect is the dilution of the water by the solids rather than increased scatter from the solids.

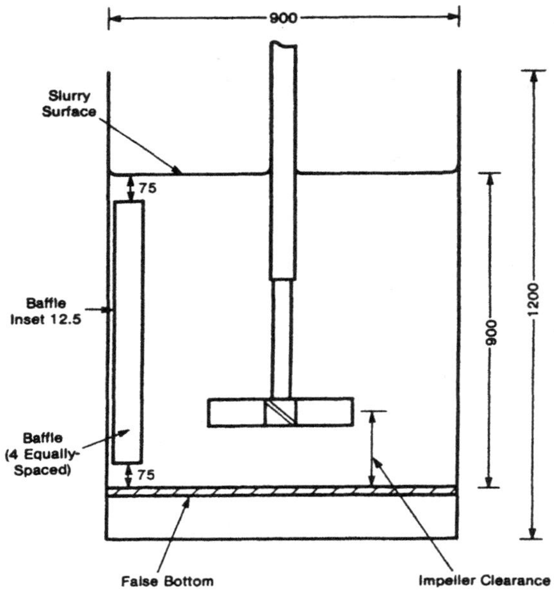

Fig. 2. Tank geometry used (All dimensions in mm)

Fig. 3. Test tank with γ-ray backscatter equipment

The γ-ray backscatter instrument was calibrated for the sand size fraction used by suspending various sand concentrations as uniformly as possible in a 300 mm diameter tank fitted with the same thickness polycarbonate window as the 900 mm diameter vessel. Because even at the closest approach to uniform suspension there is some variation in solids concentrations around the mean value, measurements were taken at several points in the tank and averaged when performing the calibration. The calibration curve obtained was fitted with a third order polynomial equation to allow the experimental software to use the calibrations in suspension profile tests.

The assumption was made both in calibration and during experiments in the 900 mm diameter vessel that the solids concentration measured at the wall was representative of the mean concentration at that horizontal plane throughout the vessel. Although this assumption was never tested in this study, Einenkel (2) has maintained that when the vessel is fitted with baffles, radial concentration differences are insignificant compared with vertical variations.

Agitator Power Measurement

Measurements of torque, impeller rotational speed and power were achieved using one of two optical sensors. Each transducer was mounted directly onto the drive shaft. The 0 - 1000 Nm range transducer was used for the coarse slurries, and the 0 - 100 Nm transducer for the fine slurries.

Slurries Used

Two sand fractions were used. The coarse fraction had a particle size range between 300 and 500 μm, with a d_{50} of 380 μm, while the finer size ranged between 50 and 200 μm, with a d_{50} of 100 μm. The size distributions obtained by sieving are shown in Figs. 4 and 5. The slurry type used will henceforth be referred to as "coarse slurry" or "fine slurry". Both size fractions were used at four concentrations: 50%, 55%, 60% and 65% by weight of solids to weight of slurry.

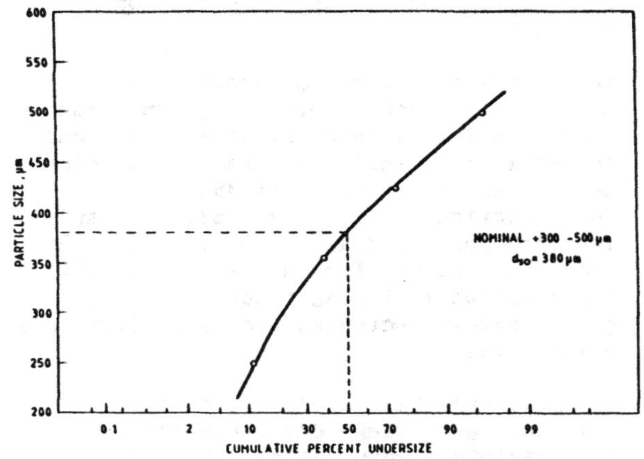

Fig. 4. Particle size distribution of coarse sand

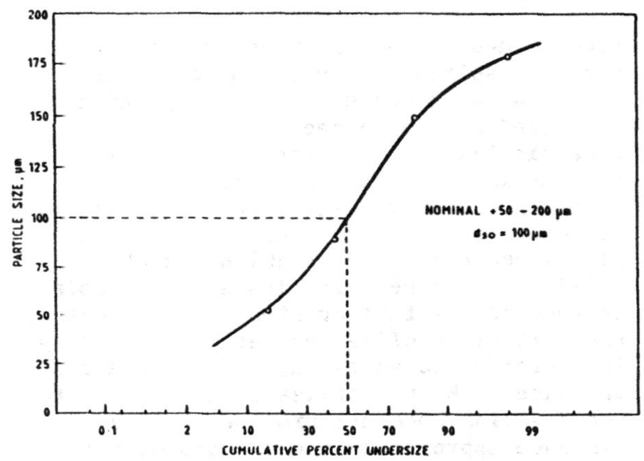

Fig. 5. Particle size distribution of fine sand

EXPERIMENTAL PROCEDURE

Each agitator was run at ten rotational speeds. The agitators were started up at their specific clearance height where possible although they were invariably fully immersed in sediment in the tank at start-up. On occasions when there was insufficient torque available for start-up in the sediment, they were started above their clearance height and the vessel then raised, by means of a hydraulic lift, until the correct agitator clearance was reached. The lowest of the ten rotational speeds used was set as that at which the agitator continued to rotate without stalling. The highest rotational speed was set at the point when slurry was beginning to be ejected from the tank.

At each speed, the γ-ray source and detector unit was moved vertically through a series of 21 measurement positions, each 37 mm apart, and the γ-ray count rate of the back-scattered radiation determined using the computerised data acquisition system. At each position, the γ-count was used to calculate the solids concentration adjacent to the measuring position using the previously determined calibration incorporated within the software. The concentration data were normalised with respect to the mean bulk solids concentration and plotted against height above vessel bottom. Simultaneous measurements of torque, and therefore power, were made at each elevation.

DATA INTERPRETATION

Although useful for some comparative purposes, the concentration profiles were of limited use for direct comparison of the five agitators on the basis of power efficiency. Instead this comparison was facilitated by reduction of individual solids concentration profiles to a single parameter. Two alternative parameters were estimated for each concentration profile: the suspension index (S_i) and the solids concentration standard deviation (σ).

Calculation of Suspension Index, S_i

The suspension index was defined as the weight of solids (or average solids concentration) in the top half of the tank divided by the weight (or average solids concentration) in the bottom half. To calculate S_i, suspension profile data were used in the definition:

$$S_i = \frac{\sum_{I=I.5}^{I=I} B_I.h}{\sum_{I=I}^{I=I.5} B_I.h} \times 100\ \% \qquad (1)$$

Eq. (1) is the summation of the areas defined by the measured points on the suspension profile and the horizontal lines to the vertical axis, for the top and bottom half of the tank. It is assumed that the solids concentration at the slurry surface was equal to that at the highest point in the slurry at which a measurement was made, and the concentration at the tank bottom the same as the lowest measured point. Typical values of S_i ranged from 20% to 50% for low agitator speeds to 90% to 105% for high speeds. The highest S_i values occasionally exceeded 100% because some agitators were capable at high speeds of lifting more solids into the top half of the tank compared with the solids remaining in the bottom half.

The lowest S_i value (S_i (st)) was achieved when the agitator speed was zero and a sediment was present. Because the bulk slurry concentrations used in the experiments were high, the initial sediment height was substantial. For the coarse slurry, S_i (st) values were estimated to be 8%, 24%, 36% and 52% for the four bulk concentrations of 50%, 55%, 60% and 65% respectively.

Calculations of Solids Concentration Standard Deviation, σ

Although suspension index is a good measure of agitator performance in terms of the quantity of solids lifted into the upper half of the tank, it does not reflect by how much the distribution of solids in the tank deviated from uniformity. Therefore, the standard deviation of the measured solids concentrations about the mean concentration in the tank was also calculated:

$$\sigma = \sqrt{\frac{1}{n}\sum_{i=1}^{n}\left(\frac{B_i}{\bar{B}}-1\right)^2} \qquad (2)$$

This was calculated using a similar expression to that used by Bohnet and Niesmak (11) except that weight concentrations, rather than volume concentrations, were used.

Bases for Agitator Comparison

The five agitators were compared using two alternative bases for comparison:

(1) for a given σ - value, the agitator requiring the least power consumption to attain that specified degree of solids suspension was noted;

(2) for a specified power consumption, the agitator providing the lowest σ-value, and the highest S_i-value, was noted.

EXPERIMENTAL RESULTS

A large amount of data was obtained but it is possible in this section to give only examples of some of the data. Full details of all experimental results may, however, be obtained from Warren Spring Laboratory. A further paper will be published providing additional details. Overall conclusions from the data are presented here when appropriate.

One-dimensional Solids Concentration Profiles

The solids concentration profile was obtained for each experimental condition of a combination of sand particle size, sand concentration, agitator type and agitator speed. In all, a total of 394 concentration profiles were measured and recorded for the coarse slurry tests (200 profiles) and the fine slurry tests (194 profiles) where the agitator speed was progressively increased from a minimum to a maximum value.

Owing to the limitations set on the agitator speed range described previously, some agitators gave relatively flat, vertical concentration profiles at their lowest speeds and increases in speed thereafter often gave only marginal improvements in the degree of solid uniformity.

Figs. 6 and 7 give examples of the ten vertical solids concentration profiles obtained for a coarse sand slurry and a fine sand slurry respectively. Dimensionless slurry height is plotted against local solids concentration, normalised with respect to the bulk slurry concentration. Both figures refer to a slurry concentration of 50% by weight solids. All other profiles are available on request for further study. As expected, the vertical profile becomes progressively less pronounced as the agitator speed is increased. Both profiles indicate that the local solids concentration peaks at a distance approximately one-third of the slurry height from the tank bottom, as observed previously (2). The profiles also indicate that once a certain maximum degree of solids homogeneity has been reached, further increases in agitator speed have little effect on the solids distribution in the tank.

Suspension Index Data (S_i)

Coarse Slurry

For any combination of agitator type and speed, and sand size and bulk concentration, S_i values were estimated using Eq. (1). In all, ten plots were constructed (5 agitator types, each with two sand sizes) from which comparisons of agitator power consumption were made; three examples plots are shown here for coarse slurry. Fig. 8 shows S_i data for the Denver 'MIL' propeller, Fig. 9 for the Ekato 'Intermig' and Fig. 10 for the 4-blade turbine.

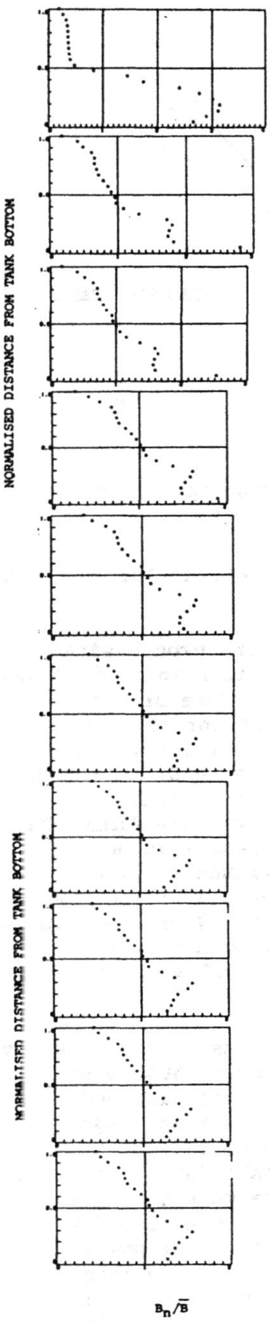

SPEED (rpm)	POWER (w)	σ	S_i (%)
97	362	1.19	19
107	520	0.62	42
118	689	0.54	47
129	904	0.43	53
139	1120	0.37	58
150	1390	0.35	60
161	1700	0.34	62
173	2030	0.34	62
185	2400	0.33	63
198	2820	0.33	63

Fig. 6. One dimensional concentration profile for 50% coarse slurry agitated using the Ekato 'Intermig'.

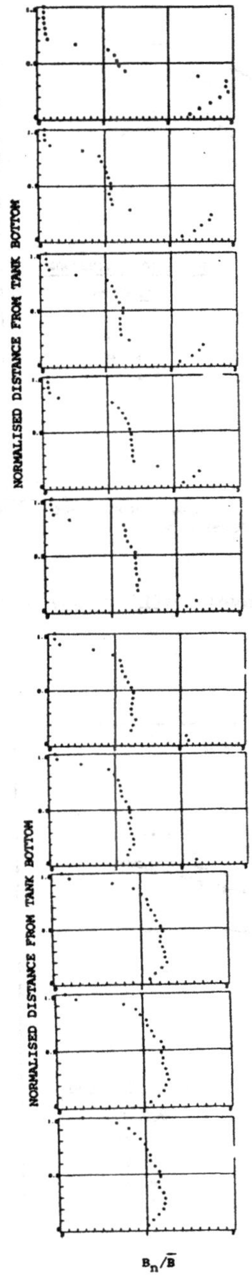

SPEED (rpm)	POWER (w)	σ	S_i (%)
67	17	1.17	20
73	26	0.82	38
82	39	0.73	44
90	48	0.71	46
99	57	0.66	49
107	71	0.52	58
118	83	0.43	65
128	102	0.32	78
131	114	0.31	78
140	134	0.21	85

Fig. 7. One-dimensional concentration profile for 50% fine slurry agitated using the 4-blade turbine (PBT).

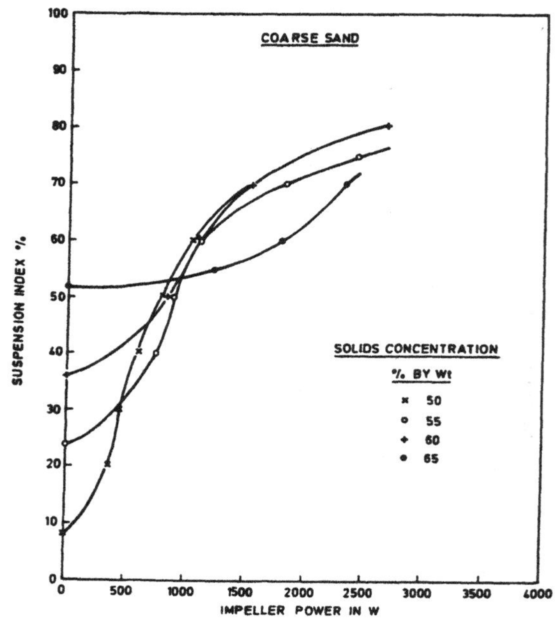

Fig. 8. S_i plot for Denver 'MIL' propeller using coarse slurry

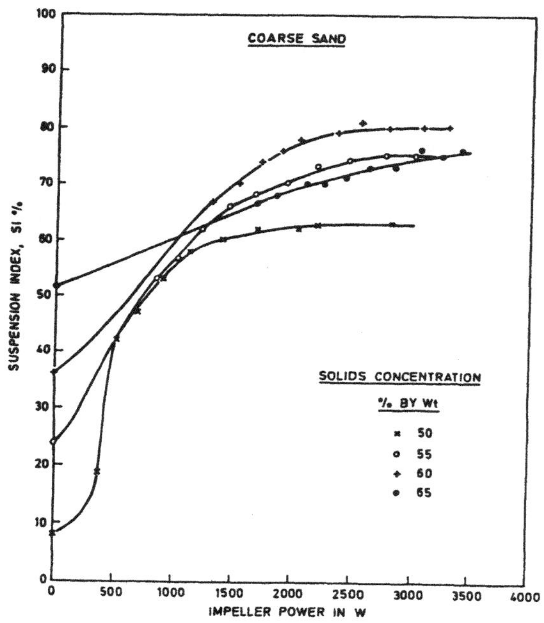

Fig. 9. S_i plot for Ekato 'Intermig' using coarse slurry

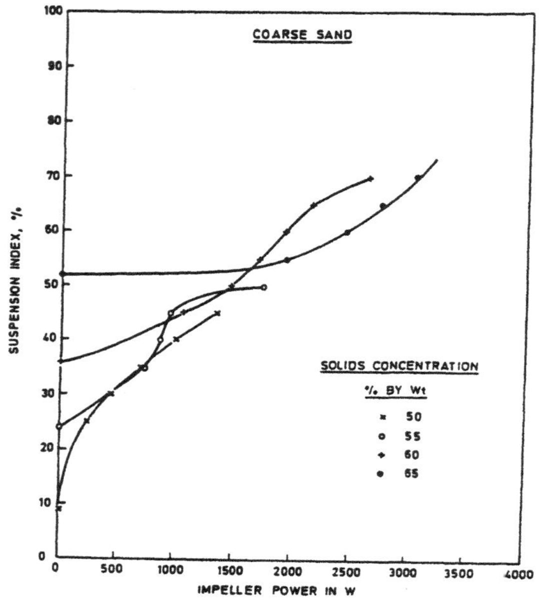

Fig. 10. S_i plot for 4-blade turbine (PBT) using coarse slurry

The plots illustrate the progressive increase in S_i as agitator speed (and hence power) is raised. The rate of increase in S_i with power falls off for the Denver and Ekato agitators for power inputs exceeding around 1500 to 2000 W once S_i values in the range 60% to 80% have been reached. Thereafter there is a rapidly diminishing return on further increases in power. However, the PBT continues to give significant improvements in S_i at power inputs above 2000 W at 60% and 65% slurry concentrations, but it was unable to achieve an S_i value of 80% at power inputs of 3000 W or more.

The maximum degree of suspension peaks over the concentration range of 50% to 65% with all three agitators. This also occurred using the marine propeller and appears to occur at a slurry concentration of about 60%. However, the Scaba 'SHP' data (not shown) gave no such peaks and S_i values in the range 80% to 90% were achieved at all four slurry concentrations at power inputs above about 3000 W. Similarly large S_i values were obtained using the marine propeller at this level of power input.

Fine Slurry

The five plots obtained for fine slurry (not shown) for each of the five agitators under test covered a power input range of 10 to 400 W. S_i values exceeding 90% were readily obtained in all cases at 50% and 55% slurry concentrations at power inputs below 200 W. All three proprietary agitator designs achieved this S_i level at lower power inputs than the marine propeller or the PBT. With 60% and 65% slurry concentrations, there was a significant concentration effect, and, unlike the coarse slurry results, a progressively increasing power requirement was observed to achieve a given S_i value in the concentration range 55% to 65%.

Standard Deviation Data (σ)

Coarse Slurry

Figs. 11, 12 and 13 show the corresponding σ-plots to Figs. 8, 9 and 10 respectively. These σ-plots provide essentially similar information. Fig. 12 shows how the 'Intermig' could rapidly achieve low σ-values (0.2 to 0.3) at more modest power inputs compared with the Denver agitator (Fig. 11) and the PBT (Fig. 13). The plots indicate again that the best suspension at around 3000 W was obtainable using a 60% slurry concentration. This trend again continued with the marine propeller, but it was possible to obtain σ-values in the range of 0.1 to 0.2 using the Scaba 'SHP' at power inputs above 2000 W.

Fine Slurry

The standard deviation plots (not shown) for all five agitators confirmed that good suspension (σ <0.2) was rapidly achieved at power inputs below 200 W and there was little slurry concentration effect for the 50% and 55% slurried. Apart from the 'Intermig' which also gave low σ-values at power levels below 200 W for the 60% slurry, the other four agitators required significantly more power when agitating the 60% or 65% slurries. However, all three proprietary designs achieved σ-values below 0.2 at these higher slurry concentrations at power inputs between 200 and 400 W. The marine propeller could not achieve a σ-value below 0.3 when agitating the 65% slurry, and the PBT was unable to reach a σ-value below 0.35 when agitating either the 60% or 65% slurries.

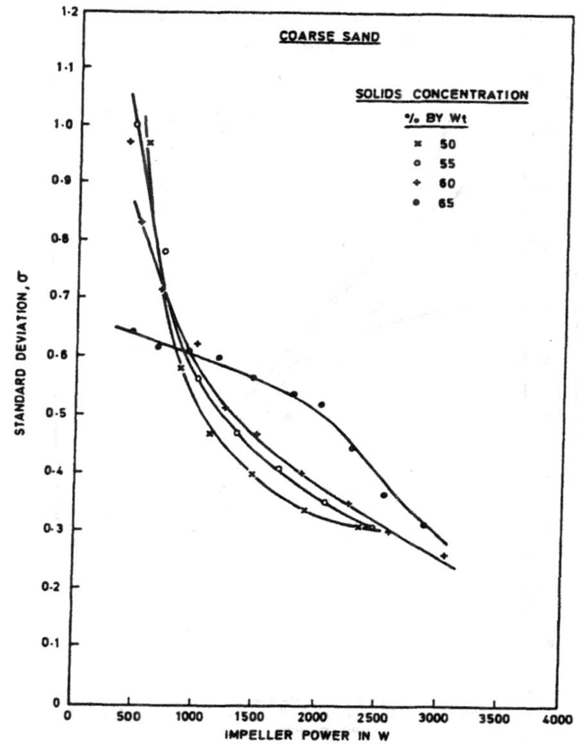

Fig. 11. σ-plot for Denver 'MIL' propeller using coarse slurry, corresponding to S_i data shown in Fig. 8.

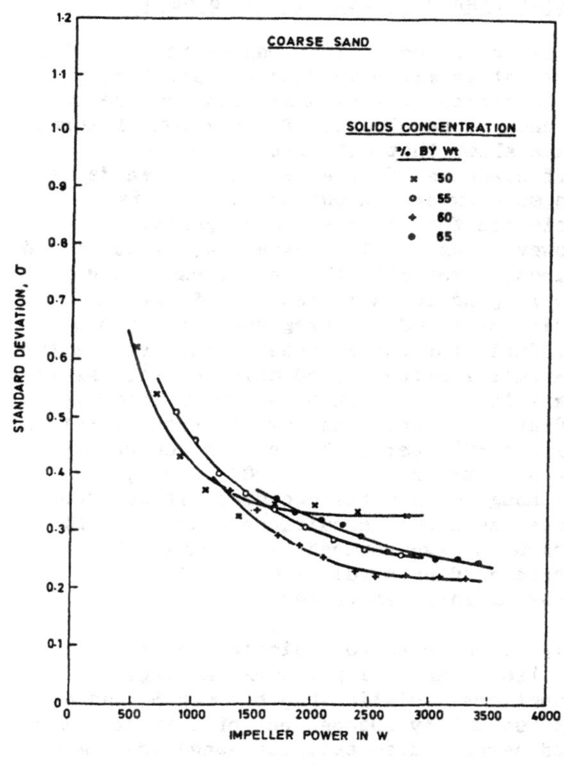

Fig. 12. σ-plot for Ekato 'Intermig' using coarse slurry, corresponding to S_i data shown in Fig. 9

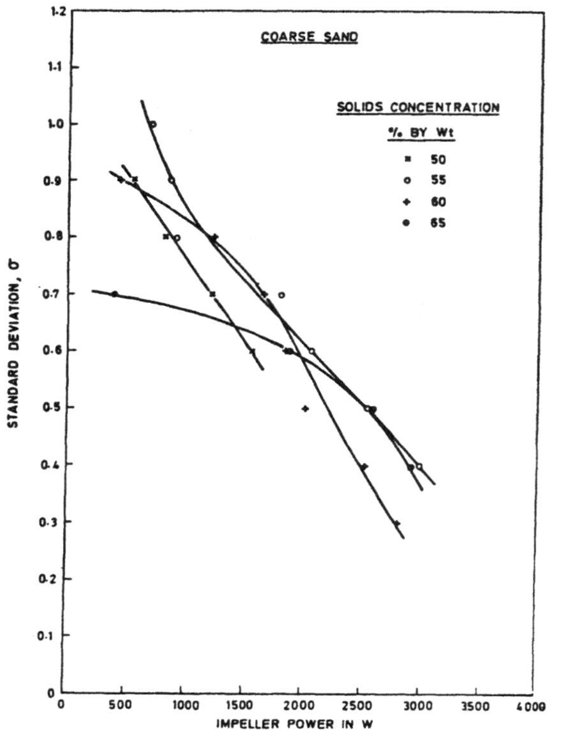

Fig. 13. σ-plot for 4-blade turbine (PBT) using coarse slurry

<u>Relationship between S_i and σ data</u>

Because of the definitions of the S_i and σ parameters given in Eqns (1) and (2), a mathematical relationship linking the two cannot be developed. The previous figures have shown that differences in the efficiencies of three agitators can depend to some extent on whether S_i or σ is selected for comparative purposes. However, as might be expected, it was found almost invariably that a decrease in σ correspond to an increase in S_i as the agitator speed was progressively raised. In fact, the correspondence between the two parameter often seemed quite close. As an example, Fig. 14 shows S_i and σ data for identical conditions for all five agitators with a 50% coarse slurry. The curve has been drawn through S = 100%, σ = 0, although on several occasions it was found under some test conditions that S_i could reach, or even exceed, 100% while the corresponding σ-values were still substantially above zero.

Fig. 14 appears to indicate that there may be, for practical purposes, a single functional relationship between S_i and σ, approximately independent of agitator type, and perhaps also agitator speed and some slurry parameters, although this was not investigated further. It is likely that the relationship will vary with slurry concentration because the initial value of S_i can be strongly dependent on slurry concentration when the sediment volume under no agitation exceeds some 50% of the total slurry volume.

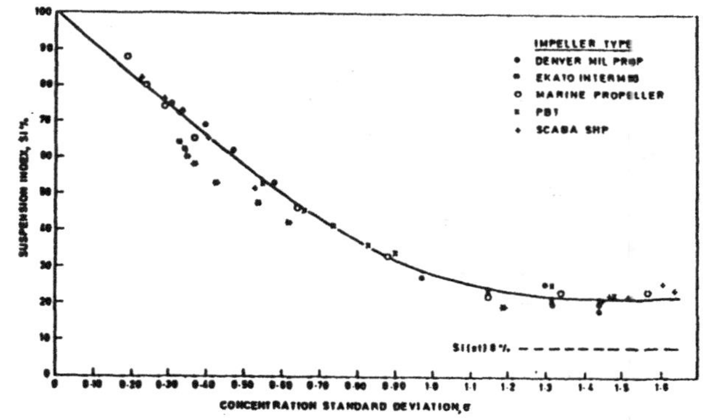

Fig. 14. Comparison between S_i and σ data for 50% coarse slurry

<u>Agitator Comparison for Coarse Slurry</u>

Figs. 15 to 18 give S_i and σ data for 50% and 60% coarse slurries. The figures show that the use of two different measures of the extent of solids suspension tends to lead to similar conclusions regarding which agitator to select if power consumption is to be minimised for a specified degree of suspension. Clearly, the PBT provides poor suspension at both 50% and 60% concentrations for the same power compared with the other four agitators.

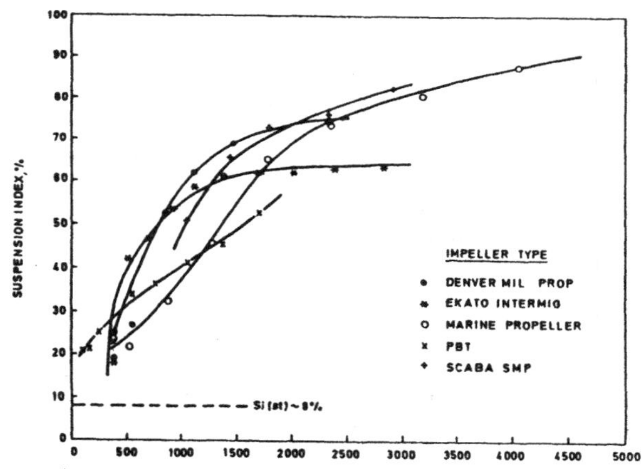

Fig. 15. S_i plot for 50% coarse slurry for all five agitators

Both the Denver 'MIL' propeller and the Ekato 'Intermig' were most power efficient at intermediate σ and S_i-values, but neither agitator was capable of achieving

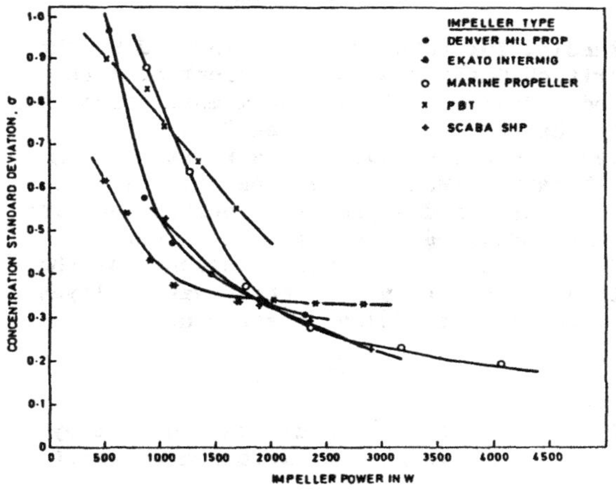

Fig. 16. σ-plot for 50% coarse slurry for all five agitators

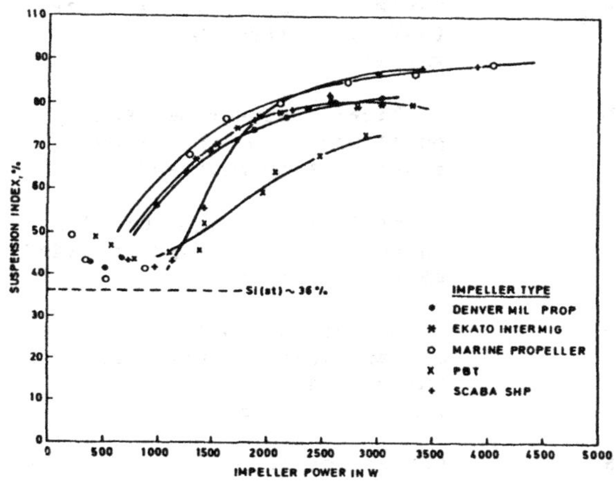

Fig. 17. S_i plot for 60% coarse slurry for all five agitators

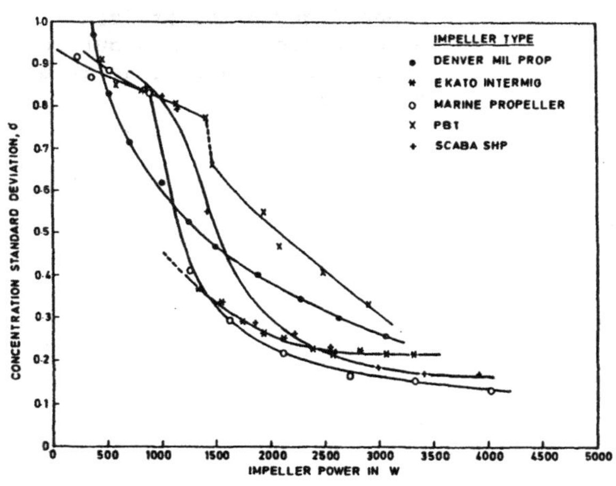

Fig. 18. σ-plot for 60% coarse slurry for all five agitators

the degrees of suspension obtainable with the marine propeller and the Scaba 'SHP', although relatively high power inputs were necessary for this. Such behaviour also occurred using the 55% and 65% slurries.

Agitator Comparison for Fine slurry

Figs. 19 and 20 show typical S_i and σ data for the five agitators using 50% fine slurry. Here there is much less of an agitator effect than for the coarse slurries. However, the PBT again performed least well at all four fine slurry concentrations in terms of either power efficiency at intermediate suspension levels or the ability to achieve high suspension levels at relatively high power inputs. Unlike their performance when using coarse slurry, all three proprietary agitators achieved high degrees of suspension (S_i > 80%, σ < 0.2). All three required progressively more power to achieve this with increasing slurry concentration. The marine propeller gave similar performance to the proprietary designs and it was difficult to identify specific trends when the four agitators were compared at different fine slurry concentrations.

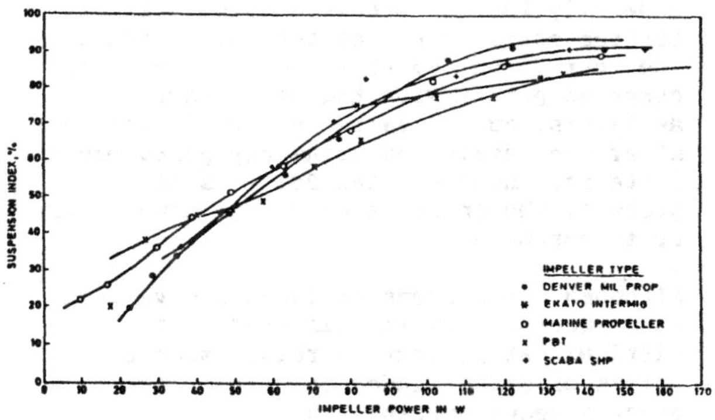

Fig. 19. S_i plot for 50% fine slurry for all five agitators

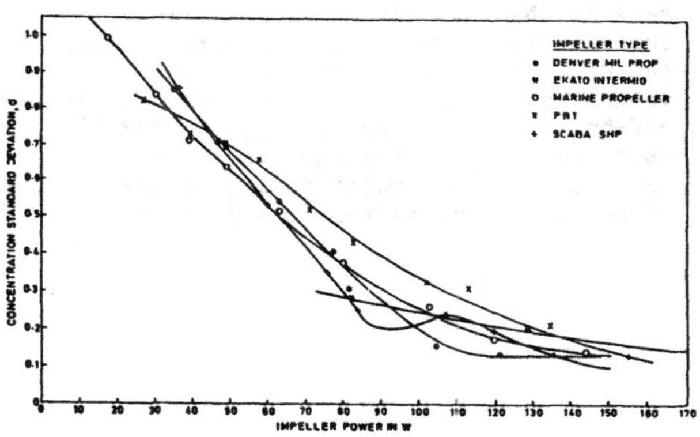

Fig. 20. σ - plot for 50% fine slurry for all five agitators

CONCLUSIONS

A novel experimental technique has been demonstrated which facilitates the comparison of different agitators on the basis of their power efficiency for suspending solids in high concentration slurries. The measured vertical solids concentration profiles obtained using a γ-ray backscatter technique may be summarised using two alternative 'indices of suspension', S_i and σ. These indices are an attempt to summarise a vertical solids concentration profile by a single number.

The results of the study have shown that, under the test conditions employed, the marine propeller and the three proprietary agitator designs can deliver similar degrees of power efficiency. Both the marine propeller and the Scaba 'SHP' achieved high degrees of suspension compared with the other three agitators at relatively high agitator power inputs when coarse slurry was used, but when using fine slurry, there was little to choose between the marine propeller and the three proprietary designs. At less demanding degrees of suspension, the Denver 'MIL' propeller and the Ekato 'Intermig' were generally found to be more power efficient for the coarse slurries tested. Overall, the four blade, 45° pitched turbine (PBT) compared poorly with the other four agitators, and appeared to have little to offer when evaluated using any performance criteria. However, the PBT is still probably the cheapest either to manufacture or to purchase.

Although two suspension indices have been used successfully in this study, it is difficult at present to relate specific values of either index to particular process requirements. In particular, positioning of a take-off point for continuous withdrawal of slurry at a specific solids concentration may require referral back to the original solids concentration profiles. However, neither the detailed profiles nor the σ and S_i values can be used to estimate the 'just-off-bottom' suspension criterion (JOBS) which in the past has been the most commonly used parameter to assess the effectiveness of different agitators.

ACKNOWLEDGEMENTS

Funding for this work was provided by the British Government's Department of Trade and Industry (DTI) and 30 member companies of the Wet Solids Handling Project, a research consortium run by Warren Spring Laboratory (WSL). The experimental test facility was designed and constructed from funds provided by DTI and the Mineral Industries Research Organisation. At the time of the study, S. Rehman was employed at WSL as a sandwich course student.

NOTATION

B_i	solids concentration in % by weight, at measurement point, n
$\bar{B}$	average solids concentration, % by weight on ratio of solids-to-water basis
d_{50}	"average" sand particle size (50% by wt less than this value)
h	vertical distance between positions at which solids concentration measured
H	dimensionless height of slurry in tank (slurry height divided by tank diameter)
n	number of horizontal locations at tank polycarbonate window at which local solids concentration measured
S_i	suspension index defined by Eqn (1)
S_i(st)	value of S_i when slurry not agitated
σ	standard deviation of local solids concentration about bulk slurry concentration, defined by Eqn (2)

REFERENCES

1. Christie, P.G. Modelling and scale-up factors of solid-liquid mixing systems. Seminar: "Mixing and Dewatering in the Minerals Industry", London, 16 April 1985. Organised by the Mineral Industries Research Organisation, UK.

2. Einenkel, W.D. Influence of physical properties and equipment design on the homogeneity of suspensions in agitated vessels. German Chem Engng, 3, 1980, 118.

3. Pavlushenko, I.S., Kostin, N.M. and Matveer, S.F. Stirrer speeds in the stirring of suspensions. J Appl Chem USSR, 30, 1957, 1235 - 1243.

4. Barresi, A. and Baldi, G. Solid suspension in an agitated vessel. Chem Eng Sci, 42, 2, 1987, 2949 - 2956.

5. Barresi, A. and Baldi, G. Solid dispersion in an agitated vessel: effect of particle shape and density. Chem Eng Sci, 42, 2, 2969 - 2972.

6. Baldi, G., Conti, R. and Gianetto, A. Concentration profiles for solids suspended in a continuous agitated reactor. AIChE J, 27, 6, 1981, 1017 - 1020.

7. Buurman, C., Resoort, G. and Plasches, A. Scaling-up rules for solids suspension in stirred vessels. Chem Eng Sci, 41, 1986, 2865 - 2871.

8. Buurman, C. Stirring of concentrated slurries: a semi-empirical model for complete suspension at high solids concentrations and 5 m^3 verification experiments. Proc Fluid Mixing IV Conference, I Chem E Symposium Series No. 121, Bradford, UK, Sept 1990, pp 343 - 350.

9. Yamazaki, H., Tojo. K. and Mujanami, K. Concentration profiles of solids suspended in a stirred tank. Powder Technol, 48, 3, 1986, 205 - 216.

10. Bos, A.S. and Heerens, J.J. Light backscattering as a technique to measure solids particle size and concentration in suspension. Chem Eng Commun, 16, 1982, 301 - 311.

11. Bohnet, M. and Niesmak, G. Distribution of solids in stirred suspensions. Chem Ing Tech, 51, 4, 1979, 314 - 315.

12. Shamlou, P.A. and Koutsakos, E. Solids suspension and distribution in liquids under turbulent agitation. Chem Eng Sci, 44, 3, 1989, 529 - 542.

13. Fajner, D., Magelli, F., Nocentini, M. and Pasquali, G. Solids concentration profiles in a mechanically stirred and staged column slurry reactor. Chem Eng Res Des, 63, 1985, 235 - 240.

14. Magelli, F., Fajner, D., Nocentini, M. and Pasquali, G. Solid distribution in vessels stirred with multiple impellers. Chem Eng Sci, 45, 3, 1990, 615 - 625.

15. Mak, A.T.C. and Ruszowski, S.W. Scaling up of solids distribution in stirred vessels. Proc Fluid Mixing IV Conference, I Chem E Symposium Series No. 121, Bradford, UK, Sept 1990, pp 397 - 412.

16. Rieger, F., Ditl, P. and Havelkova, O. Suspension of solid particles - concentration profiles and particle layer on the vessel bottom. Proc 6th European Conference on Mixing, Pavia, Italy, May 1988, pp 251 - 258. Published by BHRA Fluid Engng, Cranfield, UK.

ANNEX 1: REFEREES' QUESTIONS

1. You use d_{50} (50% by wt less than this value) as "average" sand particle size. Usually, $\bar{d}_p$ $(d_{p,16} \cdot d_{p,84})^{0.5}$. However, here $d_{50} = \bar{d}_p$.

2. What is the density of the sand particles?

3. Have you optimised the distance of each agitator (centre) from the bottom of the vessel?

4. Could you give some information concerning the rheological properties of the liquid-solid mixtures used?

5. For $N > N_{js}$ (agitator speed to just suspend particles off the tank bottom), can we assume that

$$\frac{P}{P_l} = 1 + \frac{\Sigma B_l h.(\rho_p - \rho_l)}{\Sigma h.\rho_l}$$

P_l is the impeller power consumption in the liquid density ρ_l, ρ_p is the particle density and $(\Sigma B_l h/\Sigma h)$ corresponds to the bottom half of the tank.

ANNEX 2: AUTHOR'S REPLIES TO REFEREES' QUESTIONS

1. Your suggested average particle size is a geometric mean, ours is the median of the distribution. In attempting to summarise the particle size distribution as a convenient single number, either is equally valid, although the geometric mean can be difficult to calculate if the distribution has long tails and d[illegible] and d[illegible] are difficult to estimate. However, average particle size is not made use of in any calculation in the paper, so the definition is not critical.

2. The sand density was 2650 kg m^{-3}.

3. No attempt has been made to optimise agitator clearance. All experiments were undertaken at one clearance only for a given agitator, although in an earlier unpublished study we investigated the effects of three clearances using several different designs and diameters of pitched blade turbine. The conclusion reached was that for the turbines used over the range of variables studied, there was no clear evidence that agitator clearance has any effect on the power efficiency to achieve a given degree of solids suspension as defined through the one-dimensional solids concentration profile. However, there are published data suggesting that clearance has a significant effect on the agitator speed for the "just-off-bottom" suspension condition.

   In general, the agitator clearances used in this study were set following discussion with agitator manufacturers.

4. Both sand sizes used gave rise to slurries which settled rapidly when not agitated, although there was some evidence that hindered settling effects reduced settling rates at the higher solids concentrations used. No attempt was made to measure the "viscosity" of these suspensions as the concept of "viscosity" is meaningless for a material which is unable to remain in a pseudo-homogeneous state without the need for agitation which causes turbulence in the liquid phase. However, if the agitator Reynolds number is below 10, it is possible to infer a torque-averaged agitator shear rate, and hence "viscosity", through the Metzner-Otto method. This was not carried out in the study.

5. This would appear to be a possible approximation for the agitator power consumption based on the power required for liquid agitation alone, which can be readily estimated through the agitator power number. The usefulness of this or similar expression has not, however, been assessed in the study. We may wish to do this in developing a further paper based on additional analysis of the existing experimental data.

THE DESIGN OF AGITATED DISSOLUTION TANKS

P. DITL*, F. RIEGER, I. ROUŠAR, Czech Technical University, Prague, CS

ABSTRACT

We propose a theoretical and experimental approach for optimizing agitated dissolution tanks to minimize energy requirements. Semiempirical equations for estimation of the mass transfer coefficient and the minimum impeller speed for suspension are reported. Conclusions of practical importance have been gained showing the effect of impeller speed and final solution concentration on the specific energy consumption required for preparation of storage solutions of chemicals. In addition, the efficiency of the classical dissolution tank stirred with a single axial impeller has been compared with high aspect ratio tanks agitated with multiple axial impellers.

INTRODUCTION

The dissolution of chemicals in order to prepare a storage solution of a desired concentration is important not only in the classical chemical and food industries, but also in wastewater treatment. The disproportionately high annual consumption rate in many water treatment facilities can be reduced by optimizing the dissolution process. The proper design of dissolution equipment can save energy and reduce the operating time to a minimum. Mechanical agitation is most commonly used for dissolution but air and pump mixing is also used due to their simplicity, reliability, easier maintenance and lower investment costs. Our analysis of data obtained from different water utilities proves that there are significant differences in performance between different types of dissolvers as shown in Table I. In addition each type of dissolver, depending on its specific design, can exhibit a wide range of dissolution times and energy consumption requirements. The actual energy consumption is lower than listed in Table I because these data are based on customer consumption load only and motors are usually oversized. However, these data do clearly show that stirred tanks are more efficient than the pump and air-sparged counterparts. Despite an extensive literature search we did not find any paper dealing with dissolution of chemicals from the above outlined point of view. Therefore the goal of this paper is to propose a theoretical and experimental approach for optimizing dissolution tanks, to compare the efficiency of selected geometrical configurations of high aspect ratio tanks agitated with multiple impellers, and to compare the performance of these high aspect ratio tanks, with stirred tanks of classical design. Attention also is focused on optimizing operating parameters such as impeller speed and technological parameters such as the final concentration of the storage solution.

Table I: Comparison of Different Types of Dissolution Tanks

Dissolution Tank Type	Installed Power Input [kW/m^3]	Operation Time [h]	Energy Consumption [kWh/m^3]
Agitated	0.25 - 1.5	0.5 - 4	0.2 - 1.5
Pump Mixed	0.5 - 4	0.5 - 4	0.5 - 10
Air Mixed	1.8 - 7	0.5 - 4	2 - 16

*The author is deeply grateful to Rensselaer Polytechnic Institute, Troy, N.Y, U.S.A., where he is staying on sabbatical leave, and to Prof. E.B. Nauman for the financial support to finish this paper.

THEORETICAL

CRITICAL IMPELLER SPEED FOR PARTICLE SUSPENSION

A very important parameter for suspension mixing is the minimum impeller speed needed for suspension N_s. The published literature offers more than 50 papers and recommends at least 20 different correlations for the estimation of N_s. Unfortunately, the results obtained by different authors differ substantially, especially when scaling-up. A critical review of published papers and a comparison of their results was reported by Nienow [5]. Rieger and Ditl [7] recently proposed correlations for estimation of N_s which recognize the existence of different particle-fluid regimes and are based on an extensive experimental program. The existence of different hydrodynamical regimes was first pointed out by the authors in 1984 [2]. Their results [7] for a six-pitched blade turbine are summarized in Table II.

Table II: Correlations for Estimation of Critical Impeller Speed for Particle Suspension, 6 Pitched Blade Turbine, Baffled Flat Bottom Tank, H/T = 1, C/D = 1/2, T/D = 3, Solid Concentration 2.5 vol %.

Hydrodynamical Region		Correlation
Re > 25500	d/T>0.0073	$Fr = 0.855$
Re < 25500	d/T > 0.014	$Fr = 36.5\ Re^{-0.37}$ or $Re = 1.14 Ar^{0.42}(d/T)^{-1.26}$
Re > 8600	d/T< 0.0073	$Fr = 16.4\ (d/T)^{0.6}$
Re < 8600	d/T < 0.014	$Fr = 468 (d/T)^{0.6} Re^{-0.37}$ or $Re = 3.33 Ar^{0.42}(d/T)^{-1.01}$

IMPELLER POWER INPUT

For most suspension types, it is recommended to estimate the impeller power input P by the equation

$$P = P_o \rho_{sus} N^3 D^5 \tag{1}$$

where the power number P_o is taken from the experimentally obtained graphs for liquid mixing and ρ_{sus} is an average suspension density.

DISSOLUTION

Dissolution kinetics are described by the simple equation

$$-\frac{dm}{dt} = \beta A (c_s - c) \tag{2}$$

where dm/dt represents a differential change of the solid particle mass within a differential time, β is the mass transfer coefficient, A is the surface area, and the instantaneous driving force is expressed as the difference between the saturation concentration c_s and bulk concentration c. The saturation concentration depends on the temperature, whereas the bulk concentration changes with time in accordance with the dissolution rate. Assuming that the particle surface area and the particle mass are related by $A = \alpha m^{2/3}$, one can express the dissolution time from Eq. (2) as

$$t_D = \int_0^{m_o} \frac{dm}{\beta \alpha m^{2/3} (c_s - c)} \tag{3}$$

where m_o is the initial mass of the solid particle. The integration of Eq. (3) is trivial when the driving force $(c_s - c)$ can be considered constant, e.g., when a small amount of highly soluble monodisperse particles is being dissolved. However, even more complex dissolution problems, such as dissolution of polydisperse particles at a constant driving force, can be solved analytically as proved by Ditl, et al. [1]. Introducing a dimensionless concentration $c^* = c/c_s = (m_o - m)/Vc_s$, and assuming an average value for the shape factor α during the course of dissolution, one can rewrite Eq. (3) in the form:

$$t_D = \frac{1}{\alpha} \frac{V^{1/3}}{c_s^{2/3}} \int_0^{c_o^*} \frac{dc^*}{\beta (c_o^* - c^*)^{2/3} (1 - c^*)} \tag{4}$$

where

$$c_o^* = \frac{c_o}{c_s} = \frac{m_o}{Vc_s} \tag{5}$$

A review of correlations for the estimation of the mass transfer coefficient is given by Nienow [5]. However, we recommend the correlation published by Ditl [3] which is based on a statistical treatment of 2,500 of our own experiments along with literature data for axial impellers

$$Sh = 2 + 0.43 \left(\frac{d^{4/3} \varepsilon_M^{1/3}}{\nu}\right)^{0.6} \left(\frac{T}{D}\right)^{-0.25} Sc^{0.39} \tag{6}$$

This result is valid for the following ranges of the dimensionless parameters:

$H/T = 1$	
$T/D = 4\text{-}15$	$Sh = 4\text{-}150$
$C/D = 1/3 - 2/3$	$d^{4/3}\varepsilon_M^{1/3}/\nu = 10^{-2}\text{-}10^3$
$d/T = 10^{-4}\text{-}1.55 \times 10^{-2}$	$Re = 5 \cdot 10^2 - 10^5$
$Ar = 10^{-1} - 10^3$	

The correlation was evaluated from data obtained for $N > N_s$. Neglecting the first term on the right side of Eq. (6), one can rewrite this equation in a form showing the effect of the individual parameters on β:

$$\beta \sim d^{-0.2} D^{0.4} (T/D)^{0.85} D_f^{0.61} N^{0.6} \nu^{-0.21} P_o^{0.2} \tag{7}$$

From Eqs. (4) and (6) it can be concluded that the same dissolution times in vessels of different sizes can be expected when scale-up rule ε = const is applied.

OPTIMIZATION OF DISSOLUTION TANKS

Let us assume that investment costs increase with tank volume which is proportional to the dissolution time t_D at the same production rate. The energy consumption per unit mass of the solid phase, e, accounts for a portion of the operating costs. That means that the dissolution costs are strongly influenced by both parameters t_D and e. When the dissolution time t_D and the impeller power input P are known, the specific energy consumption may be expressed as:

$$e = \frac{t_D P}{m_o} \tag{8}$$

The values of t_D and P can be either calculated using Equations (1), (4) and (6) or must be determined experimentally. Then the specific energy consumption can be easily optimized with respect to any characteristic parameter, e.g., impeller speed or relative final solution saturation $c_o^* = c_o/c_s$, or the specific consumption may be evaluated for different geometrical configurations.

In accordance with Eqs. (4) and (6) the dissolution time is proportional to $\varepsilon^{-0.2}$ and $N^{-0.6}$. Combining this with Eq. (8) one can conclude

$$e \sim \varepsilon^{0.8} \sim N^{2.4} \tag{9}$$

To express the effect of c* on e is, of course, not so trivial.

EXPERIMENTAL

Experiments have been conducted in baffled, flat bottom tanks. The geometry of various configurations tested is shown in Fig. 1 and Table III. The high aspect ratio geometry was proposed to test designs with high energy dissipation at the vessel bottom compared to tanks with uniform energy dissipation throughout the vessel ("F" and "G").

Table III: Geometry of Configurations Tested

Configu-ration	T [m]	H/T [l]	h/T [l]	i [l]	Number of Blades in Stages			
A	0.2	1	-	1	6	-	-	-
	0.3	"	-					
	0.4	"	-					
B	0.2	2.6	-	1	6	-	-	-
C	"	"	2.9	3	6	2	2	-
D	"	"	1.9	4	6	2	2	2
E	"	"	2.9	3	4	2	2	-
F	"	"	2.9	3	4	4	4	-
G	"	"	1.9	4	4	4	4	4

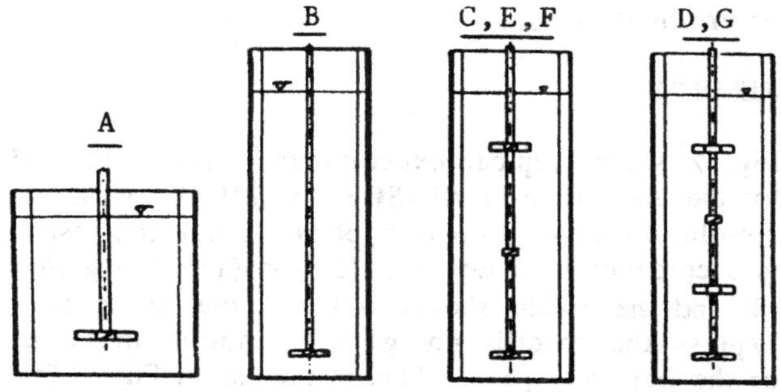

Fig. 1: Impeller to Vessel Geometry of Configurations Tested.

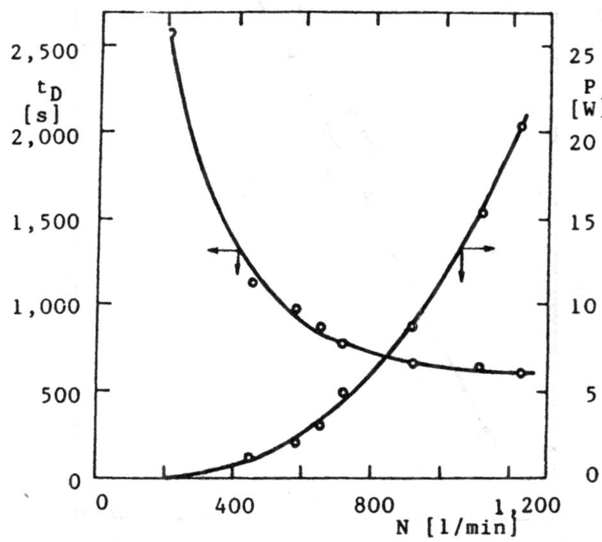

Fig. 2: Effect of Impeller Speed on Impeller Power Input and Dissolution Time. Configuration "A"; T = 0.2 m; $Al_2(SO_4)_3 \times 18H_2O$; $c_o/c_s = 0.2$.

Two different polydisperse salts, namely $Al_2(SO_4)_3$ x $18H_2O$ and $KHSO_4$, were used in the dissolution experiments. The mean particle diameters were found d = 4.53 mm for $Al_2(SO_4)_3$ x $18H_2O$ and 1.07 mm for $KHSO_4$. The $Al_2(SO_4)_3$ x $18H_2O$ mixture is used in many water treatment facilities. The saturation concentration at various temperatures has been determined experi-mentally. The dissolution experiments with $KHSO_4$ were made in the classical tank "A" having the diameter T = 0.3 m and approximately the same volume as the high aspect ratio tanks T = 0.2 m.

The impeller power input was calculated from the measured torque. The minimum impeller speed for suspension and the dissolution time were both determined visually.

RESULTS

Fig. 2 shows typical experimental results obtained for the dissolution of $Al_2(SO_4)_3$ x $18H_2O$. As the dissolution time decreases, power input increases in accordance with Eqs. (4), (6) and (1). Using Eq. (8) and the graphs shown in Fig. 2 one can easily express the specific energy consumption in terms of the impeller speed. This is plotted in Fig. 3 for two different values of the final solution saturation, $c_o^* = 0.2$ and 0.6. It is interesting that even below the minimum impeller speed for suspension, the data follow a straight line on the logarithmic plot of e vs. N. From this plot, it can be concluded that dissolution can be achieved when the solids are not fully suspended suspensions.

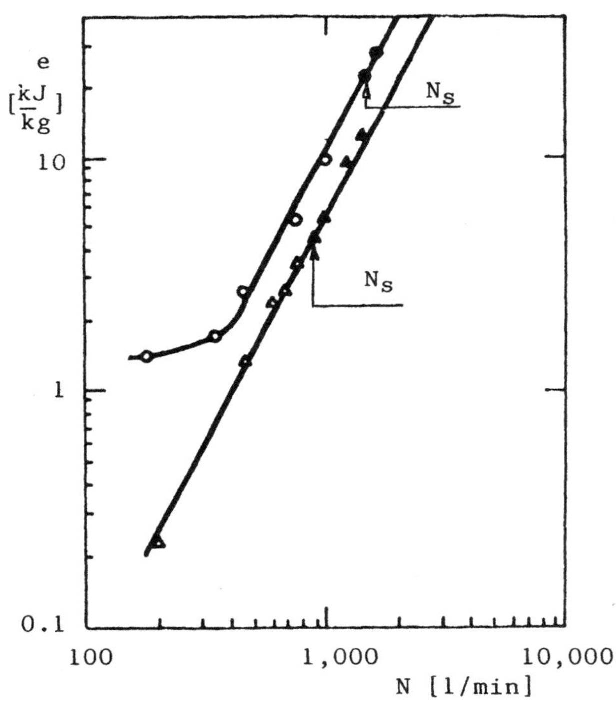

Fig. 3: Effect of Impeller Speed on Specific Energy Consumption Needed for Dissolution of $Al_2(SO_4)_3$ x $18H_2O$. Configuration "A", Δ-c_o/c = 0.2, o-c_o/c_s = 0.6

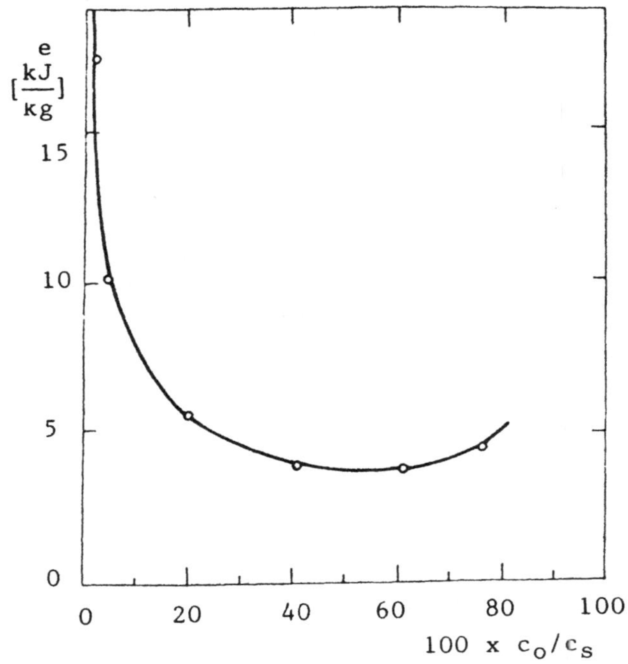

Fig. 4: Effect of Relative Saturation of the Prepared Solution c_o/c_s on Specific Energy Consumption. Configuration "A"; T = 0.2 m, N = 910 l/min, $Al_2(SO_4)_3$ x $18H_2O$.

Fig. 4 shows how the dissolution energy depends on the concentration of the storage solution. It is seen that a minimum amount of energy is expended when preparing a storage solution whose concentration is 40-60% of c_s. This conclusion offers desirable energy and also investment savings, e.g., in water treatment utilities. However the consequent homogenization of the storage solution with water must be efficient enough to allow the use a solution of higher $Al_2(SO_4)_3$ x $18H_2O$ concentration . Figs. 5 and 6 are analogous to Figs. 3 and 4, but they apply to $KHSO_4$ dissolution. The only important difference is that the curves depicted in Fig. 6 are this time based not only on experiments, but they reflect the theoretical prediction based on Eqs. (4), (6) and (8). Very good agreement between the experiments and the theoretical predictions can be observed. The curve shape of e vs. c_o/c_s is very similar to that for $Al_2(SO_4)_3$ x $18H_2O$ shown in Fig.4.

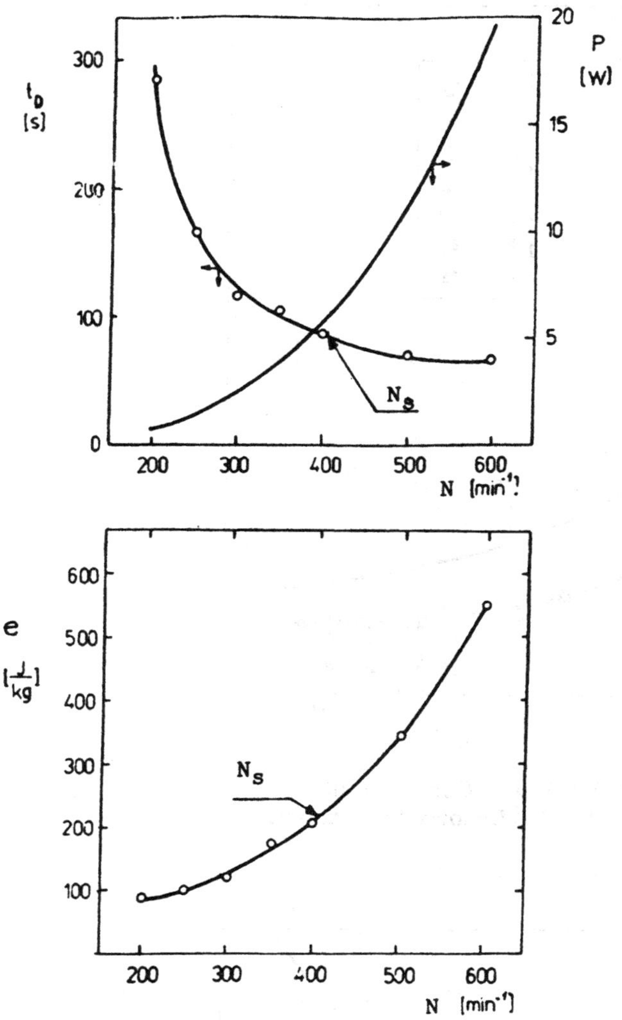

Fig. 5: Effect of Impeller Speed on Dissolution Time, Impeller Power Input and Specific Energy Consumption Needed for $KHSO_4$ Dissolution. Configuration "A", T = 0.2 m, c_o/c_s = 0.38.

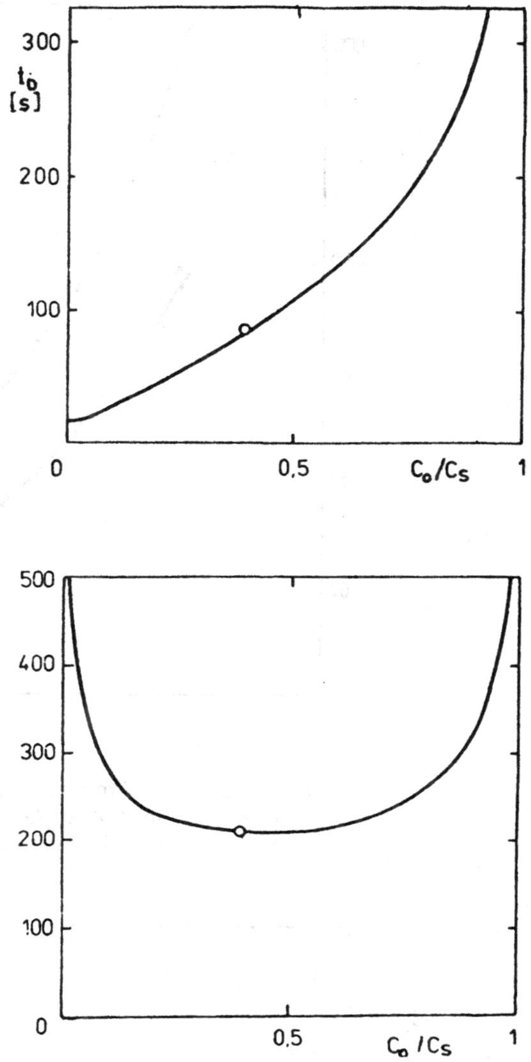

Fig. 6: Comparison of the Theoretical Prediction by Eqs. (4) and (8) (Solid Lines) with a Experiment (O). Effect of Relative Saturation of the Final Solution c_o/c_s on Dissolution Time and Specific Energy Consumption. Configuration "A", T = 0.3 m, ε = 286 W/m^3, $KHSO_4$.

Figs. 7-9 enable us to compare the efficiencies of the geometrical configurations tested. From Fig. 7, it is evident that there is no difference in dissolution time between the configurations tested (except "B") when compared at the same value of the energy input per unit of mixed volume. In configuration "B" the upper fraction of the tank volume is poorly mixed. The agitation was not sufficient to maintain a uniform concentration throughout the tank volume and consequently, the dissolution time was longer. The effect of impeller speed on energy consumption is illustrated in Fig. 8 for all impellers tested. If we transform the x-coordinate into specific impeller power input and ignore the type "B" data, we obtain only one correlation. In other words the specific energy consumption is the same regardless of the geometry of the dissolving tank, when operated at the same value of ε. However, we found significant differences in the specific impeller input required to maintain particles fully suspended as it is seen from Fig. 9. It is evident that impellers which exhibit lower values of ε_s better meet technological and operating requirements.

Further, it is worth noting that the slope of the linear part of the curve depicted in Fig. 9 equals 0.8 which is exactly the theoretical value given by Eq.. (9). The straight lines in Fig. 3 have slopes equal to 2 which is lower than the theoretical value of 2.4 predicted from Eq. (9). The difference is probably caused by the higher initial particle size.

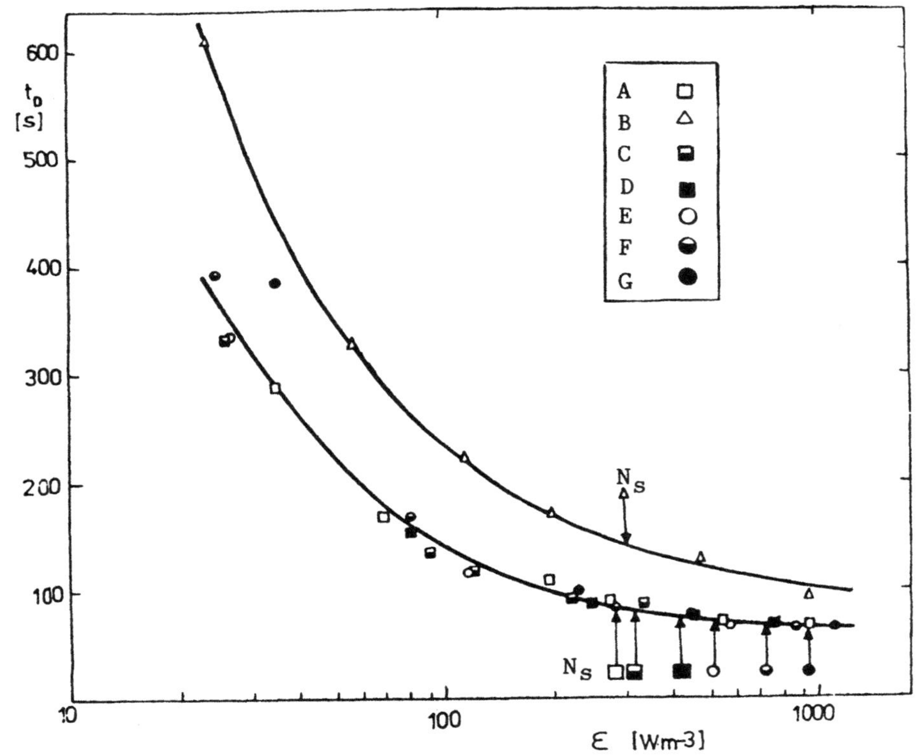

Fig. 7: Dissolution Time vs. Specific Impeller Power Consumption for All Configurations "A to G" Tested. Symbol N_S↑ Denotes Full Particle Suspension, $KHSO_4$, c_o/c_s = 0.38.

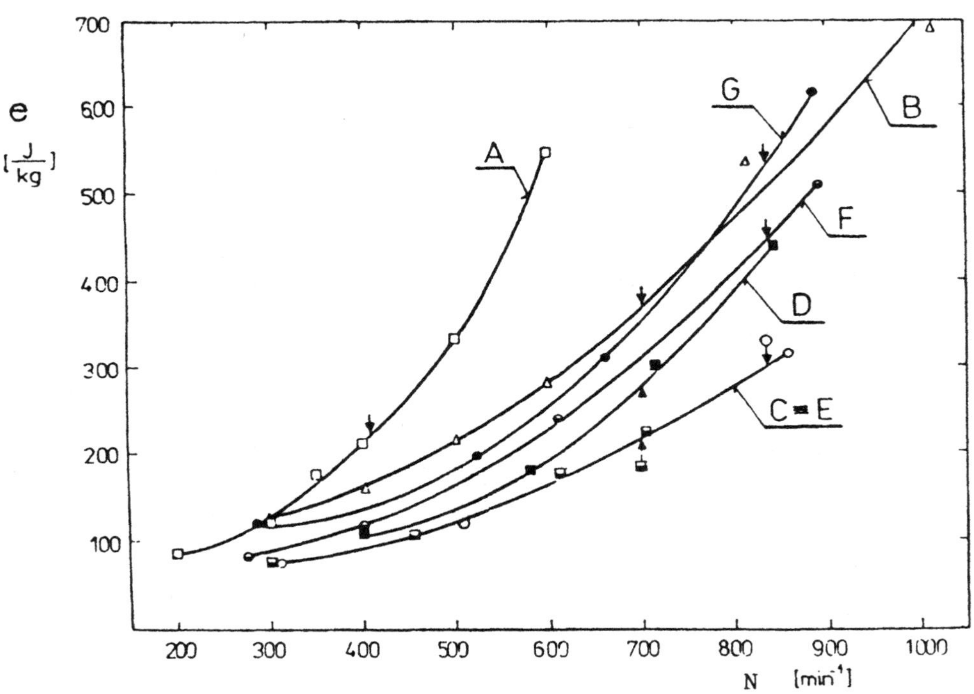

Fig. 8: Effect of Impeller Speed on Specific Energy Consumption for $KHSO_4$ Dissolution. Configurations "A to G" Tested at c_o/c_s = 0.38. Symbol ↓ Denotes Full Particle Suspension.

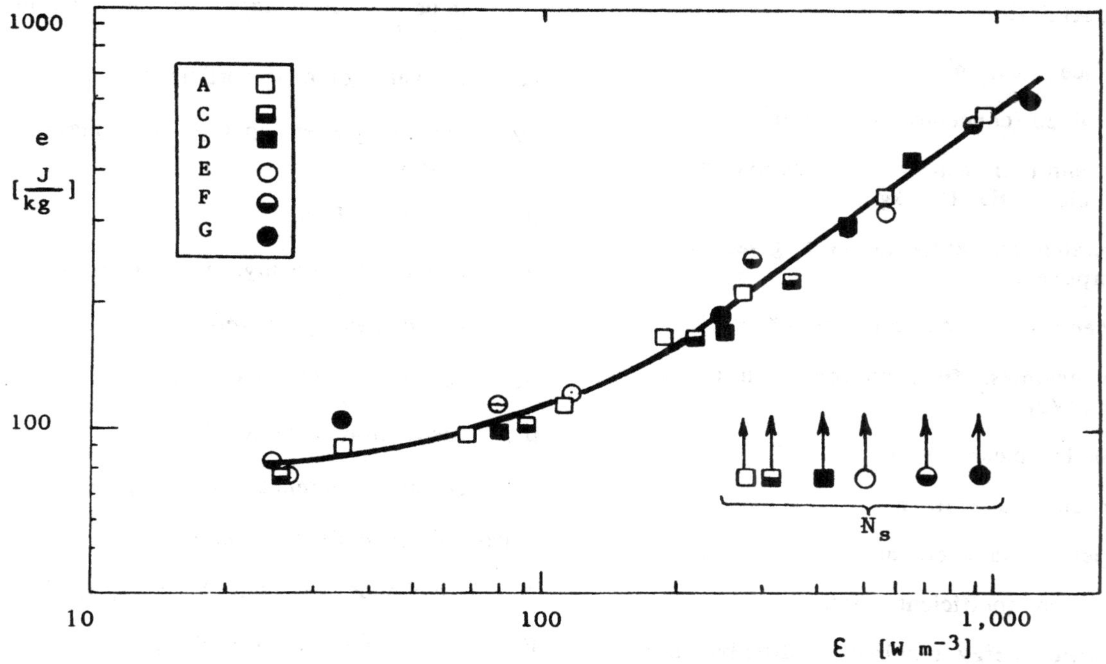

Fig. 9: Effect of Specific Power Input on Specific Energy Consumption Needed for $KHSO_4$ Dissolution. Symbol N_s↑Denotes Full Particle Suspension. Configurations "A, C - G", c_o/c_s = 0.38.

CONCLUSIONS

The following conclusions are drawn:

1. A mechanically agitated dissolution tank exhibits substantially lower energy consumption per unit mass of the solid dissolved than the other agitation schemes.

2. For the given salt dissolution rate, the size of a dissolution tank and investment costs are proportional to the dissolution time and depends on the final concentration of the solution. Energy consumption and operating costs, consequently, can be optimized with respect to parameters affecting the process, e.g., N, c_o,/c_o, geometry, etc.

3. The calculation procedure proposed in this paper to optimize agitated tanks in terms of energy consumption was experimentally verified.

4. This paper successfully utilizes the authors' equation for the estimation of the mass transfer coefficient as well as their equations for minimum impeller speed for suspension.

5. The effect of the impeller speed on specific energy consumption was found in accordance with Eq. (9), $e \sim \varepsilon^{0.7-0.8} \sim N^{2-2.4}$. It was proved that dissolution tanks can be economically operated even below the minimum impeller speed for suspension, provided there are technological limitations (e.g., particle stickining).

6. Preparation of storage solutions exhibits a minimum specific energy consumption in the range of 40-60% of their saturation concentration as seen from Figs. 4 and 6. The selection of such a concentration is limited by the efficiency of the subsequent homogenization equipment.

7. No significant difference in energy consumption was found between the classical dissolving tank "A" and the high aspect ratio tanks "C" through "G" which contained multiple axial impellers. Significant differences were, however, found with respect to the specific impeller power input required for particle suspension. Impellers achieving particle suspension at the lowest value of ε_s are preferred for dissolution tank design.

NOMENCLATURE

A	surface area, m^2
c	dissolved concentration, kg/m^3
c_o	concentration achieved by dissolving all the particles - Eq. (5), kg/m^3
c_s	saturated concentration at a given temperature
c^*	dimensionless concentration, $c^* = c/c_s$
c_o^*	dimensionless final concentration $c_o^* = c_o/c_s = m_o/Vc_s$
C	impeller clearance, m
d	particle diameter, m
D	impeller diameter, m
D_f	diffusion coefficient, m^2/s
e	specific energy required to dissolve unit mass of particles, J/kg
g	acceleration due to gravity, m/s^2
h	distance between impellers, m
H	liquid depth, m
i	number of impellers, dimensionless
m	total mass of solid particles, kg
m_o	initial mass of solid particles, kg
N	impeller speed, l/s
N_s	impeller speed required for particle suspension, l/s
P	impeller power input, W
P_o	impeller power number
t	time, s
t_o	dissolution time, s
T	tank diameter, m
V	volume, m^3
W	blade width, m
α	shape factor, $m^2/kg^{2/3}$
β	mass transfer coefficient, m/s
ε	impeller power input per unit volume, W/m^3
ε_s	the value of ε at particle suspension, W/m^3
ε_M	impeller power input per unit mass, $\varepsilon = \varepsilon_M \rho_{sus}$
μ	viscosity, Pa·s
ν	kinematic viscosity, $\nu = \mu/\rho$, m^2/s
ρ	liquid density, kg/m^3
ρ_s	density of solid particles, kg/m^3
ρ_{sus}	suspension density, kg/m^3
$\Delta\rho$	density difference, $\Delta\rho = \rho_s - \rho$, kg/m^3

Other Dimensionless Criteria

Ar	Archimedes number, $Ar = d^3\rho\Delta\rho g/\mu^2$
Fr	Froude number, $Fr = N^2D\rho/g\Delta\rho$
Re	Reynolds number, $Re = ND^2\rho/\mu$
Sc	Schmidt number, $Sc = \nu/D_f$
Sh	Sherwood number, $Sh = \beta d/D_f$

LITERATURE

[1] Ditl, P., et al.: Int. J. Heat Mass Transfer, 19, pp. 635-41 (1976).

[2] Ditl, P., Rieger, F.: Paper V3.61, CHISA Conference, Prague, August 1984.

[3] Ditl, P.: Paper 18, Proceedings of 5th European Conference on Mixing, Würzburg, June 1985, BHRA Fluid Eng.

[4] Ditl, P., Rieger, F.: Paper 8, ibid.

[5] Harnby, N., Edwards, M.F., Nienow, A.W.: Mixing in the Process Industries, Butterworths, London, 1985.

[6] Rieger, F., Ditl, P.: Proceedings of the European Conference on Mixing, Netherlands, 1982, pp. 263-269, BHRA Fluid Eng.

[7] Rieger, F., Ditl, P.: Suspension of Solid Particles, Paper presented at CHISA Conference, Prague, August 1990.

MODELLING AND CONTROL OF MULTISTAGE MIXER EXTRACTORS

C. TSOURIS, L.L. TAVLARIDES, Dept. of Chemical Engineering and Materials Science, Syracuse University, USA

ABSTRACT

A drop-population balance model is proposed for the prediction of the transient dispersed-phase fraction (holdup), drop size distribution, and solute concentration along a multistage mixer extraction column. Droplet events of breakage and coalescence and flow patterns inside the contactor are considered. The population balance model is suitable for model-based predictive control, and it is tested against experimental results of drop size distribution, holdup, and concentration profiles from a pilot plant unit. The dynamic behavior and control of multistage extraction columns is also studied and a multivariable control scheme of holdup and concentration is proposed.

INTRODUCTION

The mass transfer rate of a chemical species from one liquid phase to another is linearly related to the interfacial area between the two phases and to the concentration driving force. In a liquid-liquid extraction equipment, the two phases form a liquid dispersion where one phase is continuous and the other is dispersed in terms of droplets. Mixing provided by agitation results in droplet breakup and coalescence which control the droplet size and, thus, the interfacial area between the two phases. Furthermore, droplet coalescence tends to homogenize the concentration in the dispersed phase affecting in this way the driving force for mass transfer. Droplet interaction processes are therefore playing an important role for the mass-transfer process in liquid-liquid extraction equipment. Frequency functions of breakage and coalescence have been derived by Coulaloglou and Tavlarides [1] to describe these interaction processes based on phenomenological arguments.

The Population Balance Equation

An appropriate tool for describing particulate systems such as liquid dispersions is the population balance model which accounts for smooth changes in concentration and abrupt changes due to birth and death of particles. The birth term in liquid dispersions is a result of breakage of larger droplets and coalescence of smaller ones whereas the death term is due to coalescence to larger droplets and breakage to smaller ones. Details of the population balance model are described elsewhere [2-4]. Briefly, in a liquid dispersion, the position and the properties of every droplet can be described by means of a particle-state vector. The state vector includes all the variables which uniquely characterize a droplet, such as position, age, and solute concentration. These variables are separated into external, $\mathbf{x} = (x,y,z)$, describing position in the physical space and internal coordinates, $\mathbf{r} = (r_1, r_2, ..., r_n)$, describing droplet size, concentration, age, etc. Then, the particle state vector can be written as

$$\xi = \xi(x,r) \tag{1}$$

With this decomposition, a balance of the number of particles in a volume element of the phase space leads to the following population balance equation:

$$\frac{\partial n}{\partial t} + \nabla_x \cdot (V_e n) + \nabla_r \cdot (V_i n) + D - B = 0 \tag{2}$$

where n is the number density and denotes the average number of droplets per unit volume of the phase space; V_e is the rate of change of the external coordinates, i.e., the velocity of the droplets throughout the physical space, V_i is the rate of change of the internal coordinates; B and D are the birth and death terms and indicate the generation and loss of droplets in the control volume. The divergence terms describe the deterministic continuous variations that take place in the system. This model has been employed in the past [5-7] to explain nonuniform holdup profiles and drop size distributions along extraction columns.

Control studies of the liquid-liquid extraction process have been reported for a number of extractors including pulsed, packed, Scheibel, and Karr columns. The main objective of these studies is to control the solute concentration at the exit streams. Recent studies include also the control of the dispersed phase volume fraction [8-10].

The present work employs the population balance model to describe the performance of a multistage mixer extractor to demonstrate the capability for use in predictive control. Multivariable control of holdup and solute concentration in one of the two phases is also studied. The ultrasonic technique [11] is employed for monitoring the holdup profile.

EXPERIMENTAL

Pilot Plant

Experiments were contacted on a multistage stirred extraction column of pilot plant scale. A detailed description of the pilot plant is given by Kirou et al. (1988). The multistage extractor is of 125 mm diameter. It has four vertical baffles, and it is separated into seven interconnected stages by horizontal plates. Each plate has a hole at the center which forms 29% free area as compared to the cross-sectional area of the column. Mixing is provided by six-blade impellers of 63.5 mm diameter located at the center of each stage.

Measurement Techniques

Parameters of importance in liquid-liquid extraction are: i) the drop size distribution, ii) the dispersed phase volume fraction, and iii) the concentration of the transferred species in both phases. The drop size distribution is measured by a microphotographic technique [12], the holdup by an ultrasonic technique [11,12], and the concentration by conductivity and refraction index measurements. These parameters are measured at various operating conditions ranging from 0.2 to 1.2 L/min flow rate of each phase and from 150 to 200 rpm in the agitation speed.

Chemical System

The chemical system used in this study is water (continuous) - toluene (dispersed). For mass transfer experiments, butyric acid was added in either of the two phases, and the concentration profile along the extractor was measured in both phases.

RESULTS

Dynamic Behavior of the Multistage Column Extractor

The transient behavior of the dispersed phase volume fraction has been studied for step changes in the operating conditions. Measurements of the holdup were taken automatically by a data acquisition system developed for the ultrasonic technique [11]. Figure 1 shows the holdup response for step changes in the agitation speed, whereas Figure 2 shows the response for step changes in the dispersed-phase flow rate. As

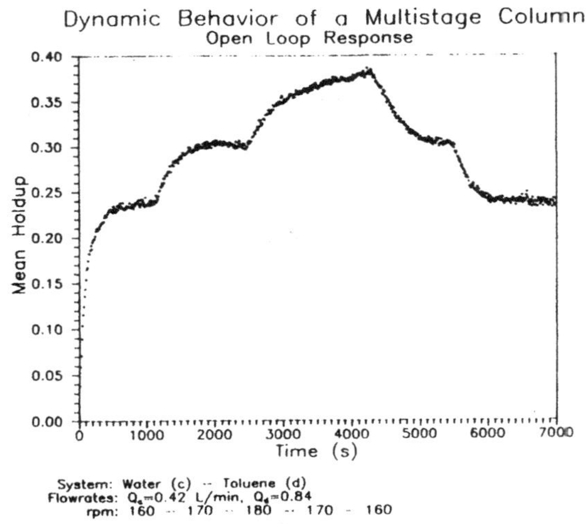

Figure 1. Holdup response for step changes in the agitation speed

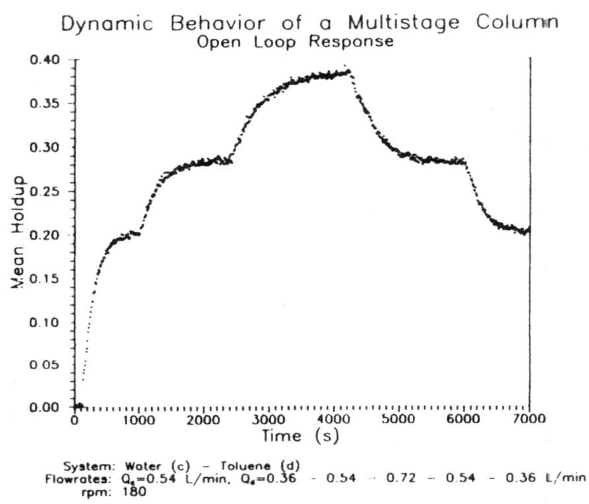

Figure 2. Holdup response for step changes in the dispersed-phase flow rate

shown in Figure 3, the response can be modelled by a first-order-plus-dead-time transfer function of the form:

$$G = \frac{K \exp(-\theta S)}{\tau S + 1} \tag{3}$$

where K = gain of the process
θ = dead time, s
τ = time constant of the process, s

Eq. (3) is an empirical model with three parameters which are normally assumed to be constant. For nonlinear processes, however, such as the liquid

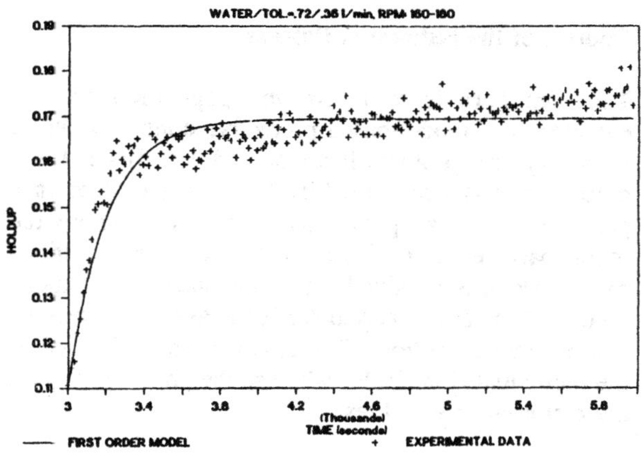

Figure 3. Comparison of the first-order-model-plus-dead-time with experimental data

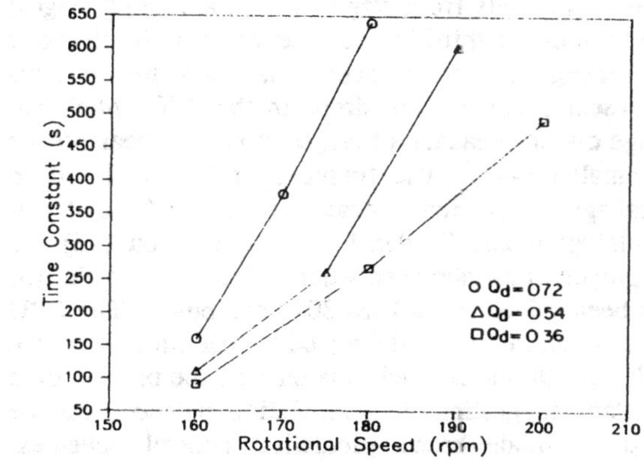

Figure 4. Time constant of the transfer function of holdup with respect to the continuous-phase flow rate versus agitation speed

extraction, the above parameters are functions of the operating conditions. As an example, the time constant of the transfer function of the holdup with respect to changes in the continuous-phase flow rate is shown in Figure 4. A more realistic approach for the modeling of the behavior of the extractor is by the population balance equation.

Solution of the Population Balance Equation for the Multistage Extractor

For only one drop internal variable, i.e., drop volume, the population balance equation for the drop population in the multistage contactor, is given as follows:

$$\frac{d[N^i(t)A^i(V,t)]}{dt} = N^{i-1}(t)A^{i-1}(v,t)F_f^{i-1}(v)+N^{i+1}(t)A^{i+1}(v,t)F_b^{i+1}(v)$$
$$-N^i(t)A^i(v,t)F_f^i(v)-N^i(t)A^i(v,t)F_b^i(v)$$
$$-N^i(t)A^i(v,t)[g^i(v)+\omega^i(v,t)]$$
$$+\int_V^{V_{max}} \beta(v',v)\nu(v')g^i(v')N^i(t)A^i(v',t)dv'$$
$$+\int_0^{V/2} \lambda^i(v-v',v')h^i(v-v',v')N^i(t)A^i(v-v',t)N^i(t)A^i(v',t)dv' \qquad (4)$$

where

$$\omega^i(v,t) = \int_0^{V_{max}-V} \lambda^i(v,v')h^i(v,v')N^i(t)A^i(v',t)dv'$$

$A^i(v,t)\,dv$ = number fraction of drops in the ith stage of the column with volume between v and v + dv at time t

$N^i(t)$ = total number of drops in the ith stage at time t

$F_f^i(v)$ = forward exit frequency of drops of volume v from the ith stage

$F_b^i(v)$ = backward exit frequency of drops of volume v from the ith stage

$g^i(v)$ = fraction of drops in the ith stage with volume v disappearing through breakage per unit time

$\nu(v)$ = average number of drops formed per breakage of drops of volume v

$\beta(v',v)$ = number fraction of drops with volume between v and v + dv formed by the breakage of a drop of volume v'

$h^i(v',v)$ = collision frequency of a drop of volume v with a drop of volume v' in stage i

$\lambda^i(v',v)$ = coalescence efficiency of a drop of volume v with a drop of volume v' in stage i.

The first two terms after the equality sign of Eq. (4) represent input of droplets to stage i from the two adjoining stages; the terms on the second line represent flow of droplets from stage i to the adjoining stages; the term on the third line represent loss of droplets due to breakage and coalescence; the last two integral terms represent production of drops in the differential size range due to breakage of larger drops and coalescence of smaller ones. The frequency functions for drop breakage, collision, coalescence, and exit of Coulaloglou and Tavlarides [1] and Kirou [13] are employed in the above equation. The resulting Eq. (3) has been solved by an IBM 3090 computer. The CPU time required for modeling 800s real time is of the order of 20s and is much smaller than the process time constant suggesting that population balances can be used in model-based predictive control schemes. Figure 5 shows a comparison of the holdup results from the population balance equations with experimental data obtained by the ultrasonic technique. It seems that the model underpredicts the effect of the agitation speed. This is shown at changes of rpm from 160 to 180. A steady-state difference of about 8 percentage units is obtained experimentally, whereas the model shows a difference of about 5 percentage units. This defficiency of the model is attributed to the breakage function.

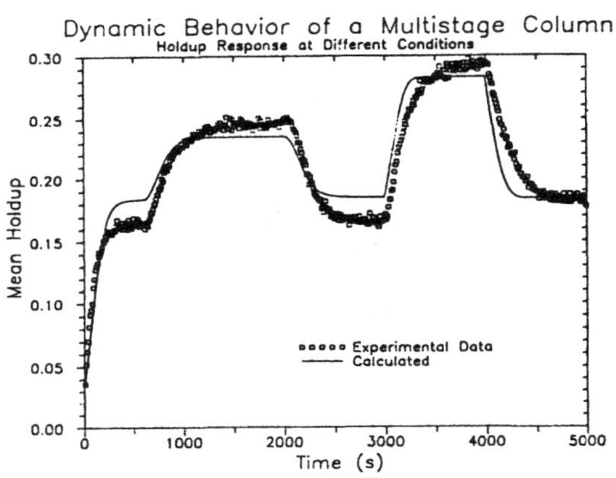

Figure 5. Solution of the PBE vs. experimental data at the following conditions:

time(s)	rpm	Q_d(L/min)	Q_c(L/min)
0	160	0.54	0.54
600	180	0.54	0.54
2000	160	0.54	0.54
3000	170	0.84	0.42
4000	170	0.42	0.84

Control of the Extraction Process

The idea here is to utilize the population balance equation in a model-based predictive control scheme. Presently, this problem is not solved yet. Instead, the empirical model presented by Eq. (3) is offered in a gain-scheduling adaptive control scheme, where the model parameters are updated with operating conditions by correlations obtained from the open loop experiments. The control parameter is the average dispersed phase volume fraction. The agitation speed is used as the manipulated variable, whereas the phase flow rates are disturbance variables.

The controller was designed with the requirement that the close-loop response be of first order with dead time. This approach is equivalent to the Dahlin algorithm in the discrete domain. The final equation of the controller has a proportional-integral form as shown below:

$$G_c = \frac{\tau}{K(\lambda + \theta)}\left(1 + \frac{1}{\tau S}\right) \tag{5}$$

where λ is the close-loop time constant. Figure 6 represents the close-loop response of the dispersed-phase volume fraction for step changes in the set point from 0.15 to 0.25, then to 0.35, and finally back to 0.15. The phase flow rates in this experiment are both 0.54 L/min. The close-loop time constant was set to be the same as the open-loop time constant. The manipulated variable is also shown in the same figure.

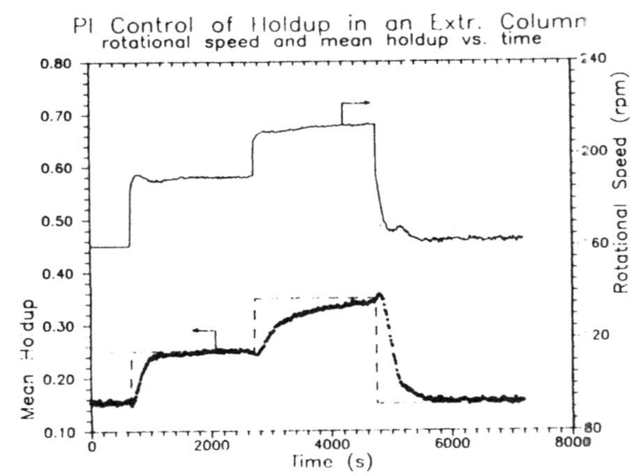

Figure 6. Servo-control of the holdup. Set point changes: 0.15 - 0.25 - 0.35 - 0.15
System: water (c) - toluene (d)
Flow rates: $Q_c = Q_d = 0.54$ L/min
Close-loop time constant: $\lambda = \tau$

Figure 7 shows the response of the holdup at exactly the same operating conditions. The only difference is that the close-loop time constant is 50% of the open-loop time constant. This difference makes the controller more aggresive as shown in the manipulated variable. Also, a slight overshoot appears even though the controller was designed for a first-order close-loop response.

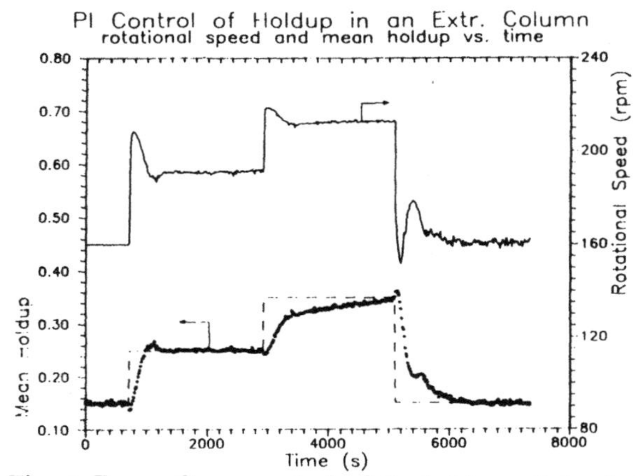

Figure 7. Servo-control of the holdup. Set point changes: 0.15 - 0.25 - 0.35 - 0.15
System: water (c) - toluene (d)
Flow rates: $Q_c = Q_d = 0.54$ L/min
Close-loop time constant: $\lambda = 0.5\ \tau$

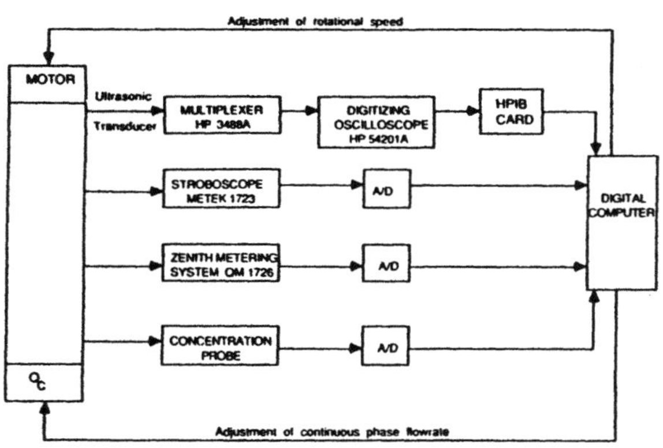

Figure 8. Multivariable control scheme

The multivariable control of holdup and concentration is studied currently. Figure 8 shows the control scheme. A conductivity probe is employed for monitoring the solute concentration in the continuous (aqueous) phase. Simultaneous control of the dispersed-phase volume fraction and solute concentration in the exit stream of the continuous phase will be studied in this program.

CONCLUSIONS

In summary, the dynamic behavior of a multistage extraction column is studied both experimentally and theoretically. The response of the dispersed-phase volume fraction is studied with variations in the operating conditions. A population balance model, which considers drop rate processes of breakage, coalescence, and exit is employed to explain the experimental results. Empirical models are also used and further utilized for the control of the holdup. Finally, servo-control of holdup has been demonstrated.

NOTATION

B	birth term in Eq. (2)
D	death term in Eq. (2)
holdup	$\frac{\textit{volume of the dispersed phase}}{\textit{total volume of the dispersion}}$
G	transfer function
K	gain of the process
θ	dead time, s
τ	open-loop time constant, s
λ	close-loop time constant, s

REFERENCES

1. Coulaloglou, C.A. and L.L. Tavlarides, "Description of Interaction Processes in Agitated Liquid-Liquid Dispersions", Chem. Eng. Sci., 32, 1289 (1977).

2. Hulburt, H.M. and S. Katz, "Some Problems in Particle Technology. A Statistical Mechanical Formulation", Chem. Eng. Sci., 19, 555 (1964).

3. Randolph, A.D. and M.A. Larson, "Theory of Particulate Processes. Analysis and Techniques of Continuous Crystallization", Academic Press, New York, 1971.

4. Ramkrishna, D., "Analysis of Population Balance - IV. The Precise Connection Between Monte Carlo Simulation and Population Balances", Chem. Eng. Sci., **36**, 1203 (1981).

5. Sovová, H., "A Model of Dispersion Hydrodynamics in a Vibrating Plate Contactor", Chem. Eng. Sci., **38**, 1863 (1983).

6. Laso, M. L. Steiner and S. Hartland, "Modeling of Axial Hold-Up and Drop Size Distribution Profiles in a Liquid-Liquid Extraction Column", paper D7.8 CHISA 84, Prague, Czechoslovakia, Sept. 1984.

7. Cassamata, G. and A. Vogelpohl, "Modeling of Fluid Dynamics and Mass Transfer in Extraction Columns", Germ. Chem. Eng., **8**, 96 (1985).

8. Taylor, P.A., M.H.I. Baird and I. Kusuma, "Computer Control of Holdup in a Reciprocating Plate Extraction Column", Can. J. Chem. Eng., **60**, 556 (1982).

9. Camurdan, M.C., "Application of Adaptive Control to a Reciprocating Plate Liquid-Liquid Solvent Extraction Column", Ph.D. Thesis, McMaster University, 1986.

10. Tsouris, C. and L.L. Tavlarides, "Control of Extraction Columns", Annual AIChE Meeting, Chicago, IL, Nov. 11-16, 1990.

11. Tsouris, C., L.L. Tavlarides and J.C. Bonnet, "Application of the Ultrasonic Technique for Real-Time Holdup Monitoring for the Control of Extraction Columns", Chem. Eng. Sci., **45**, 3055 (1990).

12. Kirou, V.I., L.L. Tavlarides, J.C. Bonnet, and C. Tsouris, "Flooding, Holdup, and Drop Size Measurements in a Multistage Column Extractor", AIChE J., **34**, 283 (1988).

13. Kirou, V.I., "Stochastic Simulation of Dispersed Phase Holdup and Drop Size Distribution in a Mixer Column Contactor", Ph.D. Thesis, Syracuse University, 1990.

ACKNOWLEDGEMENTS

The partial financial support for this work by the National Science Foundation Grant, CTS-9017138 and the general support by the Chemical Engineering and Materials Science Department of Syracuse University is gratefully acknowledged.

LOCAL MEASUREMENTS OF LIQUID AND SOLID VELOCITIES AND OF PARTICLE SIZES IN STIRRED SUSPENSIONS WITH A PHASE DOPPLER PARTICLE ANALYSER

P. GUIRAUD, J. COSTES, J. BERTRAND, Laboratoire de Génie Chimique, URA CNRS ENSIGC, Toulouse
J. BOUSQUET, SN Elf Aquitaine, Direction des Recherches, Paris la Défense, F

ABSTRACT

In many mechanically agitated solid-liquid contactors, it is of great interest to determine the local characteristics of the suspension : the local velocities of both particles and fluid, and the size distribution of the particles. This paper shows a new suitable method to do these measurements : the phase Doppler particle size technique. The first results obtained in a 20 liters fully baffled tank with glass particles are then presented.

INTRODUCTION

Mechanically agitated solid-liquid contactors are widely used in industry. Various processes can then be performed such as promoting a chemical reaction between the two phases, promoting dissolution or crystal growth or simply obtaining an uniform particle concentration (and/or size distribution) in a stream (1). In many applications, it is of great interest to determine the local characteristics of the stirred suspension, especially the local relative differences in velocity between particles and fluid and the interactions of particles against particles and against impeller or walls. These characteristics are very important to explain some phenomena like attrition during crystallization from solution, for instance.

However, in most cases the previous papers about stirred suspensions mainly deal with overall performances of such systems : power consumption, critical agitator speed to ensure the fluidization of particles, mass transfer rates, description of scale-up criteria. Only a few studies deal with local measurements in suspensions (2,3). This fact is certainly due to the difficulty of finding a suitable measurement technique for such systems.

First in this paper the needed characteristics of a measurement technique in suspensions are detailed. Then, the chosen method is described. Finally, the experimental set-up is described and results are presented.

LOCAL MEASUREMENTS IN LIQUID-SOLID SUSPENSIONS

Three characteristics are needed to describe locally liquid-solid suspensions : the velocity of the solid particles, the velocity of the liquid phase and the particle size distribution. It is now well known that the velocity fields of a homogeneous fluid and of a suspension are different. A paper by Geisler and Mersmann (3) confirms this fact in the case of a stirred vessel. So, it is necessary to know the effects of the particles on the flow field.

The use of intrusive measurement techniques is prohibited in such flows, because the probes are often larger than the particles (4).

A lot of non-intrusive methods of measuring the velocity of particles have been developed : transit-time or correlation techniques, Laser Transit Velocimetry(5), Particle Tracking Velocimetry (6), Laser Speckle Velocimetry (7) and of course Doppler techniques. Among these measurement methods, the first one does not enable the determination of the sign of the velocity, so this technique is prohibited in turbulent flows like stirred suspensions. The second and the third ones use fast camera, pulsed laser, laser analysis of pictures... They need very expansive equipments and also a great experience in this field. Doppler techniques are very interesting because Laser Doppler Velocimetry is now well known and it clearly measures particle velocity.

If signals issued from "big" particles (dispersed phase) and signals issued from "small" particles (natural seeding of the liquid phase) can be separated, then it is possible to measure both particles and liquid velocities. This is a good way to build a suitable experimental technique to study solid-liquid suspensions.

Although several methods of discrimination of the Doppler signals exist, commercial equipment is not available when a calibration is necessary at each measurement point. These methods are based on amplitude measurements of the Doppler burst occuring when a particle is crossing the probe volume. This amplitude is a function of the particles size (8,9). The main problem of these methods lies in the fact that amplitude measurements of a light signal are dependent upon a lot of optical elements, particularly when the flow is confined by walls (the tank, for instance) : calibration has to be performed at each measuring point, and measurement conditions can evolve during experiment because of fouling.

Another method of discrimination of the signals issued from small and large particles is based on the frequency of the Doppler burst. This technique uses the phase difference between Doppler signals (issued from a particle crossing the probe volume) received by photomultipliers directed towards the probe volume at different angles. The phase difference is a function of the particle size (10). The velocity and the size of each particle crossing the probe volume can be measured and stored. After data analysis, the velocity of the continuous phase (liquid) can be deduced from the smallest particles in the flow (a few µm in diameter); the velocity and the size distribution of the dispersed phase is deduced from the other particles.

This last technique has been chosen for several reasons : it is an eulerian non-intrusive technique, the validity of the velocity measurements needs no more to be proved, the size measurements do not need calibration and the equipments are now commercially available.

THE PHASE/DOPPLER SIZE AND VELOCITY MEASUREMENTS TECHNIQUE

Ten years after the works carried out by Durst and Zare (10) the first papers upon phase/Doppler equipments were published in 1984 by Bachalo and Houser (11), Bauckhage and Flogel (12) and Saffman *et al.* (13). These works have leaded to two commercially available equipments. So, it can be said that the phase /Doppler is a very recent technique.

Just like the well-known Laser Doppler anemometer (LDA), the Phase/Doppler technique also uses two Laser beams focused in order to cause a beam intersection. It produces in the beam intersection planar interference fringes parallel to each other and separated by a distance known as the fringe spacing δ :

$$\delta = \lambda_o / (2 \sin\theta/2) \quad [1]$$

where λ_o is the wavelength of the laser and θ the intersection angle between the laser beams.

While LDA measurements need only one photodetector to detect the Doppler frequency of the light scattered by a particle passing the fringes, the phase/Doppler technique needs also a second one (Fig.1) in order to detect the Doppler burst, but with a typical time difference relating to the particle velocity and to the curvature of its surface (for instance, the diameter of a spherical particle).

The velocity of the particle V_p can be deduced from the Doppler frequency F_d by the relation [2]

$$V_p = F_d \, \delta \quad [2]$$

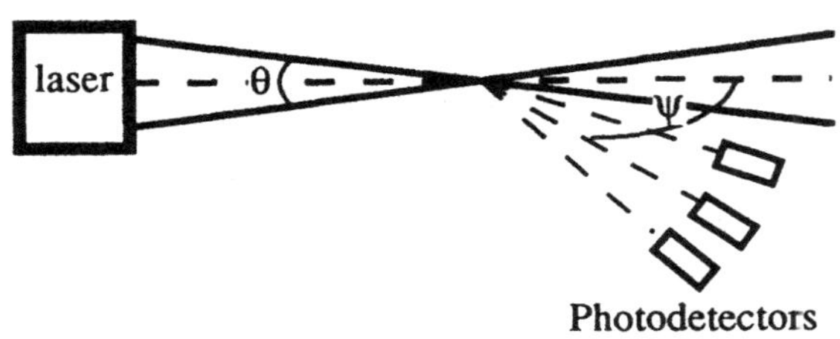

Fig.1 Phase Doppler set up

The time difference measurable as phase difference Φ between the Doppler bursts at each photodetector (Fig.2) can be correlated to the diameter D_p of spherical particles (11,12) by the following equation :

$$D_p = \Phi / 2b \, (\lambda_o / \pi \, n_l) \quad [3]$$

n_l is the refractive index of the continuous phase, and b a function of the geometrical arrangement of the experimental set up.

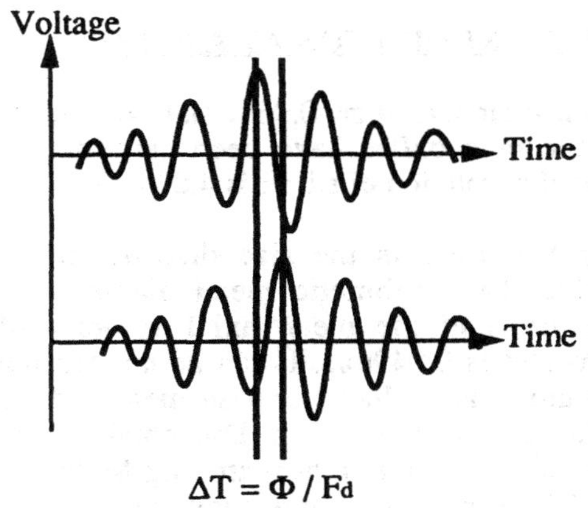

Fig. 2 : Doppler bursts at two photodetectors

The dependency of Φ versus D_p is a linear one in some special cases and when the off-axis ψ (Fig.1) is well chosen. The linearity in these special cases has been computed using Mie theory or geometrical optic laws (14) and has been checked experimentally (11).

In fact, three detectors are used to improve the accuracy and the range of the size measurements (Fig. 1).

For each particle crossing the probe volume, the Doppler frequency and the phase difference are measured, so, the velocity and the diameter are stored on a personal computer. Data analysis can be made off-line, and gives size and velocity distributions, but also the size-velocity correlation, useful to discriminate the velocity of large and small particles in the study presented here.

EXPERIMENTAL EQUIPMENTS

The experimental tests of the phase/Doppler technique have been carried out in a fully baffled cylindrical vessel (Fig.3), with a flat bottom. This tank is equipped with a disk turbine with six inclined blades.

In order to have known deflections of the laser beams, the tank is placed in an other glass tank filled with water.

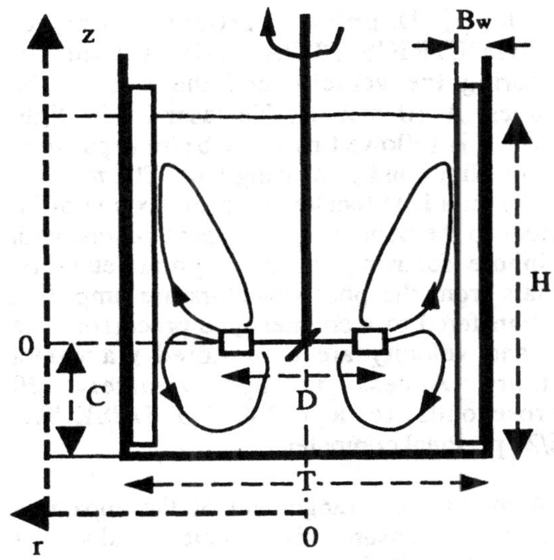

$H=T=0.3$ m ; $C=T/3$; $D/T=0.45$; $B_w=T/10$

Fig.3 : The experimental tank

The liquid phase is water. The flow pattern inside the vessel is shown on Fig.3. The velocities in discharge of the turbine are not only radial, there is also in this case an important axial component.

For the dispersed phase, spherical glass particles have been chosen. The size distribution has been measured with a laser light scattering Malvern Sizer 2600. The mean volume diameter is 252.7µm. The size distribution is reported on Table 1.

Size (µm)	% vol. in band
401	0
346	17.2
299	28.6
258	29.0
222	18.4
192	5.8
under 165	1.0

Table 1 : Size distribution of the solid phase (Malvern Sizer)

The volumic concentration of the solid phase is of course very low (0.5%). This value is necessary to enable the laser beams to go through the suspension. The limit is a function of the diameter of particles, and will be upper when the particles are larger .

A Phase Doppler Particle Analyser (AEROMETRICS PDPA 100) is used for measuring the velocity and the size of the particles. A 10 mW He-Ne laser is the light source. It is followed by a 3X beam expansion, a beamsplitter and a focusing lens. The receiver consists in a lens located at an off-axis ψ of 30 degrees to the plane of the incident beams, with a pinhole focusing on three photodetectors. Signals from the photodetectors are amplified and transfered to a counter type processor. The size and velocity are then stored via Direct Memory Access in approximately 20 microseconds to a COMPAQ DESKPRO 386/20 personal computer.

Fig.4 shows the arrangement of the apparatus necessary to measure axial velocities and size of particles at a point in the vessel.

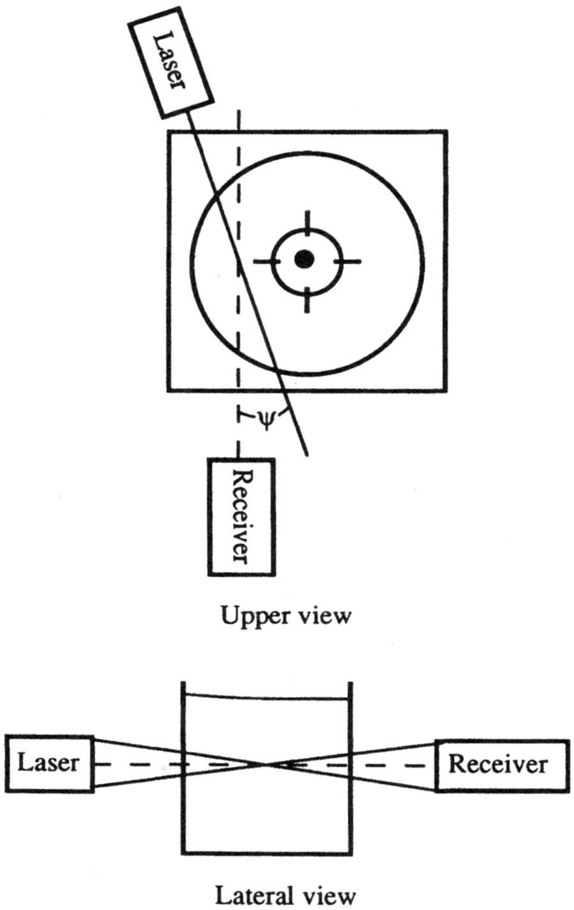

Fig.4 : Experimental arrangement of the apparatus

RESULTS AND DISCUSSION

The results presented here are the first ones obtained with the experimental equipments previously mentioned.

SIZE DISTRIBUTION AT A POINT

Measurements at z=-0.06m and r=0.08m, in the plane of a baffle have been carried out. The speed of rotation of stirrer is 4.03 (r.p.s.).

Fig.5 represents the size distribution at this point. The arithmetic mean diameter D_{10} is 189μm, and the mean axial velocity of the particles is 0.049m/s. It can be noticed that the volume mean diameter D_{30} measured here is 206.4μm. The value of D_{30} obtained for the particles with the light scattering Malvern sizer is 252.7μm. This gap may be due to the difference of measuring technique, or may be an indicator of an unhomogeneity of the suspension.

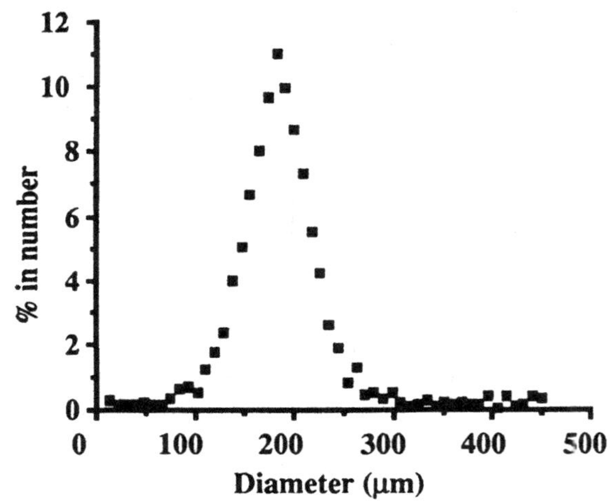

Fig.5 : Size distribution at a point

CORRELATION BETWEEN THE AXIAL VELOCITY AND THE DIAMETER OF THE PARTICLES AT A POINT

These results have been obtained at the point where the previous measurements have been made. The Fig.6 shows the mean value of the particle velocity for each particle category.

This mean velocity is constant for particle diameters between 110μm and 240μm. It is the diameter range inside of which the particles are the more numerous (see the size distribution).

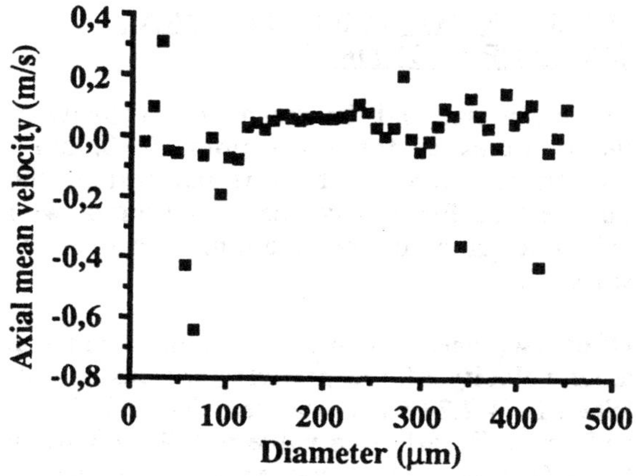

Fig.6 : Diameter - Velocity correlation at a point

After 240 µm and under 110 µm, a dispersion of the mean velocity, due to a lower number of particles in each category is observed: this number is not sufficient to calculate a correct mean value of the velocity, because the flow is here very turbulent.

However, the particles have been arbitrarily separated in two categories : the smallest ones (<60µm) and the largest ones ($130\mu m < D_p < 240\mu m$). A mean axial velocity value for both categories has been calculated. The mean value found for the small particles is -0.015m/s. The one for the large particles is 0.055m/s. This result is certainly surprising, but it must be kept in mind that the comparison is made only on mean axial values of the velocity. Largest particles (the dispersed phase) and smallest ones (supposed to represent the continuous phase) do not have the same trajectories in the vessel.

AXIAL PROFILE OF MEAN AXIAL VELOCITY

Measurements have been performed along a vertical axis at r=0.08m, in the plane of a baffle. The speed of the stirrer is also in this case 4.03 r.p.s.

Fig.7 represents the mean axial velocity along this axis. As it has been see in the previous parts of the presentation of these results, small particles are not very numerous, so the values of the velocity are those of the large particles (dispersed phase). The mean velocity is zero near the bottom of the tank, it increases under the turbine to reach a maximum value a few centimeters above. This maximum value is the axial component of the turbine discharge, due to the tilt of the blades. In the upper part of the vessel, the mean axial velocity is negative. These points lie in the downward part of the flow (see Fig.3)

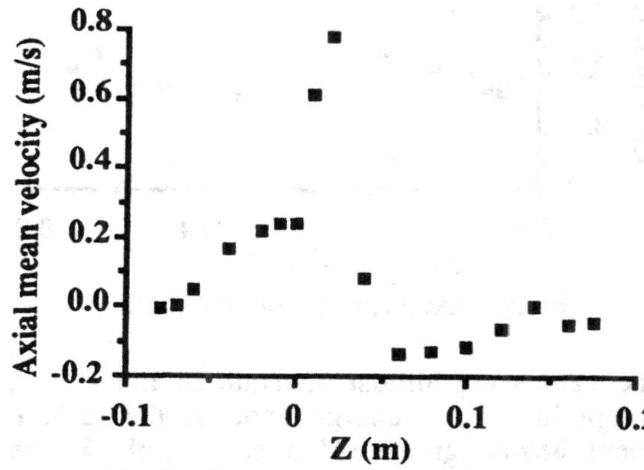

Fig.7 : Axial mean velocity profile

The axial velocity data have been treated along the axis in the same way like at a point in the previous part. Fig.8 shows the mean axial velocity profiles obtained for the smallest particles and for the largest ones. The results are here also surprising : the dispersed phase (large particles) seems to go up more quickly and to go down slower than the continuous phase (small particles). Like in the previous part, the same remarks about the small particles mean velocity calculation and about the particle trajectories in the vessel may be done.

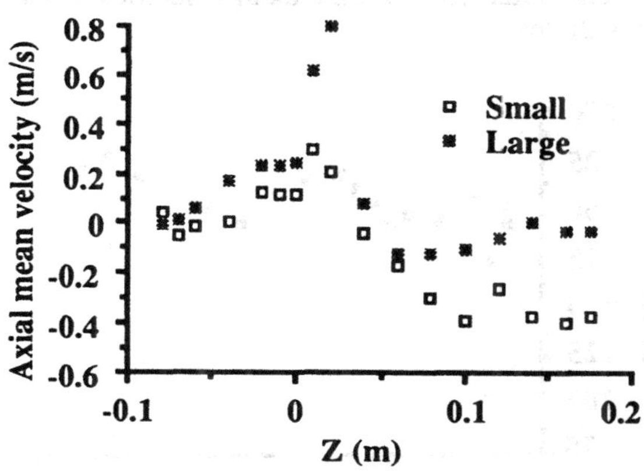

Fig. 8 : Axial velocity profile of small and large particles

AXIAL PROFILE OF R.M.S. VELOCITY

Fig.9 represents the axial profile of the root mean square axial velocity along the same vertical axis.

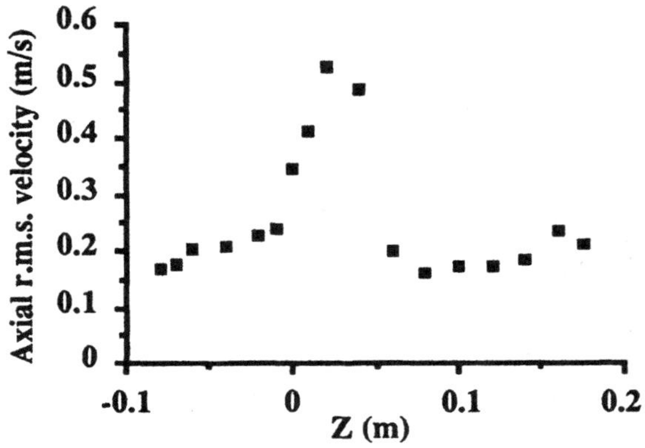

Fig.9 : Axial r.m.s. velocity profile

The values are almost constant in the vessel, except in the discharge zone of the turbine where they are greater. This results look like the one observed for a classical Rushton turbine in a single phase flow (15).

AXIAL PROFILE OF THE VOLUME DIAMETER D_{30}

The profile of D_{30} along the same axis is reported on Fig.10. The value of D_{30} slightly increases along this axis. Particles are larger in the top of the vessel than they are at the bottom, but, of course this result is only valid along this axis. Certainly, the size distribution is not uniform in the vessel, and this result confirms the one obtained about the axial velocities of the two phases.

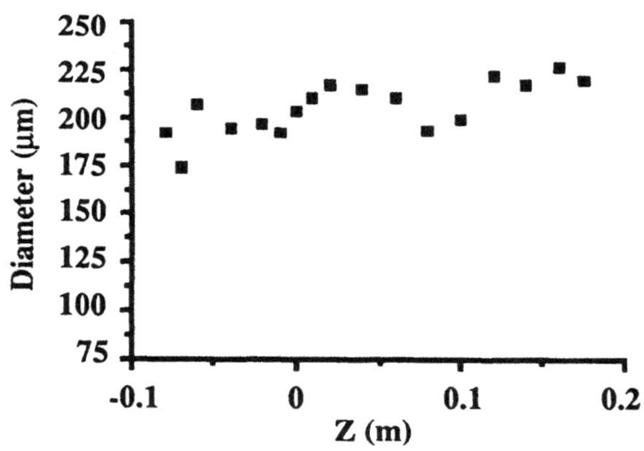

Fig.10 : Volume mean diameter profile

INFLUENCE OF THE ROTATIONAL SPEED OF THE STIRRER

The mean size and the mean axial velocity of the particles at the same point (r=0,05m ; z=-0,06m) where the flow is pumped by the turbine (see Fig.3) have been measured, with different values of the rotational speed of the stirrer N.

Of course, when N is increasing, the local mean axial velocity of the particles increases also. When N is 2.27 r.p.s., the velocity is 0.17 m/s and varies linearly to a value of 0.33 m/s when N is 4.03 r.p.s. . This result is not reported here.

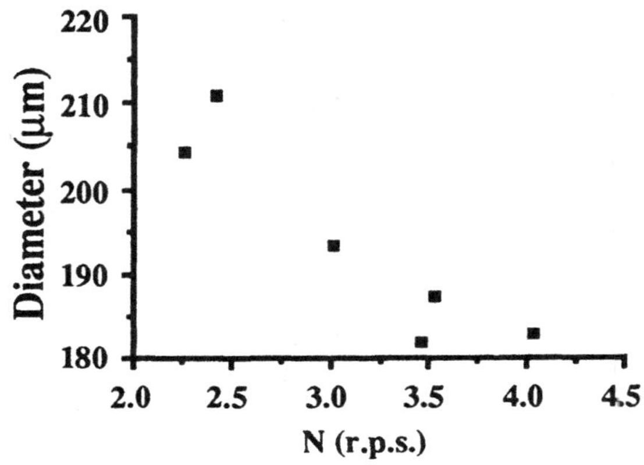

Fig. 11 : Influence of N on D_{30}

Fig.11 represents the changes in D_{30} when N is increasing : D_{30} decreases from almost 205µm when N is 2.27 r.p.s. to almost 182µm when N is 4.03 r.p.s.

This result would mean that, in this case, the largest particles are easier to lift up from the bottom than the smallest. An identic fact has been reported by only few authors (1), the majority found that the critical speed for suspension of particles increases when the mean diameter increases. However, their results have been obtained by changing the population of particles in the vessel to change the mean diameter.

In the paper presented here, an explanation of this result could be that the small particles can not be lifted up, because they are locked on the bottom wall by the largest.

CONCLUSION

In this contribution a new experimental technique to investigate solid-liquid suspensions in stirred vessels has been presented. It is the first application of the phase/Doppler method to study such a flow, the usual ones being sprays (liquid droplets in gases). Results prove that this application is successful and that it provides a lot of local data very important to describe liquid-solid suspensions. Of course these first results have to be confirmed by other measurements. Particularly, it would be interesting to add small particles in the suspension in order to improve the calculation of the mean velocity of the continuous phase.

REFERENCES

1. LYONS E.J. "Suspensions of solids", in Mixing, theory and practice; UHL V.W. and GRAY J.B.; Academic Press; 1967; Vol.2;Ch.9.

2. KUBOI R.; KOMASAWA I.; OTAKE T. "Fluid and particle in turbulent dispersion. II. Influence of turbulence of liquid on the motion of suspended particles"; Chem.Eng.Sci.,29, 1974, pp651-657.

3. GEISLER R.K.; MERSMANN A.B. "Local velocity distribution and power dissipation rate of suspensions in stirred vessels"; Pr. of the 6th European Conference on Mixing, Pavia, Italy, May 24-26th, 1988; BHRA, Bedford, England,1988.

4. YIANNESKIS M. "Velocity, particle sizing and concentration measurement techniques for multiphase flow"; Powder Technology, 49, 1987, pp261-269.

5. TSALAPATAS H.A.; RIETHMULLER M.L. "A real time personal computer-based clipped auto and cross correlator with applications in the detection of transition"; Pr. of the5th International Symposium on Application of Laser Techniques to Fluid Mechanics, Lisboa, Portugal, July 9-12th, 1990.

6. REUSS D.L.; ADRIAN R.J. "Two-dimensional velocity measurements in a laminar flame using Particle Image Velocimetry"; Chem. Phys. Processes Combust.; 1988; pp11/1-11/4.

7. PICKERING C.J.D.; HALLIWELL N.A. "Particle Image Velocimetry : data reduction and fringe analysis"; Pr. of the 3rd International Symposium on Applications of Laser to Fluid Mechanics, Lisboa, Portugal, July 7-9th, 1986.

8. LEVY Y. "LDA technique for measurement in freeboards of fluidized beds"; A.I.Ch.E. Journal, 32, 9, 1986, pp1579-1583.

9. GREHAN G.; GOUESBET G. "Corrected Laser beam techniques for simultaneous velocimetry and sizing of particles in flow and applications"; Pr. of the 3rd International Symposium on Application of Laser to Fluid Mechanics, Lisboa, Portugal, July 7-9th, 1986.

10. DURST F.; ZARE M.; "Laser Doppler Measurements in Two-Phase Flows"; Pr. of the LDA Symposium, Copenhagen, Denmark, 1975, pp403-429.

11. BACHALO W.; HOUSER M.J. "Phase/Doppler spray analyzer for simultaneous measurements of drop size and velocity distributions"; Optical Engineering, 23, 5, 1984, pp583-590.

12. BAUCKHAGE K.; FLOEGEL H.M.; "Simultaneous measurement of droplet size and velocity in nozzle sprays"; Pr. of the 2nd International Symposium on Applications of Laser Anemometry to Fluid Mechanics, Lisboa, Portugal, July 2-4th, 1984.

13. SAFFMAN M.; BUCHHAVE P.; TANGER H. "Simultaneous measurement of size, concentration and velocity of spherical particles by a Laser Doppler Method"; Pr. of the 2nd International Symposium on Applications of Laser Anemometry to Fluid Mechanics, Lisboa, Portugal, July 2-4th, 1984.

14. BACHALO W.D.; SANKAR S.V. "Analysis of the light scattering interferometry for spheres larger than the light wavelength", Pr. of the 4th International Symposium on Applications of Laser Anemometry to Fluid Mechanics, Lisboa, Portugal, July, 1988.

15. COSTES J.; COUDERC J.P. "Study by laser Doppler anemometry of the turbulent flow induced by a Rushton turbine in a stirred tank: Influence of the size of the units. Part I. Mean flow and turbulence"; Chem.Eng.Sci.,43, 1988, pp2751-2764.

Annexe :Referee's comments :

The results are astonishing, because large particles seem to go up more quickly and go down slower than small particles under the stirrer increases as the stirrer speed is increased. Published data (A. BARRESI and G. BALDI, *Solid dispersion in an agitated vessel,* Chem. Eng. Sci. 42, 2949-2956, 1987, also paper E5.6, 9th CHISA Congress, 1987 ; J.Y. OLDSHUE, *Suspending solids and dispersing gases in mixing vessels,* Ind. Eng. Chem. 61(9), 78-79, 1969), referring to wide size distribution particles, indicate a reverse trend. About the suspension of large and small particles together, the authors should examine the review by NIENOW on *Mixing of liquids by mechanical agitators* (ULBRECHT and PATTERSON Eds, Gordon and Breach, 1985) and related papers : the general conclusion is that the agitator speed to just suspend particles off vessel bottom is related to the mass mean diameter of the particles.

Reply to the comments :

Of course our results are astonishing and we have mentioned this in our paper.
BARRESI A. and BALDI .G have measured the local volumetric concentration of the particles along radial and vertical axis, but they have not presented the size distributions along these axis, so their results can not be compared with ours.
We have also mentioned in our paper that, in main papers of the literature (see reference 1), the critical speed for suspension of particles, increases as their mean diameter increases. But these results have been obtained by changing the suspension, it is not the case in this work.
OLDSHUE J.Y. indicates a reverse trend about the size distribution, but he does not explain in the paper the conditions of the measurements.
The results presented in our paper have been performed on a single vertical axis, it is possible that, along an other vertical axis, the results would be different. So, no definitive conclusions can be yet drawn about the size distribution profile within the tanks, the aim of our paper is only to present the successful application of the phase/Doppler method to the local study of suspensions.

HYDRODYNAMIC CHARACTERISTICS OF A GAS-INDUCING IMPELLER

G.M. EVANS, C.D. RIELLY, J.F. DAVIDSON, Dept. of Chemical Engineering, University of Cambridge
K.J. CARPENTER, ICI Fine Chemicals Manufacturing Organisation, Manchester, GB

ABSTRACT

Gas-inducing impellers use the relative velocity between the bulk liquid and the impeller, and the acceleration of the liquid phase over the blades, to locally reduce the pressure at an orifice on the blade and induce a flow of gas from the headspace, through a hollow shaft. In this paper a theoretical model based on a blade slip factor and a pressure coefficient is developed to predict the reduction in the pressure at the gas outlet. The model has been used to predict the critical impeller speed marking the onset of induction, and also the rate of gas induction at higher impeller speeds. The model predictions are in good agreement with experimental observations for cylindrical, aerofoil and flat paddle impeller shapes.

INTRODUCTION

Gas-inducing impellers increase the utilisation of a solute gas by recycling the gas from the headspace, without the need for an external loop or compressor. This feature is particularly desirable when recompression of the gas through an external loop is hazardous or undesirable, such as in hydrogenation, chlorination or phosgenation processes. A gas-inducing impeller uses the acceleration of the liquid phase over the blades to locally reduce the pressure at an orifice and induce a flow of gas, from the headspace, through a hollow shaft. The relative velocity of the blade and bulk liquid depends on the blade shape and tank geometry, whereas the acceleration of the liquid over the blade depends only on the impeller profile.

Figure 1 shows the design of a typical self-inducing impeller, consisting of a hollow shaft connected to hollow blades. A hole in the shaft above the free liquid surface allows gas to flow through the hollow shaft and to leave through orifices on the impeller blades. At a critical impeller speed, N_c, the reduction in pressure at the outlet orifice on each blade is sufficient to overcome the static head of liquid and a flow of gas is induced. Braginsky (1964) applied Bernoulli's equation, between the free surface and the orifice on the blade to calculate the critical impeller speed at which a gas flow is first induced, yielding

$$N_c = \sqrt{\frac{2gh}{k(\pi D)^2}} \tag{1}$$

Here D is the diameter of the impeller, h is the depth of submersion of the orifice and k is a head loss constant, which should be regarded as an empirical coefficient. Joshi & Sharma (1977) used the same form as eq.(1) and determined experimentally that k ranged from 0.9 - 1.3 for pipe and flattened cylindrical blades. Similarly, Zundelevich (1979) obtained a value for k of 1.15 for a turbo-aerator. Sawant & Joshi (1979) extended this approach to include the effects of liquid physical properties on N_c, and proposed

$$N_c = \sqrt{\frac{0.21\,gh}{D^2}\left(\frac{\mu_L}{\mu_w}\right)^{0.11}} \tag{2}$$

where μ_L and μ_w are the absolute viscosities of the liquid and water respectively. Eq. (2) shows only a weak dependence on viscosity and no dependence on surface tension or density.

At speeds greater than the critical impeller speed a flow of gas, Q, is induced, which increases with increasing impeller speed. Typically, the flow rate of gas induced increases rapidly at speeds just above the critical speed, but more slowly at higher impeller speeds. Several authors have proposed simple correlations of the form

$$Q \sim N^a D^b \tag{3}$$

where a and b are constants, for example: Raidoo *et al* (1987) report values of $a \approx 0.5$ and $b \approx 1.7$, whereas Sawant *et al* (1981) give $a = 1.5$ and $b = 3.0$. No acceptable reason for the variation in the results has been proposed.

The form of Bernoulli's equation proposed by Evans *et al* (1990) can be used to predict the pressure at the outlet orifice for impeller speeds below the critical speed (single phase system). It is the purpose of this paper to examine if this approach can be extended to predict the rate of gas induction at impeller speeds greater than the critical induction speed. The predictions for the critical impeller speed and induced gas flow rate are to be compared to experimental measurements using cylindrical, aerofoil and flat paddle blade shapes at two scales of operation.

For simplicity, the theory is developed for flow around a cylindrical blade, since inviscid flow theory should allow a

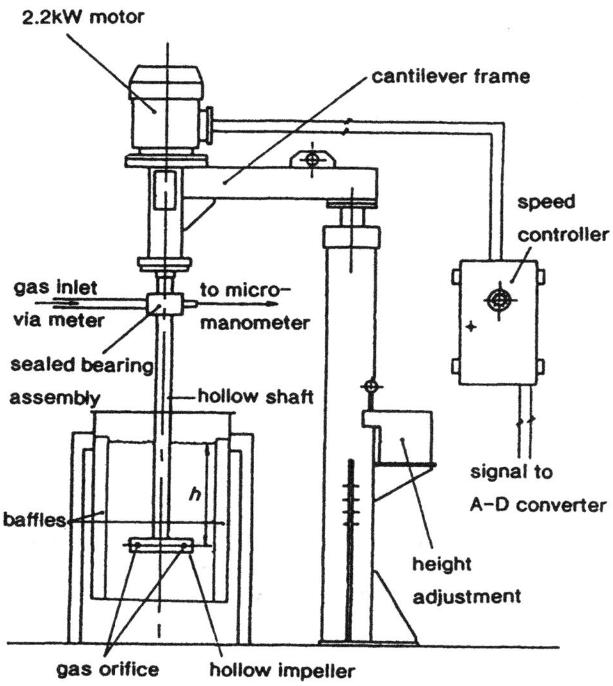

Figure 1 Gas-inducing impeller set-up

prediction of the pressure distribution on the surface of such a shape. Results from the theory can then be tested directly against experiment. However, the model is of general applicability, and Evans *et al* (1990) describe how the model constants may be derived from small scale experiments in a single phase flow, prior to the onset of gas induction.

THEORY

PRESSURE AT THE ORIFICE

The system illustrated in Figure 2 shows a cylindrical impeller blade which has a gas outlet located at a radial distance, R, from the vertical axis of the shaft, and an angle, θ, measured from the rear of the blade.

The fluid inside the vessel is set in motion as a result of the drag force exerted by the rotating impeller. Assuming the radial velocity component is small, the absolute tangential velocity of the fluid immediately upstream of the orifice can be written as

$$v_L = 2\pi KNR \tag{4}$$

where N is the impeller rotational speed and K is the blade slip factor, which is defined as the ratio of the upstream fluid velocity and the blade velocity at radius R (both relative to a stationary observer). The upstream velocity of the fluid, U, relative to the moving blade is assumed to be uniform and can be expressed as

$$U = (1-K)\,2\pi NR \tag{5}$$

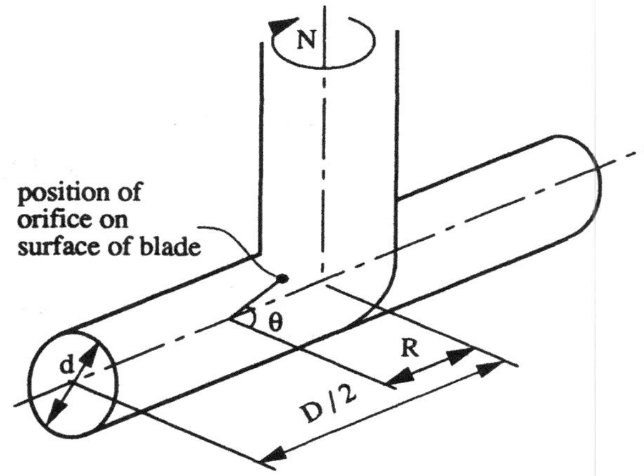

Figure 2 Cylindrical blade geometry

At any angular position of the orifice on the blade, the local relative velocity of the liquid with respect to the blade, $u_L(\theta)$, is

$$u_L(\theta) = \left\{C_p(\theta)\right\}^{1/2} 2\pi NR\,(1-K) \tag{6}$$

where $C_p(\theta)$ is a pressure coefficient (Abbot & von Doenhoff, 1959), defined by

$$C_p(\theta) = \frac{P_s - P(\theta)}{\frac{1}{2}\rho_L U^2} \tag{7}$$

In eq. (7) P_s is the stagnation pressure and $P(\theta)$ is the local absolute pressure at the orifice. In a previous publication, Evans *et al* (1990) assumed that P_s was equal to the static head of liquid for the fluid at rest. More recent measurements show that a better assumption is that the stagnation pressure is given by

$$P_s = P_0 + \rho_L g\,h + \tfrac{1}{2}\rho_L\left\{2\pi NR\,(1-K)\right\}^2 \tag{8}$$

where P_0 is the headspace pressure. The absolute pressure, $P(\theta)$, on the surface of the blade is then equal to

$$P(\theta) = P_0 + \rho_L gh + \tfrac{1}{2}\rho_L\left\{2\pi NR\,(1-K)\right\}^2\left\{1 - C_p(\theta)\right\} \tag{9}$$

This is of the same form as the equation for $P(\theta)$ as proposed by Evans *et al* (1990), but with a more rigorous definition of the pressure coefficient, which can be compared with inviscid flow calculations.

CRITICAL IMPELLER SPEED

The critical induction speed can be obtained from eq. (9) by realizing that induction will begin when the pressure at the gas outlet is equal to the headspace pressure. For the case where $P(\theta) = P_0$, eq. (9) can be rearranged to give

$$N_c = \sqrt{\frac{gh}{2\left\{(1-K)\,\pi R\right\}^2\left\{(C_p(\theta)-1)\right\}}} \tag{10}$$

RATE OF GAS INDUCTION

The rate of gas induction for a gas inducing impeller is a function of the total pressure drop ΔP_T across the system *i.e.* the difference between the absolute pressure at the gas outlet on the blade, $P(\theta)$, and the inlet pressure, P_0. From eq. (9)

$$\Delta P_T = \tfrac{1}{2}\rho_L\{2\pi NR(1-K)\}^2\{1-C_p(\theta)\} + \rho_L g h \qquad (11)$$

This pressure difference is equal to the sum of: (*i*) frictional pressure losses along the gas pathway through the hollow shaft and blades ΔP_P; (*ii*) frictional losses through the gas orifice ΔP_o; (*iii*) pressure losses due to imparting kinetic energy to the liquid during the bubble formation process, ΔP_{KE}; and, (*iv*) due to the work done against surface tension forces in forming a bubble, ΔP_σ.

The frictional pressure losses along the gas pathway are a function of the type of flow. For *laminar* gas flow ΔP_P is directly proportional to the total rate of gas induction, Q, *i.e.* the Hagen-Poiseuille law may be written as

$$\Delta P_P \propto Q \qquad (12)$$

Similarly, for *turbulent* gas flow the pathway pressure losses are proportional to the square of the induced gas flow rate (assuming an approximately constant friction factor), *i.e.*

$$\Delta P_P \propto Q^2 \qquad (13)$$

The proportionality constants in eqns. (12) and (13) depend on the physical properties of the gas and also the geometry of the gas pathway (the inside diameter of the hollow shaft and loss coefficients for any sudden expansions or contractions in the system). The values for the proportionality constants are best determined experimentally.

In all applications for gas-inducing impellers the gas flow initially will be laminar at low rates of gas induction for impeller speeds just above the induction speed, but turbulent at higher impeller speeds. Furthermore, in many small scale experiments the gas flow through the system may be laminar over the entire impeller speed range examined. Since eq.(11) shows that the total pressure driving force, ΔP_T, is a function of N^2 plus a constant term, then a simple power law expression, such as eq.(3) would not be able to correlate the induced gas flow rate against impeller speed. This difficulty would be further compounded if in a small scale experiment there was a transition from laminar to turbulent flow of the gas within the hollow pathway, with increasing impeller speed. In this case the dependence of the induced gas flow rate on the pressure driving force would change from eq.(12) to eq.(13). The actual shape of the rate of gas induction curve will depend on the magnitude of the static head and the dynamic pressure, which are determined by the radial distance of the gas orifice from the axis of rotation, the pressure coefficient and blade slip factor, and the rotational speed of the impeller.

The orifice pressure drop ΔP_{KE} associated with the kinetic energy imparted to the liquid by the expansion of a bubble forming underwater can be estimated from the expression obtained by Davidson and Schüler (1960)

$$\Delta P_{KE} = \frac{3Q_o^2\rho_L}{32\pi^2 r_b^4} \qquad (14)$$

where Q_o is the volumetric gas flow rate through a single orifice and r_b is the radius of the bubble at the point of detachment. Eq. (14) is based on the assumptions that the bubble is formed in a stagnant liquid and also that the volumetric gas flow rate is constant with time.

For a stationary orifice Davidson and Schüler estimated that ΔP_{KE} was negligible, since in their experiments the bubble size was large; eq.(14) shows that ΔP_{KE} is inversely proportional to the bubble radius so that smaller bubbles would impart more kinetic energy to the liquid. Ghosh and Ulbrecht (1989) found that for a moving orifice the bubble radius decreased with increasing cross-flow liquid velocity. Under these conditions the pressure associated with the kinetic energy imparted to the liquid may become significant.

To obtain an estimate of the bubble size at the moment of detachment, consider a horizontal orifice of radius r_o in a liquid moving with a relative cross-flow velocity of u_L. Assuming there is no slip between the bubble and the liquid, the bubble centre will have moved a distance equal to $u_L t$ in the cross-flow direction, at a time t after the bubble has begun to form. If the bubble remains spherical, and the volumetric flow rate into the bubble is steady with time, the radius of the bubble at any time is given by

$$r_b(t) = \left[\frac{3Q_o t}{4\pi}\right]^{1/3} \qquad (15)$$

Using the assumption that the bubble is detached from the orifice when it has travelled a distance $r_o + r_b$ then the time for growth, t, can be eliminated from eq. (15), to give

$$\frac{4}{3}\pi r_b^3 - \frac{Q_o(r_b + r_o)}{u_L} = 0 \qquad (16)$$

In the case of a bubble forming at a moving orifice the local relative velocity, u_L, is given by eq. (6) and Q_o is the induced gas flow rate from a single orifice. Eq. (16) may be solved to give the bubble radius at detachment and hence the pressure drop ΔP_{KE} may be obtained.

The pressure drop necessary to overcome surface tension forces at the orifice may also be calculated, knowing r_b, from

$$\Delta P_\sigma = 2\frac{\sigma}{r_b} \qquad (17)$$

The total pressure difference generated by the liquid flow over the blade, may then be equated to pressure losses through the system, *i.e.*

$$\Delta P_T = \Delta P_P + \Delta P_o + \Delta P_{KE} + \Delta P_\sigma \qquad (18)$$

Eq. (17) may then be solved iteratively, in conjunction with eqs.(11) to (17) to yield the variation in gas flow rate with impeller speed, knowing the model constants, K and C_p.

EXPERIMENTAL

The experimental apparatus, shown in Figure 1, consisted of a flat bottomed cylindrical perspex vessel fitted with four evenly spaced baffles of width $0.1T$, where T is the diameter of the vessel. In all of the experiments the working fluids were air and water, and in the majority of cases the experiments were performed in a 0.45 m diameter vessel. A 0.29 m diameter vessel was also used for some experiments to observe the effect of scale on the critical impeller speed. The impeller was driven by a variable speed 2.2 kW motor mounted on a cantilever frame which could be raised or lowered using a hydraulic jack to change the submersion depth of the blades (Figure 1). The speed range of the motor was 50 - 1500 rpm; the signal from the speed controller was sent to a CED 1400 analogue-to-digital converter and logged to a personal computer. The drive unit was fitted with a hollow shaft which allowed gas to be drawn through a wet gas test meter and sealed bearing assembly. The wet gas meter had been previously calibrated against a soap-bubble meter, and contributed a pressure drop of only a few mm water gauge to the induced flow of gas. The impeller designs consisted of two blades, spaced 180° apart. A 0.005 m diameter hole was drilled along the axis of each blade to allow the induced gas to pass and exit through a 0.002 m diameter orifice on the surface of each blade. The blade shapes investigated were a cylinder, an aerofoil section and a flat paddle. All three impellers were 0.225 m in overall diameter, with an orifice in the middle of each blade, at a radial distance of 0.075 m. The aerofoil had a chord length of 0.043 m, a height of 0.0075 m and an angle of attack of 0°. The width, W, of the flat paddle was 0.03 m and the orifice was on the back face of the blade. The cross-sectional diameter, d, of the cylindrical blade was 0.03 m and the orifice angular position was fixed at 90° measured from the rear of the blade, *i.e.* the orifice was facing upwards. According to inviscid flow theory this should give the largest pressure coefficient and hence the lowest induction speed, N_c.

The experimental procedure consisted of setting the submersion depth (measured from the horizontal axis of the blade) and clearance of the blade. The critical impeller speed was determined visually to be the speed at which bubbles were first seen to leave the orifice; at speeds higher than the induction speed the flow rate of gas induction was also measured using the wet gas meter. The reduction in pressure at the orifice was obtained, as a function of impeller speed, by closing the gas inlet from the meter and measuring the change in pressure inside the sealed shaft using a micro-manometer. The pressure from the micro-manometer was not a direct reading of the pressure at the orifice when there was liquid inside the shaft or impeller. The procedure used to obtain the pressure at the orifice from the micro-manometer pressure readings is given in an Appendix.

The model constants K and C_p were calculated from measurements of the pressure at the orifice as a function of impeller speed, prior to gas induction. Extrapolation of these results using eq. (10) allowed prediction of the critical impeller speed for gas induction. These model constants could were also used to predict the induced gas flow rate at speeds greater than N_c. However, before this was possible it was necessary to experimentally determine ΔP_P and ΔP_o. The set-up used to calibrate the gas flow rate *vs* pressure drop through the 0.030 m diameter cylindrical impeller was essentially the same as that described above. Instead of submerging the impeller and rotating it to induce a flow of gas, the impeller remained stationary in air and was enclosed inside an air-tight vessel. The air inside the vessel was drawn off using a vacuum pump resulting in a constant flow of gas through the system and the flow rate was measured using the wet gas meter. The micro-manometer was used to measure the pressure difference between the gas inlet and (*i*) a point inside the blade, just before the orifice, ΔP_P, to demonstrate the transition from laminar to turbulent flow inside the shaft and gas supply lines; and, (*ii*) a point downstream of the orifice, which could be used to calculate ΔP_o. In addition to the pressure drop measurements for the impeller in air, the stationary blade was submerged under water in the sealed vessel. A vacuum pressure was applied in the enclosed headspace above the liquid surface and the pressure was measured as a function of gas flow rate through the system at the same positions mentioned above. These values were compared to the air-only measurements.

RESULTS AND DISCUSSION

STATIONARY IMPELLER PRESSURE DROP

In Figure 3 the pressure drop through the gas pathway from the gas inlet to a position just upstream of the orifice, ΔP_P, has been plotted against the total gas flow rate for the stationary blade in both air, and water. At low gas flow rates the pressure drop varied linearly with Q, but at larger gas flow rates varied approximately as Q^2. For this arrangement the transition from laminar to turbulent flow was expected to occur at about 200×10^{-6} m^3/s. The small positive pressure drop at zero gas flow rate is the static head of the wet gas flow meter.

Figure 4 shows the pressure drop across a single orifice

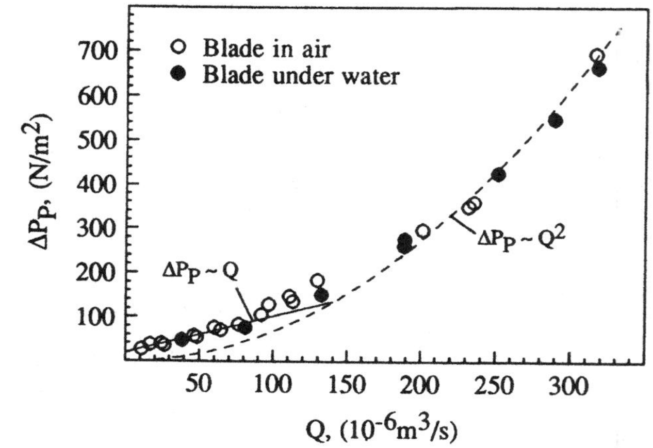

Figure 3 Gas pathway pressure losses through the hollow shaft and blades vs total gas flow rate

(the impeller had two orifices, one on each blade) plotted against the gas flow rate through a single orifice, for the same conditions as Figure 3. Comparison of figures 3 & 4 show that at speeds well above N_c the predominant pressure loss is due to the flow through the orifice. Figure 4 shows that when the impeller was in air the pressure drop was proportional to the square of the gas flow rate, in accordance with turbulent flow through an orifice; the orifice discharge coefficient was calculated to be 0.72. When the impeller was submerged under water a finite pressure difference across the orifice at zero gas flow rate was obtained which was due to the surface tension forces at the orifice (the pressure losses due to imparting kinetic energy to the liquid was negligible for the size of bubbles formed). The pressure, ΔP_σ, required to overcome the surface tension forces, such that a bubble might form at a flat circular orifice is given by $2\sigma / r_o$ (equal to 144 N/m^2, based on $r_o = 0.001$ m and $\sigma = 0.072$ N/m). Even though the orifice on the cylindrical blade is not flat the estimated value for ΔP_σ, is in reasonable agreement with the value of 159 N/m^2, obtained by extrapolating the data values to zero gas flow rate.

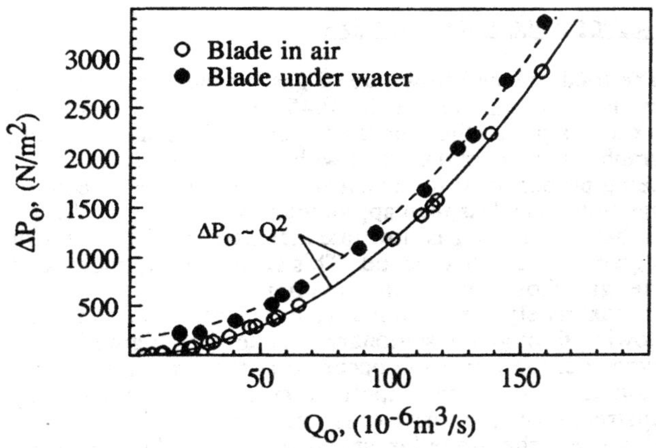

Figure 4 Pressure drop vs gas flow rate through a single diameter gas orifice on a cylindrical blade.

PRESSURE AT THE ORIFICE

In Figure 5 the pressure difference ΔP_T (obtained by closing the gas inlet and measuring the pressure inside the shaft — see Appendix) has been plotted against the impeller speed for a cylindrical blade in the 0.45 m diameter vessel. At zero impeller speed, the pressure difference is equal to the liquid submersion depth — equivalent to 2205 N/m^2. As the impeller speed is increased $P(\theta)$ decreases and the pressure difference between the orifice and the headspace reaches zero just above 3 rps (for the cylindrical blade) which was in good agreement with the minimum speed that bubbles were first observed escaping from the orifice (see Figure 5). At higher impeller speeds the pressure difference across the orifice becomes negative and gas induction would take place if the gas inlet was opened to atmosphere. Figure 5 also shows the close agreement between a parabolic least squares fit and the data, in accordance with eq. (11), which demonstrates that both the pressure coefficient $C_p(\theta)$ and blade slip factor K are independent of the impeller speed — one of the initial assumptions in deriving eq. (11). This was to be expected since, the flow around the cylinder is at

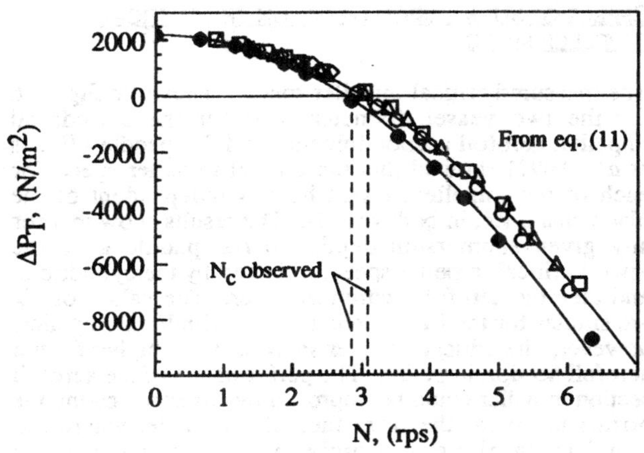

Figure 5 Total pressure driving force, generated by the impeller rotation, vs impeller speed.

$T = 0.45$ m, $h = C = 0.225$ m, $D = 0.225$ m, $R = 0.075$ m

Cylindrical blade: $d = 0.030$ m, $\theta = 90°$
$Q_s = 0$ (○), 90 (□), 160 (△), 270 (◇) $\times 10^{-6}$ m^3/s

Flat paddle: $W = 0.030$ m, $Q_s = 0$ (●) m^3/s

a high Reynolds number, in a regime where the drag and pressure coefficients are approximately constant.

Watanabe *et al* (1990) reported a large effect of the presence of a small voidage fraction of bubbles on the pressure and drag coefficients for a cylinder in cross-flow. Their results showed that for a voidage fraction of 0.05 % of bubbles there was a decrease in the drag coefficient by about 50 %. To determine if the presence of bubbles in the bulk liquid altered the pressure at the orifice of a gas-inducing impeller (by changing either the blade slip factor or the pressure coefficient), gas was introduced into the bottom of the vessel through a pipe sparger with 1 mm diameter holes drilled along its length. The inlet to the impeller shaft was sealed so that no gas flow rate was induced. Figure 5 also shows that the presence of bubbles altered the pressure difference, ΔP_T, by less than 5 % over the range of gas flow rates examined. This effect could be accounted for by the slight reduction in the mean density of the fluid flowing over the impeller blades. There was no indication that either K or $C_p(\theta)$ were affected by the presence of the bubbles.

Watanabe *et al's* (1990) results were for a uniform stream flowing over a stationary cylinder. A small number of bubbles within the uniform stream alters the turbulence intensity in the upstream fluid and causes the boundary layer to undergo a transition to supercritical flow at a lower cylinder Reynolds number than would be expected in single-phase flow. By comparison, in a stirred vessel, each impeller blade moves in the wake of the preceding blade, such that the stream ahead of each blade is highly turbulent. In this case the transition to supercritical flow may occur at a much lower cylinder Reynolds number than in a 2-D cross-flow experiment. The presence of bubbles in the vicinity of the impeller would then have a negligible further effect on the pressure distribution around the blade.

EFFECT OF SUBMERSION DEPTH ON THE CRITICAL IMPELLER SPEED

The measured critical impeller speed is shown in Figure 6 for the two vessel diameters and for the cylindrical impeller, aerofoil section impeller and flat paddle. Evans *et al* (1991) showed that the critical impeller speed for each of the impellers tested here is independent of the blade clearance, in both vessels. The results show that for any given submersion depth the flat paddle gave the lowest critical impeller speed, followed by the cylindrical, and then the aerofoil section impeller. The values of N_c are similar for the flat paddle and the cylindrical impeller, however, the critical impeller speed is much higher for the aerofoil section impeller. The performance of the aerofoil section impeller could be improved by either relocating the orifice nearer to the front face of the blade, where the liquid is travelling at a higher velocity relative to the blade, or by increasing the angle of attack of the blade.

The results also show that the critical impeller speed was approximately proportional to the square root of the submersion depth for both impeller set-ups (shown by the lines in Figure 6, which were calculated using constant values of K and C_p for each impeller). The deviation at small submersion depths was due to a slight increase in the blade slip factor as the impeller was positioned closer to the liquid free surface. These are useful results since the effect of varying both the blade clearance and depth of liquid in the vessel on the induction speed or gas flow rate, can then be calculated from the same set of model constants, K and C_p.

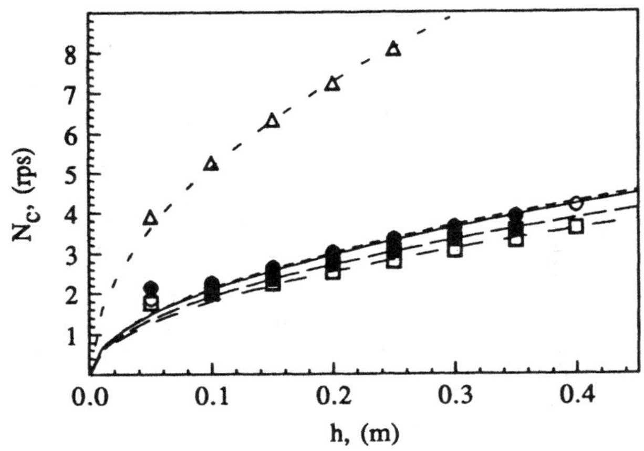

Figure 6 Effect of submersion depth and blade clearance on the critical impeller speed for gas induction. The lines show predicted induction speeds calculated from eq.(10) using the model constants given below.
$C = 0.1$ m, $D = 0.225$ m, $R = 0.075$ m

			$(1-K)^2(C_p(\theta)-1)$	
$T = 0.45$ m	○	Cylinder	1.99	———
	□	Flat paddle	2.70	— — —
	△	Aerofoil	0.33	- - - - -
$T = 0.29$ m	●	Cylinder	1.92	———·
	■	Flat paddle	2.37	--------

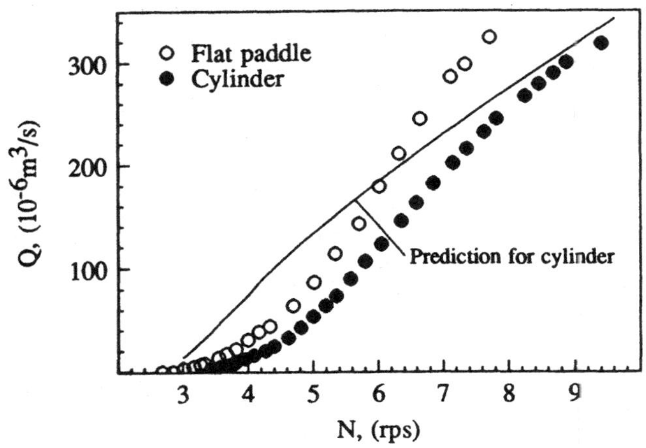

Figure 7 Total gas flow rate vs impeller speed for flat paddle and cylindrical impellers.
$\theta = 90°$, $T = 0.45$ m, $h = C = 0.225$ m, $R = 0.225$ m, $W = 0.03$ m, $d = 0.03$ m

INDUCED FLOW RATE OF GAS

The total induced flow rate of gas for the flat paddle and cylindrical impeller in the 0.45 m diameter vessel are plotted against the impeller speed in Figure 7. Both graphs show a similar trend with the rate of gas induction being proportional to the square of the impeller speed for gas flow rates less than approximately 80×10^{-6} m^3/s. At higher speeds the gas flow rate increases steadily, but with a gradual decrease of slope. This change of dependence of the gas flow rate with impeller speed corresponded approximately to the laminar-to-turbulent transition for gas flowing through the stationary impeller (see Figure 3). As expected, the transition occurred at the same rate of gas induction for both impellers since the gas pathway upstream of the orifice is the same in both cases. However, the impeller speed at which the transition occurred was lower for the flat paddle than for the cylindrical blade, due to the greater reduction in pressure in the liquid. The flat paddle also induced a greater gas flow rate, at the same impeller speed, than the cylindrical impeller.

Also shown in Figure 7 is a prediction of the rate of gas induction for the cylindrical impeller, which is in reasonable agreement with experimental measurements. The predicted gas flow rate was calculated using the model constants K and C_p calculated from pressure measurements in single-phase flow. It has already been shown that these may be extended beyond the critical induction speed and apply in the presence of gas bubbles. In addition to the pressure loss terms listed in eq. (18), the prediction also took into account the reduction in the pressure driving force due to the variation of liquid velocity on the surface of the bubble, as it expanded away from the orifice (see Evans *et al*, 1991). The method used to obtain the increase in radius of the bubble as it grew was based on an assumption of a constant gas flow rate into the bubble during formation.

IMPELLER BLADE SHAPE

The model shows that the gas induction speed and gas

flow rate depend on the constants K and C_p, determined by vessel geometry and blade shape. Blades with a large pressure coefficient and which have a large relative velocity with the fluid would give the largest pressure reduction at the orifice. These impellers would also have the lowest critical gas induction speed and the greatest gas flow rates. From the designs which have been examined so far, (*i.e.* the cylinder, flat paddle and the aerofoil), the flat paddle gave the lowest induction speed (Figures. 5 & 6) and gave the largest gas flow rate (Figure. 7). Although not considered in this study, gas cavities can form on the surface of impeller blades at high gas flow rates. Impellers at which flow separation occurs from a leading edge, such as the flat paddle, are particularly prone to flooding; this would significantly reduce the pressure driving force for gas flow. Under these conditions it is possible that a streamlined impeller design, such as an aerofoil, would give a greater rate of gas induction than a bluff impeller. However, flat-bladed impellers would provide better liquid mixing and higher rates of energy dissipation than streamlined impeller shapes. Therefore, the flat paddle could be considered as a good benchmark with which to compare other impeller designs .

CONCLUSIONS

The model proposed to predict the gas induction speed and gas flow rate has been shown to be in good agreement with experimental results. The model uses two constants which may be determined experimentally from single phase flow experiments. It has been shown that these same constants are independent of impeller Reynolds number and that they allow prediction of the pressure at the gas orifice at impeller speeds well above N_c. They are also unaffected by the presence of small bubbles in the flow upstream of each impeller blade, and so may be applied after the point of gas induction.

At gas induction speeds only slightly greater than the critical induction speed, the flow of gas in some parts of the pathway through the hollow shaft is laminar. At larger impeller speeds the flow will become turbulent on large scale equipment, and so the dependence of gas flow rate on pressure driving force is likely to change. Simple power law expressions would not be successful in correlating these data, and this partly explains the variation in the literature values of exponents in eq. (3).

The bubbles from a gas-inducing impeller form at a moving orifice, with a relative cross-flow of the order of 1 m/s. Thus the bubbles formed would be small; work is currently underway to measure bubble sizes and compare these with the predictions of eq. (16). In this case the kinetic energy imparted to the liquid during bubble formation would be significant, resulting in an additional pressure loss. The pressure difference between the orifice and the headspace, by the flow over liquid over the impeller blades, is lost due to frictional effects in the gas pathway and across the orifice, work done in imparting kinetic energy to the fluid and work done in overcoming surface tension effects at the orifice. All of these pressure losses have been taken into account in modelling the gas flow rate induced.

Several different design of gas-inducing impeller have been discussed and more blade shapes are currently being examined. The critical impeller speed for gas induction serves as a good indicator of an impellers performance as a gas-inducing device. Blade shapes on which flow separation occurs from a leading edge appear to have lower induction speeds and induce larger gas flow rates than do more rounded shapes.

APPENDIX

MEASURING THE PRESSURE AT THE ORIFICE USING THE MICRO-MANOMETER

When the gas inlet at the top of the hollow shaft is open to the atmosphere the pressure at the orifice on the rotating blade is given by (Evans *et al*, 1990)

$$P(\theta) = \tfrac{1}{2}\rho_L \left\{2\pi NR\right\}^2 + \rho_L g(h - \Delta h_0) + P_0 \qquad \text{(A1)}$$

where Δh_0 is the drop in liquid height inside the shaft from the free surface level within the vessel.

To measure the pressure at the orifice using the micro-manometer it is necessary to close the gas inlet at the top of the hollow shaft. When the impeller is stationary the drop in the liquid level inside the shaft is zero and the enclosed volume of gas inside the impeller is at atmospheric pressure. However, when the impeller is rotated the pressure inside the shaft, P_c, is no longer at atmospheric pressure as a result of the change in the enclosed volume as the liquid level inside the shaft falls by Δh_c. The expression for the pressure at the orifice becomes

$$P(\theta) = \tfrac{1}{2}\rho_L \left\{2\pi NR\right\}^2 + \rho_L g(h - \Delta h_c) + P_c \qquad \text{(A2)}$$

The micro-manometer was used to measure P_c directly, and by applying the ideal gas law to the enclosed volume, the change in height Δh_c can also be related to P_c, i.e.

$$\Delta h_c = \frac{4V_i}{\pi d_s^2}\left\{\frac{P_0 - P_c}{P_c}\right\} \qquad \text{(A3)}$$

where d_s is the inside diameter of the hollow shaft and V_i is the volume occupied by the enclosed gas when the impeller is stationary.

The *initial* enclosed volume (*i.e.* when the shaft is stationary) includes the space inside the shaft above the liquid surface, the volume of the sealed bearing assembly at the top of the shaft (see Figure 1), and also the volume of the pressure lines to the micro-manometer. A displacement method was used to determine the volume, V_i, by inserting a rod, with a rubber seal on its end, inside the stationary hollow shaft to a position equivalent to the initial height of liquid. The gas inlet at the top of the shaft was closed and the rod was moved a known length, ΔL, inside the shaft. The pressure inside the shaft was measured using the micro-manometer, and applying the ideal gas law the *i*nitial enclosed volume of the system was calculated from the expression

$$V_i = \frac{\pi d_s^2 \Delta L}{4}\left\{\frac{P_c}{P_c - P_o}\right\} \qquad \text{(A4)}$$

A number of different displacements ΔL were used to obtain the initial enclosed volume, with agreement to within $\pm 2\,\%$.

Acknowledgements: This work was carried out with support from an SERC / ICI Fine Chemicals Manufacturing Organisation collaborative award, which GME gratefully acknowledges.

REFERENCES

ABBOTT, I.H. & VON DOENHOFF, A.E. (1959) "Theory of Wing Sections", p.42, Dover Publications, New York.

BRAGINSKY L.N. (1964) "The Study of Aeration Characteristics of Impellers for Autoclave Leaching Processes", Sci. Research Report, R. and D. Institute, NIIKhIMMASH, Leningrad, USSR.

DAVIDSON, J.F. & SCHÜLER, B.O.G. (1960) "Bubble Formation at an Orifice in a Viscous Liquid", *Trans. Inst. Chem. Eng.* **38**, 144-154.

EVANS, G.M., RIELLY, C.D., DAVIDSON, J.F. & CARPENTER, K.J. (1990) "A Fundamental Study of a Gas-Inducing Impeller", *I.Chem.E. Symposium Series No. 121*, 137-152.

EVANS, G.M., RIELLY, C.D., DAVIDSON, J.F. & CARPENTER, K.J. (1991) "Modelling & Design of a Gas-Inducing Reactor", *A.I.Ch.E. Spring National Meeting, Houston, April 1991*, paper No. 33e.

GHOSH, A.K. & ULBRECHT, J.J. (1989) "Bubble Formation from a Sparger in Polymer Solutions-2. Moving Liquid", *Chem. Eng. Sci.*, **44**, 969-977.

JOSHI, J.B. & SHARMA, M.M. (1977) "Mass Transfer & Hydrodynamic Characteristics of Gas-Inducing Type of Agitated Contactors", *Can. J. Chem. Eng.*, **55**, 683-695.

RAIDOO, A.D., RAGHAV RAO, K.S.M.S, SAWANT, S.B. & JOSHI, J.B. (1987) "Improvements in Gas-Inducing Impeller Design", *Chem. Eng. Commun.*, **54**, 241-264.

SAWANT, S.B. & JOSHI, J.B. (1979) "Critical Impeller Speed for the Onset of Gas Induction in Gas-Inducing types of Agitated Contactor", *The Chem. Eng. J.*, **18**, 87-91.

SAWANT, S.B., JOSHI, J.B., PANGARKAR, V.G. & MHASKAR, R.D. (1981) "Mass Transfer & Hydrodynamic Characteristics of the Denver Type of Flotation Cells", *The Chem. Eng. J.*, **21**, 11-19.

WATANABE, H., IHARA, A. & ONUMA, S. (1989) "Effects of a Few Small Air Bubbles on the Performance of Circular Cylinder at Critical Flow Range in Water", *ASME J. Fluids Eng.*, **112**, 67-72.

ZUNDELEVICH, Y. (1979) "Power Consumption and Gas Capacity of Self-Inducing Turbo Aerators", *A.I.Ch.E.J.*, **25**, 763-773.

NOMENCLATURE

C_p	pressure coefficient	
d	cylindrical blade cross-section diameter	m
d_s	inside diameter of shaft	m
g	gravitational acceleration	m/s^2
h	submersion depth of horizontal axis of the impeller blade	m
K	blade slip factor	
k	head loss coefficient in eq.(2)	
N	impeller speed	rps
N_c	critical impeller speed for gas induction	rps
$P(\theta)$	surface pressure outside the orifice	N/m^2
P_0	pressure in the head space	N/m^2
P_c	pressure in the shaft with gas inlet closed	N/m^2
P_s	stagnation pressure	N/m^2
Q_o	gas volumetric gas flow rate per orifice	m^3/s
Q_s	sparged gas volumetric gas flow rate	m^3/s
Q	total gas volumetric gas flow rate	m^3/s
R	radial position of orifice	m
r_b	radius of bubble at detachment	m
r_o	radius of orifice	m
T	diameter of vessel	m
t	time	s
U	relative velocity of fluid upstream of blade	m/s
u_L	relative velocity of blade and fluid over the orifice	m/s
v_L	liquid absolute tangential velocity	m/s
Wf	flat paddle width	m

Greek Symbols

Δh_0	change in liquid height when pressure inside shaft is at atmospheric pressure	m
Δh_c	change in liquid height inside shaft with gas inlet closed	m
ΔL	length	m
ΔP_{KE}	pressure drop due to increase in kinetic energy of liquid	N/m^2
ΔP_o	pressure drop across gas orifice	N/m^2
ΔP_P	pressure drop through gas pathway	N/m^2
ΔP_σ	pressure drop due to surface tension	N/m^2
ΔP_T	total pressure driving force	N/m^2
μ_L	liquid viscosity	Pa s
μ_ω	viscosity of water	Pa s
θ	angle from the rear stagnation point	°
σ	surface tension	N/m

REFEREES' QUESTIONS AND AUTHORS' ANSWERS

Question. For a given configuration of the gas-inducing impeller, what is the relation between the total gas flow rate Q, impeller diameter D and the rotational speed N ?

Reply. It is not possible to give an explicit expression relating the total gas flow rate to the impeller diameter and the rotational speed – the reason being that the gas flow rate through the system is a function of the detached bubble radius, which is determined by the rate of growth of the bubble, and the time it takes to be detached from the orifice. The procedure used to calculate the gas flow rate requires that the pressure driving force be balanced by the overall pressure drop across the system. This requires solving simultaneously the expressions for each of the ΔP terms in accordance with eq.(18).

An approximation for the total gas flow rate can be obtained by excluding those pressure terms which are insignificant. For our system the pressure drop due to surface tension, plus the pressure losses upstream of the orifice, accounted for less than 8% of the total pressure drop. This result is typical of most gas-inducing impeller systems, where the overall pressure drop is mainly due to the pressure drop across the orifice, and also the bubbles imparting kinetic energy to the surrounding liquid.

If only the kinetic energy and orifice pressure drop terms are considered, and using the pressure driving force given by eq.(11), the total gas flow rate is given by

$$Q = n_o C_o \pi r_o^2 \sqrt{\frac{\rho_L}{2\rho_G}[2\pi NR(1-K)]^2 \left\{ C_P(\theta) \left[1 - \frac{1}{3\left(1+\frac{r_o}{r_b}\right)^2} \right] - 1 \right\} - \frac{\rho_L}{\rho_G} gh} \quad \ldots(19)$$

where n_o is the number of orifices. Assuming that the blade slip factor and the pressure coefficient are known, eq.(19) is solved numerically, along with eq.(16) to give the induced gas flow rate and detached bubble radius.

Question. At constant vvm, what scale up rule do you advise?

Reply. The basis of constant *vvm*, is that

$$\frac{Q_1}{V_1} = \frac{Q_2}{V_2} = \text{constant} \tag{20}$$

where Q is the induced gas flow rate, V is the volume of the vessel, and the subscripts *1, 2* refer to the two scales of operation. We assume exact geometric similarity between the two scales, such that

$$\frac{Q_2}{Q_1} = \frac{V_2}{V_1} = \lambda^3 \tag{21}$$

where λ is the scale factor. Geometric similarity implies that the liquid flow pattern inside the vessel, and hence the blade slip factor and pressure coefficient, are the same at both scales of operation. Moreover, if the impeller operates well above the minimum speed for gas induction, where

$$\frac{\rho_L}{2}[2\pi NR(1-K)]^2 \{C_P(\theta) - 1\} >> \rho_L gh \tag{22}$$

and the detached bubble radius is much larger than the orifice, i.e.

$$\frac{1}{3\left(1+\frac{r_o}{r_b}\right)^2} << 1 \tag{23}$$

then, it follows from eqs.(19) and (21) that

$$\frac{N_2}{N_1} \approx 1 \tag{24}$$

This type of scale up would be difficult to achieve in practice as the impeller speed at large scale would be limited by the power requirement. A more appropriate process objective would be to maintain the same value of $k_L a$, and this type of comparison will be the subject of a future publication.

HEAT TRANSFER IN MECHANICALLY STIRRED GAS - LIQUID SYSTEM

J. KARCZ, F. STREK, Technical University of Szczecin, PL

ABSTRACT

In the paper, results of experimental studies of heat transfer for a gas - liquid system in a stirred tank, equipped with baffles and two disc turbines are presented. The stirred tank with inner diameter T = 0.3m was filled with liquid up to $H_L = 2T$. Distributions of the heat transfer coefficient were measured by means of thermal and electrochemical methods. Results of the experiments were presented in the form of equations (19) and (20). Experimental values of h, obtained by thermal method were compared with results, calculated on the basis of the local values of the hydrodynamic parameters.

INTRODUCTION

Heat transfer in mechanically stirred gas - liquid systems can be important, because temperature is one of very significant factors, decisive for correct runs of chemical and biochemical processes. Heat transfer in the reactor with stirrer may be carried on through the heating (or cooling) jacket or coil. Heat transfer coefficients depend on the geometry of the tank - stirrer system. Analysis performed in paper [3], pointed out that theoretical calculation of the local heat transfer coefficient demands knowledge of hydrodynamic parameters which are characteristic for boundary layer. Up till now information about the above magnitudes has not been sufficient in literature.

Measurements of the heat transfer coefficient are usually carried out with application of thermal method [4,7,8]. Nonconventional measuring methods, such as the electrochemical method, are also used [1,5]. Measurements conducted by means of thermal method are troublesome because there occur heat losses and difficulties with evaluation of the wall temperature. For the above reason, accuracy of the measurement decreases. Disadvantages of the thermal method do not occur with electrochemical method in which the concentration value on the surface of electrode is known

Mean value of heat transfer coefficient $\bar{h}$ on the transfer surface can be evaluated using one of two procedures

a) direct measurement of $\bar{h}$

b) integration of the measured local values h

Knowledge of the distribution of heat transfer coefficient deserves special interest, because it allows to find the correct location of the short jacket, coil or baffles in the tank.

Slender stirred tank (with geometrical parameter $H_L/T>1$) should be used in mixing gas - liquid systems. The stirred tank should also be equipped with more than one stirrer. Up till now distributions of heat transfer coefficient for such stirred tank have not been known in literature.

The purpose of the experimental studies in mechanically stirred gas - liquid systems, presented in the paper, was to obtain

a) distributions of heat transfer coefficient along the wall of the nonstandard stirred tank, equipped with two disc turbines ($H_L/T= 2$; $i = 2$), by means of thermal method.

b) distributions of the hydrodynamic parameters in the wall region with use of the electrochemical method.

Moreover, comparison of the experimental values of heat transfer coefficient with that, calculated from theoretical equation has been intended.

THEORETICAL CALCULATION OF THE LOCAL HEAT TRANSFER COEFFICIENT

Heat flux $\dot{q}$ in the "y" direction, perpendicular to wall surface, can be described for turbulent fluid flow in the stirred tank by equation

$$\dot{q} = -\rho \cdot c_p \cdot (a + a_t) \cdot \frac{d\Theta}{dy} \qquad (1)$$

where a denotes thermal diffusivity and a_t - turbulent thermal diffusivity.

Heat transfer coefficient h is defined by Newton equation

$$h = \frac{\dot{q}}{\Delta\Theta} \qquad (2)$$

which, after substituting of equation (1), is as follows

$$h = - \frac{c_p \cdot \rho}{\int_0^\infty (a + a_t)^{-1} \cdot dy} \qquad (3)$$

On the basis of the integration of equation (3) under assumption

$$a_t \sim y^3 \qquad (4)$$

and after substitution

$$Pr = \nu / a \qquad (5)$$

we can obtain

$$h \sim \lambda \cdot Pr^{1/3} \cdot v^* / \nu \qquad (6)$$

where

$$v^* = \sqrt{\tau_w / \rho} \qquad (7)$$

is so called dynamic velocity [6] and τ_w denotes shear stress on the wall. Equation (6) can be transformed in the dimensionless form

$$\frac{h \cdot T}{\lambda} \sim Pr^{1/3} \cdot \frac{v^* \cdot T}{\nu} \qquad (8)$$

where

$$\frac{v^* \cdot T}{\nu} = Re^* \qquad (9)$$

denotes local value of Reynolds number.
Substituting the definition of the local friction coefficient f [11]

$$f = \frac{\tau_w \cdot T^2}{\rho \cdot n^2 \cdot D^4} \qquad (10)$$

we may express Re^* as

$$Re^* = \sqrt{f} \cdot n \cdot D^2 / \nu \qquad (11)$$

or

$$Re^* = Re \cdot \sqrt{f} \qquad (12)$$

where $Re = nD^2 \rho / \eta$ is the conventional definition of Reynolds number for mixing process.
Assuming validity of equation (8) for liquid and gas - liquid systems and substituting in both cases physical properties for laminar sublayer we can calculate the local value of heat transfer coefficient for gas - liquid system from interrelation

$$h_{G-L} = h_L \cdot \frac{v^*_{G-L}}{v^*_L} \qquad (13)$$

or

$$h_{G-L} = h_L \cdot \sqrt{f_{G-L} / f_L} \qquad (14)$$

Such way of the calculation of heat transfer coefficient h_{G-L} is possible if the parameters f_L and f_{G-L} are known.

EXPERIMENTAL

Measurements of heat transfer distributions with use of thermal method were carried out in the stirred tank with inner diameter T=0.3m and liquid height H_L=2T (Fig.1). Stirred tank had four baffles (J=4; B/T=0.1)

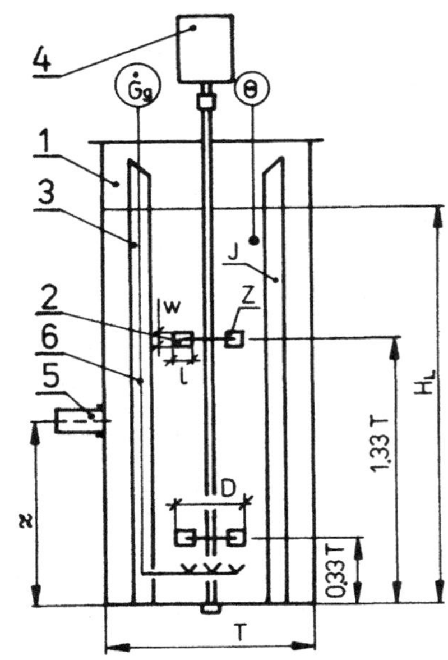

Fig. 1. Sketch of experimental equipment (thermal method)
1 - stirred tank (T=0.3m; H_L/T=2); 2 - disc turbine; 3 - baffle; 4- electromotor; 5 - local heat source; 6 - gas sparger

and two disc turbines (D/T=0.33; w/D=0.2; l/D=0.25; Z=6). The distances from the bottom of the tank to the lower and upper stirrers were h_1=T/3 and h_2=4T/3, respectively. A ring with a diameter d_r=0.07 m was applied as a gas sparger. The measurements of the local heat transfer coefficients at the wall of the stirred tank were carried out by means of a special heat flux meter [10] with diameter d_m=0.06 m, for eight various dimensionless values of coordinate z/H_L (z - distance of measuring point to the tank bottom). Experiments were performed for distilled water - air system in the following range of the variables: stirrer speed $n[1/s] \in \langle 5; 10.8 \rangle$ and gas flow rate $\dot{G}_G [m^3/s] \in \langle 0; 4 \cdot 10^{-4} \rangle$ (superficial gas velocity $v_{og} = 4 \cdot \dot{G}_g / \pi \cdot T^2 [m/s] \in \langle 0; 5.1 \cdot 10^{-3} \rangle$). The local values of the heat transfer coefficient h were calculated on the basis of directly measured local heat flux $\dot{q}$ and the temperature difference $\Delta\Theta$ between a wall and the stirred liquid

$$h = \frac{\dot{q}}{\Delta\Theta} = \frac{\dot{q}}{\Theta_w - \Theta_m} \qquad (15)$$

Measurements with use of electrochemical method were carried out in the standard stirred tank (1) with inner diameter T = 0.3 m and liquid height H_L= T, equipped with single disc turbine (2) and four baffles (3) (Fig. 2). The cathodes (5) with radius R = 2 mm (in number k), made from nickel were built-in in the wall of the stirred tank. The nickel anode (6), located on the bottom of the tank, had dimensions 10 x 10 cm^2. The voltage feeding the electrodes circuit was regulated by feeder (7). The measuring circuit included also precise, highly stable resistor (8). Voltage drop on the resistor, as the input voltage, was entered, via the analog-to-digital converter (9) (12 bit card ADDA), to the microcomputer IBM (10), where sampling of the actual voltage values and further calculations of their mean value were performed. Measurements of diffusion current I_d were carried out in the range of plateau of the voltage-current characteristics. Solution of potassium ferricyanide, potassium ferrocyanide and sodium hydroxide with concentrations 0.01; 0.05 and 0.5 kmol/m^3, respectively, was used as active electrolyte. Two series of experiments were made. In the first of them, distributions of diffusion current I_d were measured, in order to obtain local heat transfer coefficients h in the stirrer region, i.e. for parameter $z/H_L \in \langle 0.30; 0.37 \rangle$. The measurements were performed by means of k=10 cathodes for the following range of variables: stirrer speed $n[1/s] \in \langle 4.2; 7.5 \rangle$ and superficial gas velocity $v_{og} [m/s] \in \langle 0; 5.1 \cdot 10^{-3} \rangle$. The values of h were calculated from equation [5]

$$h = \frac{I_d}{z_e \cdot F \cdot A \cdot c_A} \cdot c_p \cdot \rho \cdot \left(\frac{Sc}{Pr}\right)^{2/3} \qquad (16)$$

Distributions of hydrodynamic parameters along the height of stirred tank wall were measured in a second series of experiments, in which k=32 cathodes were used. The local values of shear stress τ_w was calculated from equation [11]

$$\tau_w = \dot{\gamma} \cdot \eta = \left[\frac{I_d}{2.156 \cdot z_e \cdot F \cdot D_A^{2/3} \cdot c_A \cdot R^{5/3}}\right]^3 \cdot \eta \qquad (17)$$

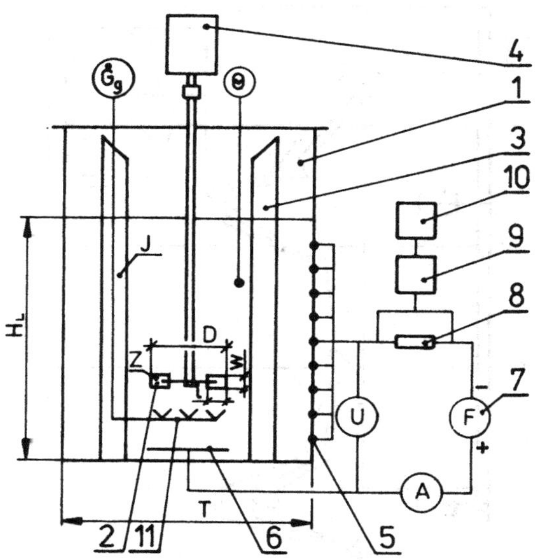

Fig. 2. Sketch of experimental equipment (electrochemical method) 1 - stirred tank (T=0.3m; H_L/T=1); 2 - disc turbine; 3 - baffle; 4 - electromotor; 5 - cathode; 6 - anode; 7 - feeder; 8 - resistor; 9 - card ADDA; 10 - microcomputer IBM; 11 - gas sparger

RESULTS AND DISCUSSION

On the basis of measurements, performed by means of thermal method, distributions of the heat transfer coefficients along the wall of the stirred tank with two disc turbines

($H_L/T=2$; $i=2$) were obtained. They are presented graphically as function $h = f(z/H_L)$ for selected series of measurements (coordinate z is calculated from the tank bottom). In generally, experimental curves have maximum in the region of both stirrers, where heat transfer is most intensive. Furthermore, heat transfer coefficients increase with the rise of the stirrer speed.

For example, distributions of function $h = f(z/H_L)$ for distilled water-air system and superficial gas velocity $v_{og} = 3.9 \cdot 10^{-3}$ m/s are presented in Fig. 3. The graph shows certain asymmetry between local maximum. Namely, values of maximum for lower stirrer are reduced. It is doubtlessly conditioned by the fact that gas was fed into stirred tank under the lower stirrer and accumulated behind the blades of stirrer in the form of gas cavities [2].

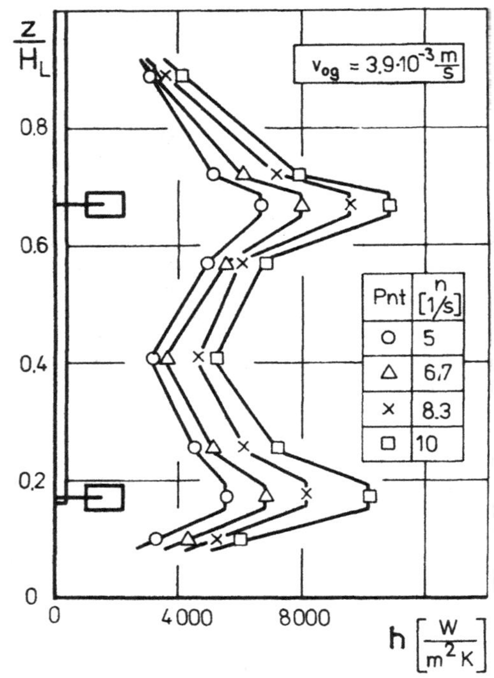

Fig. 3. Distribution $h = f(z/H_L)$ for distilled water - air system, $v_{og} = 3.9 \cdot 10^{-3}$ m/s; $H_L/T=2$; $i=2$

Influence of gas phase on the distribution of heat transfer coefficient along the height of stirred tank wall can be analysed on the basis of runs for function $h = f(z/H_L)$, plotted for constant value of stirrer speed and various superficial gas velocity v_{og}. Fig. 4 illustrates that gas phase in the distilled water - air system affects the decrease of local heat transfer coefficient in the following zones: above both lower and upper stirrers, in the region of upper stirrer and, first of all, in the region of lower stirrer.

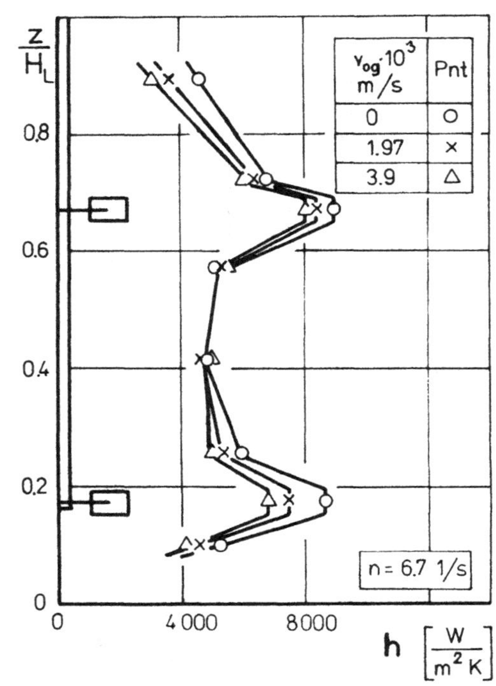

Fig. 4. Dependence $h = f(z/H_L)$ for distilled water - air system, $H_L/T=2$; $i=2$; $n = 6.7$ 1/s

Distribution of heat transfer coefficient for gas - liquid system in relation to that for liquid ($n=6.7$ 1/s; $v_{og}=3.9 \cdot 10^{-3}$ m/s) is shown in Fig. 5 as function $h_{G-L}/h_L = f(z/H_L)$. The decrease of the heat transfer coefficient in the region of upper stirrer, approximately 10% in comparison with that for liquid, is distinctive for this diagram. Analogous dependence appears in the region of lower stirrer, but there, coefficient h decreases more (approximately 20%).

Measurements with use of electrochemical method, performed for stirred tank with single disc turbine ($H_L/T=1$; $i=1$), confirmed the above conclusion. Results of experiments, obtained for stirrer region ($n=6.7$ 1/s; $v_{og}=3.9 \cdot 10^{-3}$ m/s), are presented in Fig. 6. As it is seen, the distribution of heat transfer coefficient along the width of stirrer blade w is approximately constant and it is equal to 0.8.

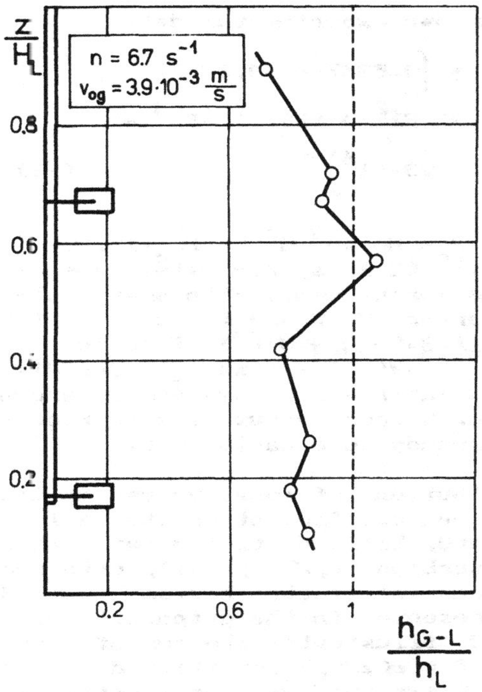

Fig. 5. Relationship $h_{G-L}/h_L = f(z/H_L)$ for distilled water - air system; n=6.7 1/s; v_{og} = 3.9·10^{-3} m/s; H_L/T=2; i=2

Mean values of heat transfer coefficient $\bar{h}$ for assumed operating variables n and v_{og} have been obtained on the basis of numerical integration of local values. Influence of the superficial gas velocity on mean value of heat transfer coefficient is presented graphically. As Fig. 7 illustrates, the coefficient h decreases with the increase of the superficial gas velocity v_{og}.

Dependence of mean heat transfer coefficient on stirrer speed for liquid phase and gas - liquid system have been evaluated from the calculation of the exponent A in the equation

$$\frac{Nu}{Pr^{0.33} \circ Vi^{0.14}} = C \circ Re^{A} \qquad (18)$$

Physical properties of liquid phase were used in the definitions of dimensionless numbers in equation (18). Exponents at Prandtl number and viscosity simplex were assumed on the basis of literature data [8,9]. Mean value of the exponent A for whole height of the wall, calculated from experimental data is equal to A = 0.668 ± 0.015. This exponent is in

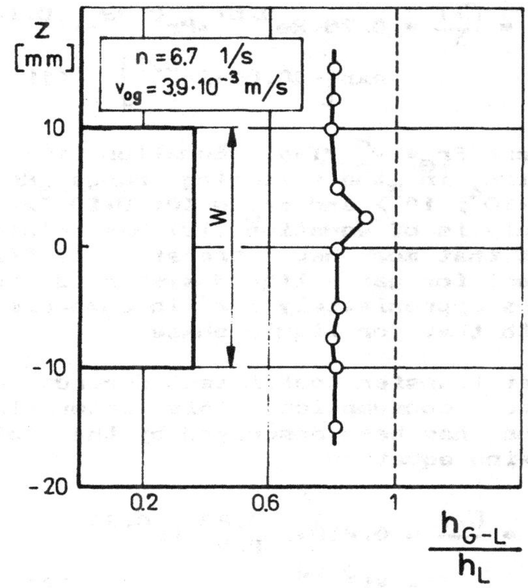

Fig. 6. Relationship $h_{G-L}/h_L = f(z/H_L)$ for region of stirrer, n=6.7 1/s; v_{og} =3.9∘10^{-3}; H_L/T=1; i=1

agreement with literature data [8, 9], which concern standard, baffled stirred tank, equipped with disc turbine.

Results of the experimental studies for slender stirred tank with two turbines (H_L/T=2; i=2) and distilled water - air system have been approximated with mean relative error ±2% in the form of dimensionless equation

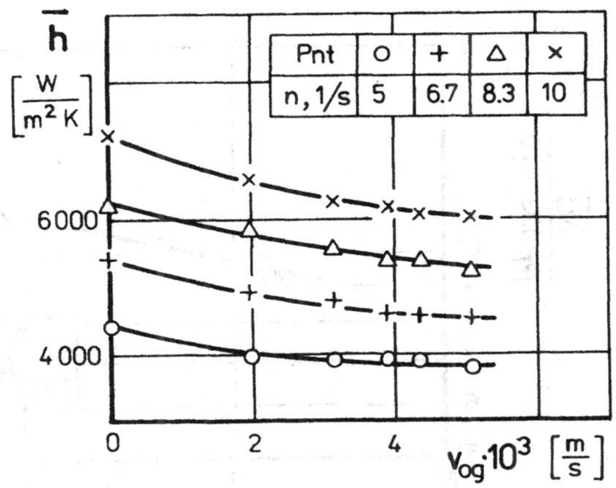

Fig. 7. Dependence $\bar{h} = f(v_{og})$ for distilled water - air system, H_L/T =2; i=2

$$\overline{Nu} = \frac{\bar{h} \cdot T}{\lambda} = 0.76 \cdot Re^{0.67} \cdot Pr^{0.33} \cdot Vi^{0.14} \cdot \exp\left[-40.65 \cdot Fr_G^{0.5}\right] \quad (19)$$

where $Fr_G = v_{og}^2 / T \cdot g$. Equation (19) is valid in the following range Re $\in$ $\langle 5 \cdot 10^4;\ 10^5 \rangle$ and $Fr_G \in \langle 0;\ 9 \cdot 10^{-6} \rangle$. Analysis of equation (19) has pointed out that mean heat transfer coefficient for gas - liquid system decreases approximately 12% in comparison with that for liquid phase.

Heat transfer coefficient depends on power consumption. This interrelation has been described by the following equation

$$\overline{Nu} = \frac{\bar{h} \cdot T}{\lambda} = 0.419 \cdot Re_{P,V}^{0.22} \cdot Pr^{0.33} \cdot Vi^{0.14} \quad (20)$$

$Re_{P,V}$ number in equation (20) is defined as

$$Re_{P,V} = \frac{(P_G / V_L) \cdot T^4 \cdot \rho^2}{\eta^3} \quad (21)$$

where ρ and η denote properties of continuous phase. Equation (20) is valid in the following range $v_{og} \langle 0;\ 5.1 \cdot 10^{-3} \rangle$ and $Re_{P,V} \in \langle 1.6 \cdot 10^{15};\ 2 \cdot 10^{16} \rangle$ and it approximates results of experiments with mean relative error ±3.5%. Power consumption P_G for distilled water - air system and stirred tank with two disc turbines ($H_L/T = 2$; $i = 2$) has been described by equation (22), obtained on the basis of our own experimental data

$$P_G / P_o = \left[0.6367 + 40.30 \cdot Fl - 560.39 \cdot Fl^2 + 1237.37 \cdot Fl^3 + 15220.93 \cdot Fl^4\right]^{-1} \quad (22)$$

where $P_o = Ne_o \cdot n^3 \cdot d^5 \cdot \rho$; $Ne_o = 10.8$. Equation (22) approximates results of the measurements with mean relative error ±3.5% and it is valid for $Fl \in \langle 1.3 \cdot 10^{-2};\ 8 \cdot 10^{-2} \rangle$. Function $Nu/Pr^{0.33} \cdot Vi^{0.14} = f(Re_{P,V})$ for distilled water - air system is shown in Fig. 8 (continuous line in Fig. 8 corresponds to equation (20)).

Distributions of local values of the friction coefficient on the wall of standard, baffled stirred tank with disc turbine ($H_L/T = 1$; $i = 1$), obtained by means of electrochemical method are presented in the graphical form. Fig. 9 illustrates the run of function $f = \varphi(z/H_L)$ for stirred speed $n = 6.7$ 1/s and various superficial gas velocity v_{og}, whereas distributions of f function for $v_{og} = 3.9 \cdot 10^{-3}$ m/s and various stirrer speed n are shown in Fig. 10.

Coefficient h in gas - liquid system can be calculated from equation (14), if values of the friction coefficient on the stirred tank wall are known. The computations were carried out for slender stirred tank equipped with two disc turbines $H_L/T = 2$) in the range of the parameter z/H_L (0; 0.5). It has been assumed that differences between values

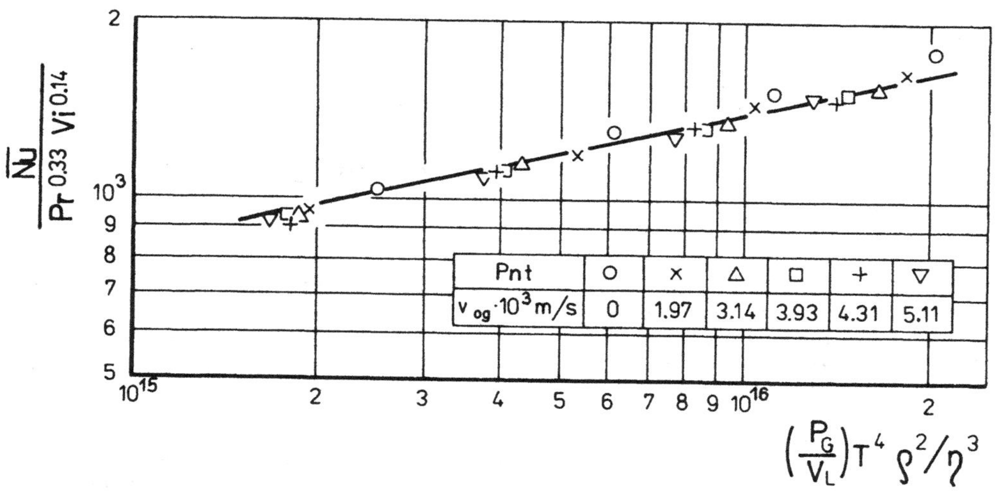

Fig. 8. Function $\overline{Nu}/Pr^{0.33} \cdot Vi^{0.14} = f(Re_{P,V})$ for distilled water - air system and $H_L/T = 2$; $i = 2$

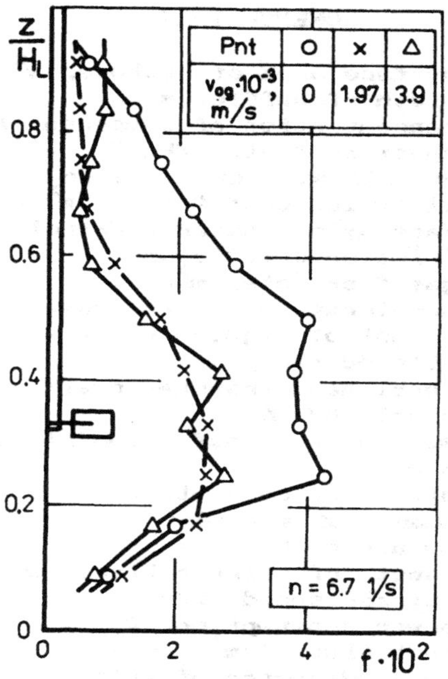

Fig. 9. Dependence $f = \varphi(z/H_L)$ for n=6.7 1/s; $H_L/T=1$; i=1

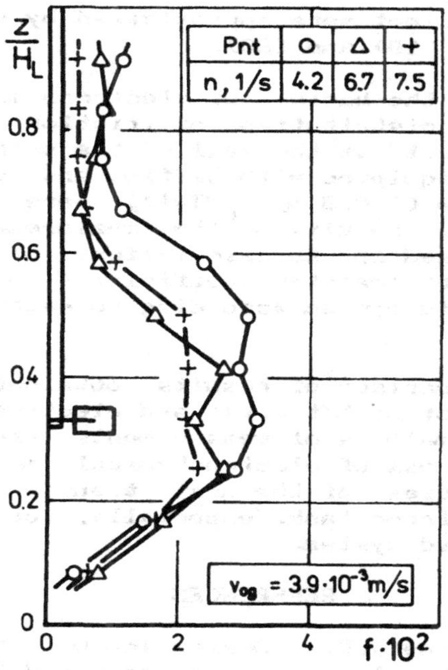

Fig. 10. Dependence $f = \varphi(z/H_L)$ for $v_{og}=3.9\cdot10^{-3}$ m/s; $H_L/T=1$; i=1

f for such stirred tank (in the range z/H_L (0; 0.5)) and for standard stirred tank can be neglected. Therefore, the following experimental results of the measurements were used in the calculations: values f obtained by means of electrochemical method and values of the heat transfer coefficient for a liquid, evaluated by means of the thermal method. Comparison of the experimental and calculated from equation (14) values of the heat transfer coefficient is shown in Fig. 11. As it is shown in Fig. 11, these values are consistent (mean relative error of approximation is ±6.5%).

Comparative analysis presented above pointed out that local heat transfer coefficient in the stirred gas - liquid system can be computed on the basis of hydrodynamic parameters (f, v^*, τ_w). The above parameters depend on the stirred tank geometry, but their values can be measured in the relatively simple way, using eletrochemical method.

CONCLUSIONS

1. Distributions of heat transfer coefficient along the stirred tank wall for gas - liquid system were obtained. Measurements were carried out in distilled water - air system, for stirred tank (T=0.3 m, $H_L/T=2$) equipped with baffles and two disc turbines.

2. Two different measuring methods (thermal and electrochemical) were used for estimation of the local values of heat transfer coefficient.

3. Mean values of the heat transfer

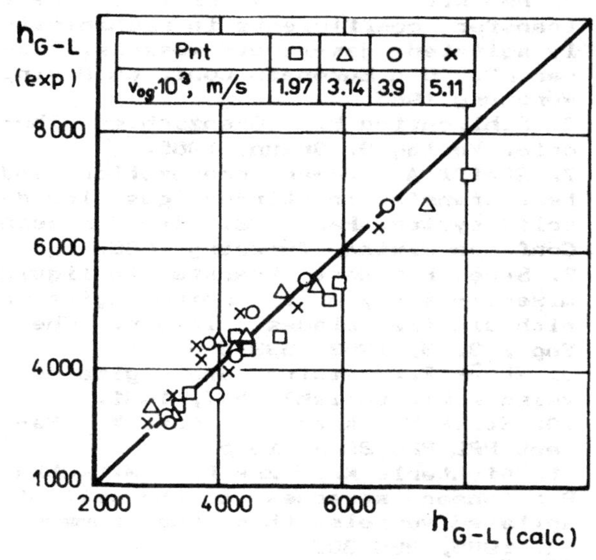

Fig. 11. Comparison of experimental and calculated from Eq. (14) values of h

coefficient were approximated by equations (19) and (20).

4. On the basis of electrochemical method distributions of friction coefficient at the wall of the stirred tank, equipped with baffles and disc turbine ($T=0.3$ m, $H_L/T=1$), were obtained. Results of the measurements were used in the calculations of local heat transfer coefficient in gas - liquid system according to equation (14).

5. Comparison of results obtained by means of both thermal and electrochemical methods of measurements proved usefulness of electrochemical method to studies of the heat transfer in the stirred tank, especially, for gas - liquid system.

REFERENCES

1. Bourne J.R., Dossenbach O., Post T.A.: Local and average mass and heat transfer due to turbine impellers. Paper 21. 5th European Conf. on Mixing, Würzburg 1985.
2. Harnby N., Edwards M.F., Nienow A.W.: Mixing in the process industries, Buterworth Co. Ltd., 1985.
3. Karcz J.: Mathematical modelling of local heat transfer for stirred tank. Paper J9.50. Summ. p. 129. 10th Congress CHISA, Prague 1990.
4. Kurpiers P.: Wärmeübergang in Ein- und Mehrphasenreaktoren. VCH, 1985.
5. Man K.L.: A study of local heat transfer coefficients in mechanically agitated gas-liquid vessels. Paper 23. 5th European Conf. on Mixing Würzburg 1985.
6. Schlichting H.: Grenzschicht-Theorie. Verlag G. Braun, 1965.
7. Steiff A.: Power consumption and heat transfer in stirred gas-liquid-solid system. Paper 22. 5th European Conf. on Mixing, Würzburg 1985.
8. Stręk F.: Heat transfer in liquid mixers - study of a turbine agitator with six flat blades. Intern. Chem. Eng., 3, 3, 1963, 533-556.
9. Stręk F.: Mixing and agitated vessels (in polish). WNT, 1981.
10. Stręk F., Karcz J., Góra W.: Patent PRL P26 2943, 1986.
11. Wichterle K., Žak L., Mitschka P.: Shear stresses on the wall of agitated vessels. Chem. Eng. Commun. 32, 1985, 289-305.
12. Žak L.: Elektrochemicka diagnostika přenosovych jevu na stene michanych nadob. Thesis. ČSAV, Prague 1986.

NOMENCLATURE

A - surface area of cathode, m^2
B - width of baffle, m
c_A - concentration of component A
c^P - specific heat, J/kg·K
D - overall stirrer diameter, m
D_A - diffusion coefficient, m^2/s
F - Faraday's constant, C/kmol
f - friction coefficient
$\dot{G}_G$ - gas flow rate, m^3/s
g - acceleration due to grav. m^2/s
H_L - height of liquid from bottom of stirred tank, m
h - local heat transfer coefficient, W/m^2K
$\bar{h}$ - mean heat transfer coefficient, W/m^2K
I_d - diffusion current, A
i - number of stirrers
J - number of baffles
l - length of stirrer blade, m
n - stirrer speed, 1/s
P - power consumption, W
$\dot{q}$ - heat flux, W/m^2
T - inner diameter of stirred tank, m
v^* - velocity def. by Eq. (7), m/s
v_{og} - superficial gas velocity, m/s
V_L - volume of liquid, m^3
w - width of stirrer blade, m
y - coordinate, perpendicular to the stirred tank wall, m
Z - number of blades on stirrer
z - axial coordinate, m
z_e - number of electrons
γ - shear rate, 1/s
η - dynamic viscosity of liquid, Pa·s
Θ - temperature, ^oC
λ - thermal conductivity, W/mK
ν - kinematic viscosity of liquid, m^2/s
ρ - density of liquid, kg/m^3
τ_w - shear stress, N/m^2

Indicies

G - gas L - liquid
m - mean w - wall

Dimensionless numbers

$$Fr_G = \frac{v_{og}^2}{T \cdot g}; \quad Fl = \frac{\dot{G}_G}{n \cdot D^3}; \quad Nu = \frac{h \cdot T}{\lambda}$$

$$\overline{Nu} = \frac{\bar{h} \cdot T}{\lambda}; \quad Pr = \frac{c_p \cdot \eta}{\lambda}; \quad Re^* = \frac{v^* \cdot T}{\nu}$$

$$Re = \frac{n \cdot D^2 \cdot \rho}{\eta}; \quad Re_{P,V} = \frac{(P_G/V_L) \cdot T^4 \cdot \rho^2}{\eta^3}$$

$$Sc = \frac{\nu}{D_A}; \quad Vi = \frac{\eta}{\eta_w}$$

<u>Referee 2:</u>

1. From Fig. 1 it looks as though the liquid temperature (θ_m) was measured a long way from the wall and heat source. Did the authors assume the liquid to be perfectly mixed? Is there any data to show the presence of temperature gradients in the liquid?

A: We did not assume the liquid to be perfectly mixed, but the distribution of the temperature can be neglected for turbulent flow of the liquid in the stirred tank. In our experiments, we have measured the temperature at three representative points in order to evaluate the bulk temperature of the liquid more exactly. This measurement procedure is described in paper [8].

2. How does the electrochemical method operate?

A: The details of the physical backgrounds for the electrochemical method are given in Ref. [5]. For this reason we have not described the procedure of the measurement more exactly.

<u>Referee 3:</u>

1. What is the clearance of baffles from wall?

A: The clearance of baffles from wall is equal 0.

2. What kind of cavities are behind the lower turbine as a function of working conditions?

A: The dispersion of the gas below the lower turbine was observed for ranges of n and v_{og} used in the measurements. The regimes of bubbles flow corresponded to states (d) and (e) shown in Fig. 17.9 (p. 334, [2]).

3. Could you give a correlation for P_G/P_O for your monoagitated system $H_L = T$?

A: Our experiments were carried out only for nonstandard geometry of the stirred tank ($H_L/T=2$). Therefore we have not the experimental correlation P_G/P_O for monoagitated system ($H_L/T=1$).

4. Desplanches and Bruxelmane ("Characteristic variable, prediction and scale-up for heat transfer to coil in agitated vessels - Part 1: Newtonian fluids in the turbulent region", Chem. Eng. Res. Des. 61(1),3-12,1983) have proved that:

- in a given vessel, v_c, the mean (circulation) fluid velocity is the main characteristic variable for h_c, the process heat transfer with a coil (agitated liquid non aerated) i.e. v_c is the main parameter which fixes the value of h_c.
- in systems of different size and eventually of different shape, $Re_c = \rho_L N_{qc} n D^3 d_c/\eta_L$ is the main characteristic for $Nu_c = h_c d_c/\lambda_L$

This "circulation hypothesis" can be proved for heat transfer to the vessel wall in agitated tanks (at least in non aerated solutions). Therefore, we don't agree with the use of P_G/V_L ("turbulence hypothesis") in your correlation (20) unless that $N_{qpG}/N_{qpo} = Q_{pG}/Q_{po} = Ne_G/Ne_o = P_G/P_o$

We have observed that in a given tank N_{qpG}/N_{qpo} is in reality lower than Ne_G/Ne_o.

Moreover, when you used your electrochemical measurement in order to compare your experimental and calculated values of h_{G-L}, implicitly you consider the "circulation hypothesis" for the heat transfer with the vessel wall!

A: The definition of Re_c is as folows (Eq. 9, p. 5 in above mentioned Ref.)

$$Re_c = \frac{\rho v_c d_c}{\eta} = \frac{\rho N_{qc} n D^3 d_c}{\eta T^2}$$

if we assume

$D,\ d_c,\ T \sim l$ (l - linear dimension) and N_{qc} = const then

$$Re_c \sim \frac{\rho N_{qc} n l^3 l}{\eta l^2} \sim n l^2$$

$$Nu \sim (Re_c)^A \sim (n l^2)^A$$

On the other hand

$$Re_{P,V} = \frac{(P/V_L) T^4 \rho^2}{\eta^3}$$

if $D/T=k_1,\ V_L=\pi T^2 H_L/4,\ H_L/T=k_2,$ $P=Ne n^3 D^5 \rho$

and $T \sim D,\ V \sim D^3,\ P \sim n^3 D^5$

then

$$Re_{P,V} \sim \frac{n^3 D^5 D^4}{D^3} \sim n^3 D^6$$

for $D \sim l$

$$Re_{P,V} \sim n^3 l^6, \quad (Re_{P,V})^{1/3} \sim n l^2$$

and

$$Nu \sim \left[(Re_{P,V})^{1/3}\right]^A \sim (n l^2)^A$$

Therefore the interrelation

$$Re_c \sim (Re_{P,V})^{1/3}$$

can be obtained.

HEAT AND MASS TRANSFER EFFICIENCIES OF MULTI-IMPELLER GAS-VISCOUS LIQUIDS CONTACTORS

H. LINDEN, H. DESPLANCHES, Y. GASTON-BONHOMME, J.L. CHEVALIER,
ENSSPICAM, Marseille, F
M. BRUXELMANE, V. DUFOUR, Faculté Polytechnique de Mons, B

ABSTRACT

Measurements of impeller power P_g, heat transfer coefficients stirred solution-coil h_{cg} and oxygen transfer coefficients k_La are carried out on two gas-viscous liquid contactors stirred by axial and radial impeller couples.

The lower impeller imposes the hydrodynamic regime especially n_{CD}, minimum of $P_g/P_o = f(Re)$ curves and beginning of the working zone. For the most important combinations, in particular DT-PBT, n_{CD}, h_{cg} and k_La correlations ($n \geq n_{CD}$) are proposed.

Two new efficiency criteria have been defined $E_H = h_{cg}/P_g$ and $E_M = k_La/P_g$ in order to quantify the multiple impellers heat and mass transfer performances.

For all the studied systems, the complete unflooding point (close to the loading point) represents the working optimum for h_{cg} and k_La too.

A great care must be paid for the multi-impellers choice because their relative performances estimated by E_H and E_M, can reverse when the comparisons are not carried on in n_{CD} optima but with imposed h_{cg} or k_La.

INTRODUCTION

Many processes, especially in the biochemical industry, involve multi-impeller gas-liquid contactors.

In view of the liquid phase complexity, whose viscosity can strongly increase or can become pseudoplastic, a particular attention must be paid to the oxygen transfer, but also to the heat transfer towards the cooling wall which can limit the process.

This study deals with the optimization of the impeller combination : disc turbine - 45° pitched-blade turbine, making allowance for the impeller power consumption P_g, for the heat transfer coefficient from the solution to the helical coil h_{cg}, and for the oxygen transfer coefficient k_La in highly viscous liquids.

EXPERIMENTAL

Two pilot scale fittings (T=0.63m) are used respectively for heat transfer (vessel n°1) and for mass transfer (vessel n°2). Geometrical characteristics of these two installations are indicated in Table 1.

		Vessel n°1	Vessel n°2
	H_L/T	2	1.27
	T/b	12	12
impellers position	h/T	0.33	0.38
	e/T	1	0.63
coil	d_c/T	0.019	no coil !
	p_c/d_c	3	
	D_c/T	0.8	
	C_c/T	0.2	
	H_c/T	1.6	

Table 1 : Installations geometry

The tested combinations of disc turbine - DT(D/L/W : 20/5/4; z=6) - and pitched-blade turbine pumping down or up : PBT↓ or ↑ (D/W : 20/4; α = 45°; z = 6) - are shown in Table 2.

Aeration takes place under the lower impeller, in the vessel axis (G/V_L : 0-0,26 vvm).

Vessel n°1	Vessel n°2
DT050 - DT033	DT039 - DT032
DT050 - PBT050↓	DT039 - PBT039↓
PBT050↓- PBT033↓	DT039 - PBT039↑
PBT050↓- PBT050↓	DT039
DT050	
PBT050↓	

Table 2 : Tested impeller combinations (DT050 - PBT050↓ : corresponds to a disc turbine with a relative diameter $D_1/T = 0.50$ as the lower impeller coupled with a pitched-blade turbine, pumping down, with a relative diameter $D_2/T = 0.50$ as the upper impeller).

The mixing power is measured with a Lebow strain-gauge torquemeter put in between the shaft and the engine (n : 0.5 to 3.3 s^{-1}). In the vessel n°1, two aqueous diluted solutions with pseudoplastic behavior are used : a highly pseudoplastic 0.6 % weight solution of guar gum (guar IRX 23337 E412 Sté. Iranex; N = 0.57; mean η_a : 0.29 Pa.s) and a **nearly** newtonian 1.5 % solution of carboxy-methyl-cellulose (CMC 7MF Pharmac.grade Sté Aqualon). In the vessel n°2, a weakly pseudoplastic 0.48% solution of methyl-cellulose (Methocel HE90-400 cPo Sté Dow Chem. ; N = 0.9; mean η_a = 0.12 Pa.s) is used.

The rheological parameters are measured with a Rheomat 135 (Sté Contraves) over a shear rate range from 1 to 100 s^{-1} and the apparent viscosity in stirred vessel is defined owing to the Metzner and Otto model.

On the installation n°1, the temperatures of the stirred medium (6 probes) and of the water which passes through the coil are sequentially scanned during the cooling cycles. The data acquisition and processing enable the determination of the heat transfer coefficient stirred two-phased solution-coil, h_{cg}, for each impeller combinations, as a function of the operational parameters (n, G) and of the rheological properties. The method for the calculation of h_{cg} has been described elsewhere (1).

In the vessel n°2, the liquid phase oxygen concentrations are measured with a probe (model 621 Sté Dow Chem. EMBH) which is plunged 5 cm under the liquid surface near a 25 s^{-1} rotating disc. Deoxygenation and oxygenation cycles exploitation through the dynamic method allows the determination of the volumetric mass transfer coefficient (k_La) of the oxygen concentration evolution as a function of time.

$$k_La = - \operatorname{Ln}\left[(C^* - C_L)/(C^* - C_o)\right] . 1/t \quad (1)$$

The G and n effects on k_La are studied for the different impeller combinations.

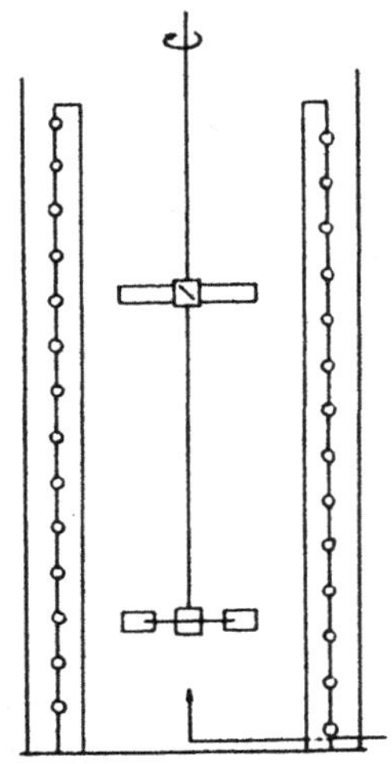

Fig. 1 Experimental set-up

RESULTS AND DISCUSSION

HYDRODYNAMICS AND POWER CONSUMPTION

Viscous **Newtonian** and pseudoplastic liquid-gas systems hydrodynamic is complex. Moreover, the solution cloudiness makes the visual observations problematic.

The evolution of the power curves, with variable n and constant G, allows the most objective characterization of the different hydrodynamic regimes. It's the lower impeller which logically dictates the contactor working, specially the position of the complete unflooding point (beginning of the effective zone) n_{CD}. The upper impeller only provides the liquid recirculation. This particular point corresponds to the surface "blurps" vanishing and to the gas bubbles incidence against the vessel wall (it's difficult to speak of gas complete dispersion with viscous **media**).

	CD point	$n \geq n_{CD}$	Non aerated medium
DT050-PBT050↓	$n_{CD} = 36.1\ G^{0.375}\ \eta_a^{0.145}$ (2)	$P_g = 4.63\ G^{-0.146}\ n^{3.05}\ \eta_a^{0.056}$ (3)	$P_o = 19.7\ n^{3.08}\ \eta^{0.035}$ (4)
PBT050↓-PBT050↓	$n_{CD} = 40{,}6\ G^{0{,}341}\ \eta_a^{0.090}$ (5)	$P_g = 0.452\ G^{-0.291}\ n^{3.20}\ \eta_a^{-0.237}$ (6)	$P_o = 13.2\ n^{3.06}\ \eta_a^{0.056}$ (7)
DT039-PBT039↑ ($\eta_a \gtrsim 0.1$)	$n_{CD} = 17{,}0\ G^{0.265}$ (8)	$P_g = 0.517\ G^{-0.249}\ n^{3.25}$ (9)	
DT039-DT032	$n_{CD} = 9{,}21\ G^{0.179}$ (10)	$P_g = 0.449\ G^{-0.231}\ n^{3.44}$ (11)	

Table 3 : Hydrodynamic correlations

Like in the single impeller systems, n_{CD} corresponds to the minimum of the P_g/P (or N_{pg}) = f(Re) curves, at constant G, for all the studied combinations.

The n_{CD} correlations and the correlations which allow to calculate P_g over the contactor effective working zone are indicated in Table 3.

HEAT TRANSFER

Like in the single impeller systems (2) the h_{cg}/h_{co} and P_g/P_o curves as a function of Re, with constant G, have a similar trend and pass through a minimum for the same abscissa whatever the studied impeller combination may be (fig. 2).

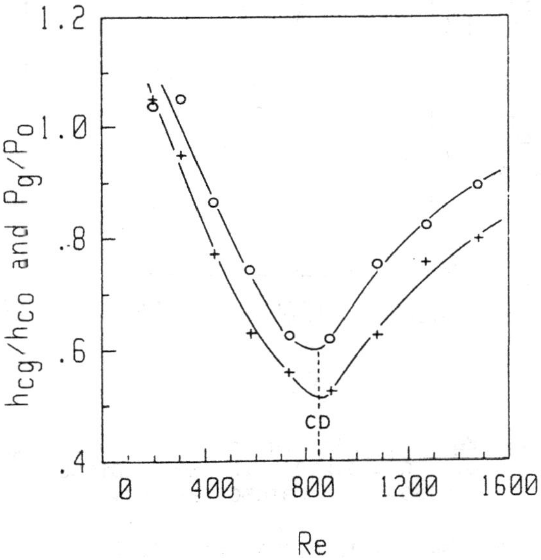

Fig. 2 P_g/P_o = f(Re) and h_{cg}/h_{co} = f(Re); PBT050↓-PBT050↓; G = 0.278 x 10^{-3} m^3/s; liquid : guar, N = 0.57 and $\bar{\eta}_a$ = 0.29 Pas P_g/P_o : + ; h_{cg}/h_{co} : o

Those minimums correspond to the complete unflooding point (n_{CD}) where full development of the stable gas cavities at the near of the blades is reached.

In the working zone, which corresponds to $n \geq n_{CD}$, general equations, including the effects of n, G, η_a on h_{cg}, are proposed concerning the following impeller combinations : DT050-PBT050↓ and PBT050↓-PBT050↓

Equations 12 and 15 have the heat transfer equation general form in one phase medium, whereas, taking into account the hereabove described analogy, the equations 13 and 16 correlate the transfer coefficient reduction to the impeller power's one. These correlations are obtained over the following ranges :

Re : 200-6000; N : 0.5-1; G/V : 0.04-0.26 vvm
-DT050-PBT050↓ combination

$$Nu_{cg} = 0.0987\ Re^{0.531}(P_g/P_o)^{0.146} Pr^{0.3} Vi^{0.2} \quad (12)$$
$$(\bar{\Delta} = 0.2\ \%)$$

$$Nu_{cg}/Nu_{co} = (P_g/P_o)^{0.55} \quad (\bar{\Delta} = 3.0\ \%) \quad (13)$$

$$Nu_{co} = 0.0755\ Re^{0.565} Pr^{0.3} Vi^{0.2} \quad (14)$$

-PBT050 -PBT050↓ combination

$$Nu_{cg} = 0.132\ Re^{0.499}(P_g/P_o)^{0.487} Pr^{0.3} Vi^{0.2} \quad (15)$$
$$(\bar{\Delta} = 5.4\ \%)$$

$$Nu_{cg}/Nu_{co} = (P_g/P_o)^{0.62} \quad (\bar{\Delta} = 3.5\ \%) \quad (16)$$

$$Nu_{co} = 0.0893\ Re^{0.544} Pr^{0.3} Vi^{0.2} \quad (17)$$

In order to be able to optimize the contactor working and to compare the different impeller couples heat transfer performances, a new con-

concept is being proposed : the heat efficiency E_H : $E_H = h_{cg}/P_g$ (18)

Relating to the working of a given single impeller system with constant G, the E_H efficiency is maximal at the n_{CD} point thus representing the working optimum and then monotonously decreases while Re increases whatever the impeller combination may be (fig. 3).

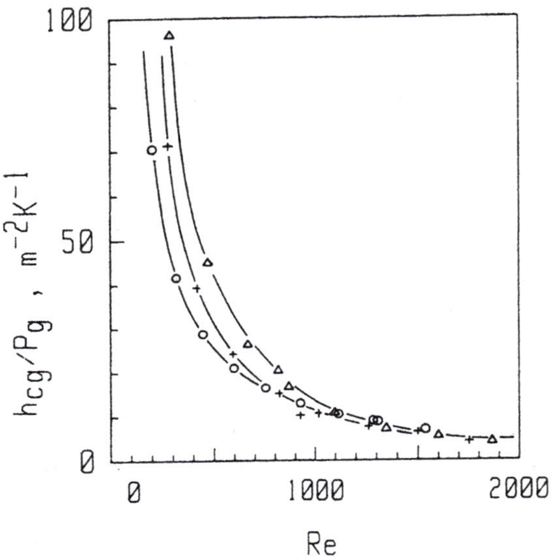

Fig. 3 Thermal efficiency : $G = 0.833.10^{-3}\ m^3/s$ liquid : guar, N = 0.57 and $\bar{\eta}_a = 0.29$ Pas; DT050-PBT050↓ : + ; PBT050↓- PBT 050↓ : o ; DT050-DT033 : Δ

On the other hand, at the optimum n_{CD}, E_H decreases when the gas flow increases and this for all the impeller couples tested (fig. 4).

These results show that the greatest care must be taken in the comparison of multiple impeller systems.

For a given G, if the heat transfer is the limiting phenomenon, multiple impeller efficiencies must be compared for higher speeds than n_{CD} which enable to reach the imposed h_{cg}.

In this case, with constant h_{cg}, the PBT↓-PBT↓ combination is more effective than the DT-PBT↓ couple (fig. 5).

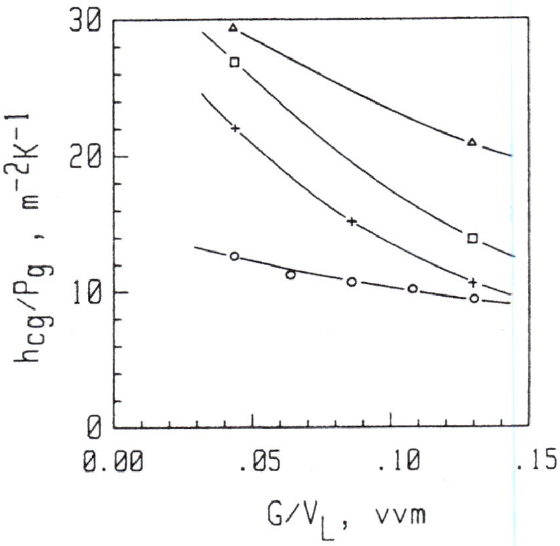

Fig. 4 $E_H = f(G/V_L)$; liquid : guar, N = 0.57 and $\bar{\eta}_a = 0.29$ Pas; DT050-DT033 : Δ ; DT050 : □ ; DT050-PBT050↓ : + ; PBT050↓-PBT050↓ : o ; $n = n_{CD}$

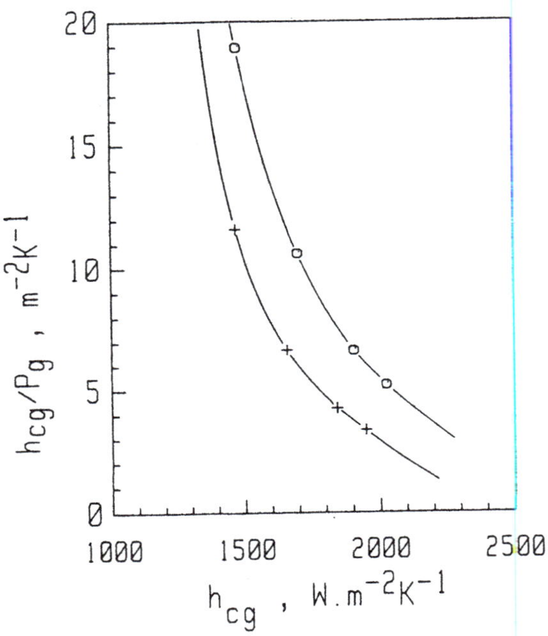

Fig. 5 $E_H = f(h_{cg})$; liquid : N = 0.9 and $\bar{\eta}_a = 0.1$; $G = 0.278 \times 10^{-3} m^3/s$; DT050-PBT050↓ : + ; PBT050↓-PBT050↓ : o

On the other hand, for a given G, if the different impeller combinations give sufficient h_{cg} values at their own CD , efficiency comparisons have to be done at this point. As a matter of fact, the different multiple impeller systems are then in the same hydrodynamic conditions and works in an optimal way which permits more objective comparisons.

According to this last criterion, the whole tested combinations are classified with respect to their efficiencies; the DT050-DT033 system seems to provide the best performances.

E_H(DT050-DT033) > E_H(DT050) > E_H(DT050-PBT050↓) > E_H (PBT050↓-PBT033↓) > E_H(PBT050↓) > E_H(PBT050↑-PBT050↑)

It's to be noted that the double impeller PBT↓-PBT↓, more effective than the DT-PBT↓ combination, becomes, according to this new criterion less effective.

Nethertheless, the $E_H = f(G)$ curves don't have the same die-away factor (fig. 4). This could mean that a classification inversion may occur for high G values.

This agrees with previous comparisons contactors stirred by a single disc pitched-blade turbine (2).

MASS TRANSFER

This study is only outlined since the impeller combinations are compared for a sole gas flow G and a sole liquid. The G effect on k_La has only been obtained for the DT-PBT↓ couple ($k_La \div G^{0.50}$). Contrary to the heat transfer in which the two PBT pumping ways correspond to equivalent performances in the case of mass transfer, the DT-PBT↑ combination leads to higher k_La values than the DT-PBT↓ for an equal Re.

Concerning two impeller couples, k_La estimation correlations as a function of the specific power P_g/V_L and of the gas superficial velocity v_s are proposed for the following experimental conditions : Re : 500-2000, N = 0.9; mean η_a : 0.12 Pa.s. This last parameter is important since k_La highly decreases when the viscosity increases ($k_La \div \eta^{-0.6}$; (3))

-DT039-PBT039↑ combination

$$k_La = 0.00461\ (P_g/V_L)^{0.55} v_s^{0.64} \quad (19)$$

-DT039-DT032 combination

$$k_La = 0.00491\ (P_g/V_L)^{0.48} v_s^{0.61} \quad (20)$$

The P_g/V_L and v_s exponents are coherent with those proposed in the litterature (3, 4). Moreover the k_La values predicted by the second equation are very close to those obtained by Machon et coll (5) for an identical stirring system and for the same liquid viscosity.

By analogy with E_H, a mass transfer efficiency E_M has been defined.

$$E_M = k_La/P_g \quad (21)$$

Like in heat transfer, E_M is maximal in the n_{CD} point and decreases when Re increases for a given G (fig. 6).

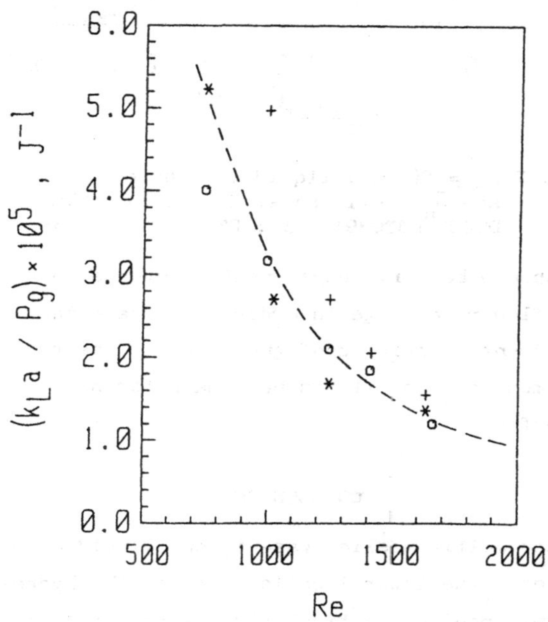

Fig. 6 $k_La/P_g = f(Re)$; liquid : methocel , N = 0.90 and $\bar{\eta}_a$ = 0.12 Pa.s; G = 0.417 m^3/s; DT039-DT032 : ✱ ; DT039-PBT039↑ : o ; DT039 : +

The impeller combinations are to be compared with an imposed k_La and a given G beyond the n_{CD} point.

The DT-PBT↑ couple seems to be lightly more effective than the DT-DT combination (fig. 7). Nethertheless, the differences are very weak and it can be said that the volumetric mass transfer coefficients, k_La, obtained with the different stirring systems are similar for given P_g/V_L and G.

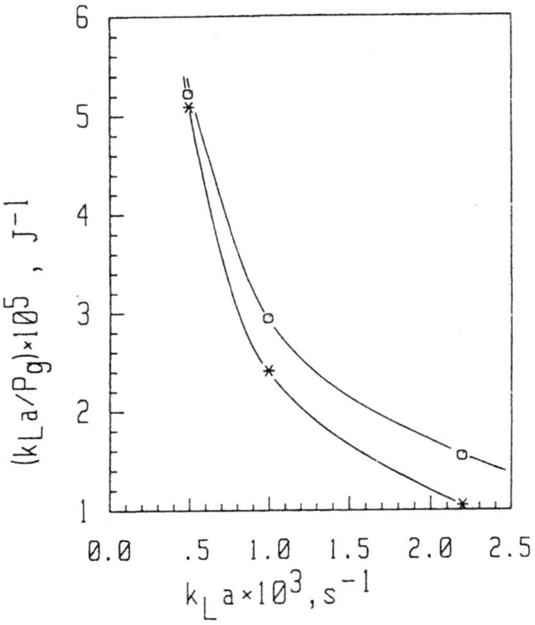

Fig. 7 $E_M = f(k_L a)$; liquid : methocel, N = 0.9 and $\bar{\eta}_a$ = 0.12 Pa.s; G = 0.417 m^3/s; DT039-PBT039↑ : o ; DT039-DT032 : *

It should be worth experimenting over a large gas flow rate range in order to compare the different stirring configurations at their optimums n_{CD}, as it has been made for heat transfer.

CONCLUSION

In the multi-impeller gas-viscous liquid contactors, the lower impeller imposes the hydrodynamic regime, particularly the complete unflooding point n_{CD} which corresponds to the minimum of the P_g/P_o = f(Re) curves.

For the different radial and axial impeller combinations, correlations are proposed for the n_{CD} point calculation and for the P_g, h_{cg} and $k_L a$ prediction in the reactor working zone ($n \geqslant n_{CD}$).

Two new efficiency criterions are proposed $E_H = h_{cg}/P_g$ and $E_M = k_L a/P_g$ in heat and mass transfer.

Whatever the studied impeller combination may be, the E_H and E_M efficiencies are maximal in n_{CD} point, then decrease with the increasing stirring speed at constant G.

A great care must be paid with the multiple impellers performances comparisons which can't always be done in the optimal working point n_{CD}; as a matter of fact some processes impose the h_{cg} or $k_L a$ values. Thus, the best combination in the n_{CD} point doesn't necessarily remain the best with given h_{cg} or $k_L a$.

Nethertheless, this study is only outlined, more particularly in mass transfer; it's then advisable to carry on the experimentation in order to propose more general conclusions which can be valid whatever the installation dimensions may be.

NOMENCLATURE

A	vessel section (m^2)
b	baffle width (m)
c_p	liquid specific heat (J/kg.K)
c_c	clearance of the lower coil turn from vessel bottom (m)
c_o	initial dissolved oxygen concentration (mol/m^3)
c_L	dissolved oxygen concentration (mol/m^3)
c*	equilibrium dissolved oxygen concentration (mol/m^3)
D	impeller diameter (m)
D_c	mean coil diameter (m)
d_c	outside coil tube diameter (m)
DT	disc turbine
e	clearance between the two impellers (m)
E_H	heat transfer efficiency
E_M	mass transfer efficiency
G	gas flow rate (m^3/s)
H_L	height of liquid from bottom of vessel (m)
h	distance of agitator centre from bottom of vessel (m)
h_c	process side heat transfer coefficient for a coil ($W/m^2.K$)
$k_L a$	volumetric mass transfer coefficient (s^{-1})
L	length of static mixer (m)
N	power law index
n	impeller speed (s^{-1})
n_{CD}	complete unflooding speed (s^{-1})
$N_p = P_g/\rho n^3 D^5$	dimensionless power number
$Nu_{cg} = h_{cg} d_c/\lambda$	: Nusselt number
p_c	coil tube pitch (m)
P	mixing power (W)

$Pr = c_p \eta_a / \lambda$: Prandtl number

PBT pitched-blade-turbine

$Re = \rho nD^2/\eta_a$ lowed impeller Reynolds number

t time (s)

T inside diameter of stirred vessel (m)

V_L volume of liquid in vessel (m^3)

v_s = G/A superficial gas velocity (m/s)

$Vi = \eta_a/\eta_{ac}$ correction of Sieder and Tate

w width of impeller blade (m)

z number of blades on impeller

α blade pitch angle to the horizontal

λ thermal conductivity (W/m.K)

η_a apparent viscosity (Pa.s)

ρ density (kg/m^3)

Subscripts

a apparent
c coil
g aerated medium
o non-aerated medium
CD complete unflooding

REFERENCES

(1) Bruxelmane M. and Desplanches H., in Cheremesinoff N.P. (ed) "Handbook Heat & Mass Transfer", pp. 965-1008, Gulf Pub.Cy. Houston, 1986

(2) Gaston-Bonhomme Y., Desplanches H. and Chevalier J.L., "Etudes hydrodynamiques et thermiques de la turbine à 6 pales inclinées à 45° en pompage bas en milieux gaz-liquides newtonien et non newtonien", Proc. 3° Congrès Français Génie Procédés, Compiègne, septembre 1991

(3) Hickman A.D., "Gas-liquid oxygen transfer and scale-up, a novel experimental technique with results for mass transfer in aerated agitated vessels", Proc. 6th Europ. Conf. Mixing, Pavia, may 1988, p. 369

(4) Jurecic R., Berovic M. and coll., "Mass transfer in aerated fermentation broths in a stirred tank reactor", Canadian J. Chem.Eng., 62, 3, p 334, 1984

(5) Machon V., Vlcek J., Hudcova V., "Multi-impeller gas-liquid contactors", Proc. 6th Europ. Conf. Mixing, Pavia, may 1988, p. 351.

SECTION 3: EXPERIMENTAL TECHNIQUES

MEASUREMENT OF CONCENTRATION FLUCTUATIONS DURING MIXING IN A STIRRED TANK AND IN A CHANNEL FLOW BY LASER INDUCED FLUORESCENCE SPECTROSCOPY

C. ANDRE, R. DAVID, J. VILLERMAUX, Laboratoire des Sciences du Génie du Chimique, CNRS-ENSIC/INPL, Nancy
J.N. GENCE, Laboratoire de Mécanique des Fluides et d'Acoustique, Ecole Centrale de Lyon, Ecully, F

ABSTRACT

The mixing of two feed streams in a reactor, one with a fluorescent tracer, the other without, results in a fluctuating concentration field, due to the turbulent flow. Laser induced fluorescence spectroscopy was used. This allows concentration fluctuations to be detected locally in a volume of characteristic size about 30 μm and up to a frequency of 5 kHz (Gaskey et al., 1988 and 1990); the measurement volume can be positioned at will in the tank. The reduced variance β^2 (or intensity of segregation) and the segregation spectra $S_c(f)$ were determined from the concentration fluctuations. The spectra can be used to calculate the macroscales and microscales of the fluctuations.

New experiments were carried out in a 1125 cm^3 continuous stirred tank. The mixing power is supplied by a six-bladed Rushton disc turbine and by the inlet fluid. The values of the two power sources can be adjusted at will by varying the rotation speed and the space time. The effects of stirring speed N, space time τ, fluid viscosity (in water/glycerol mixtures or with polymeric solutions), probe and feed positions were investigated.

The method was also applied to concentration fluctuations in a channel flow with grid turbulence and yielded results in excellent agreement with previous studies (Gence et al. 1989).

INTRODUCTION

Molecular fluorescence spectroscopy is an excellent method for the measurement of local concentrations in continuous flow reactors such as a stirred tank. It is particulary attractive for the study of turbulent mixing where it is necessary to rapidly measure fluctuating concentration at a very small size scale. The absorption and subsequent reemission of photons by a fluorescent dye makes it possible to follow very rapid concentration fluctuations. Moreover, measurements can be carried out without the intrusion of a probe which would disturb the flow pattern. And, by using a laser beam as the excitation source, the measurement can be made in a very small volume; the point of measurement is easily changed by simple adjustments of the optical set-up.

Previous methods for the measurement of concentration fluctuations relied on marker nephelometry (Becker, 1667), microconductivity probes (Gibson and Schwartz, 1963 and Barthole et al., 1982), and classical light absorption (Rao and Brodkey, 1972). The first attempt at using fluorescence spectroscopy in a stirred tank was done by Patterson et al. (1982) with a conventional light source. Papanicolaou and List (1988) used a laser method for the investigation of buoyant jets, but the spatial resolution of their system was about 400 μm. Gaskey et al. (1988, 1990) were the first to reach spatial and temporal resolutions corresponding approximately to the size and time scales of Kolmogoroff microeddies.

DESCRIPTION OF EXPERIMENT

A schematic diagram of the experimental system concerning the stirred tank is shown in fig. 1. It is similar to that used by Gaskey et al. (1988, 1990) with a scale-up factor of $5^{1/3}$ in tank size.

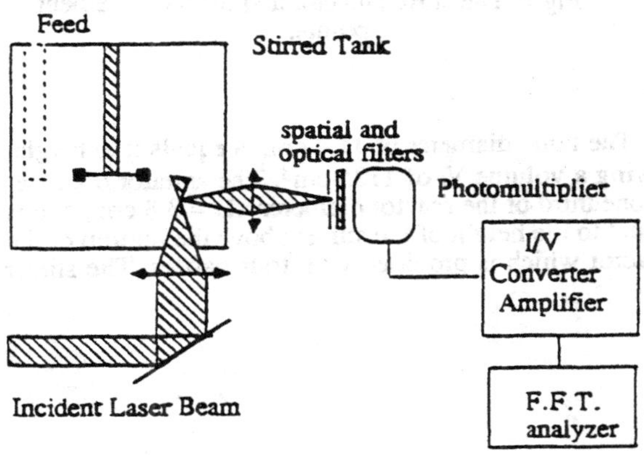

Fig. 1: Experimental set-up in the stirred tank.

This system has three main components: the reactor, the optical system (excitation and detection), and the measurement system. The reactor is a stirred tank having the same geometry as that of Holland and Chapman (1966). This reactor and the positions of the measurement points are shown in fig. 2.

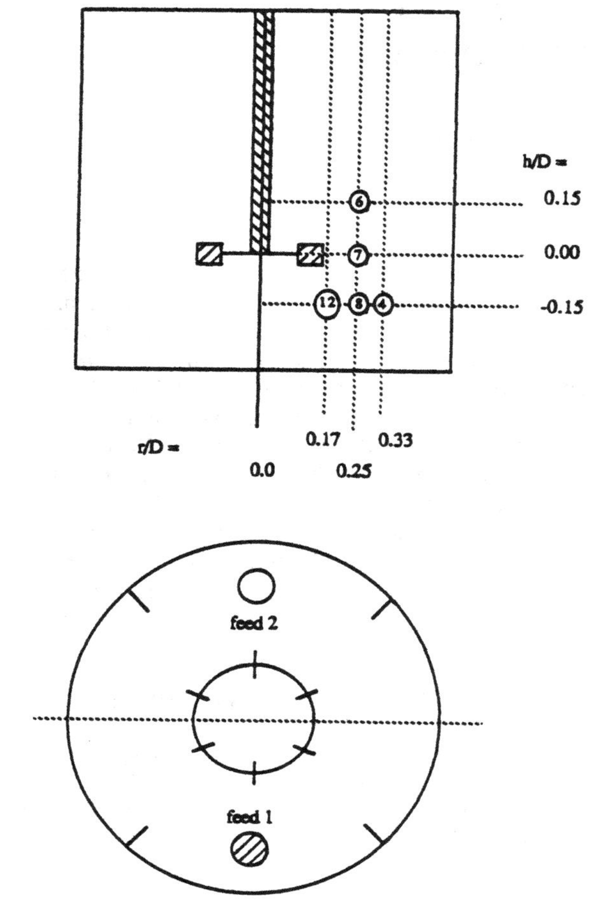

Fig 2: The stirred reactor and the measurement points.

The tank diameter of 11.4 cm, is equals it to height, giving a volume V_r of 1125 cm^3. The agitator diameter is one third of the reactor diameter (D = 3.8 cm) and is equal to the height of the stirrer above the bottom of the reactor which is provided with four baffles. The stirrer is a six bladed Rushton turbine. There are two feed tubes separated by 180°, at 90° from the measurement plane and 45° from the baffles. The tank is fed by a Rhodamine 6 G solution (feed 1) and by pure water (feed 2) with equal flowrates in the base case. Both feed tubes have an internal diameter of d = 4 mm and the liquid is injected at the level of the agitator towards the tank bottom. Two optical quality quartz windows, one in the bottom, the other in the wall, allow the excitation and analysis beams to be precisely focused in the measurement plane. The excitation beam (wave length 488 nm) is passed through a beam expanding telescope, which results in a beam of 3 cm diameter. The expanded beam is reflected by a prism and passed through a lens which focuses it within the reactor.

This measurement method was also tested in simple situations of mixing by turbulence in which it was possible to compare the measurement results and theoretical predictions. A typical example of such a simple case is given by a concentration plume of rhodamine developing from a line-source downstream a grid which generates a quasi-homogeneous and isotropic turbulence in a water-tunnel, as it is schemed on fig.3.

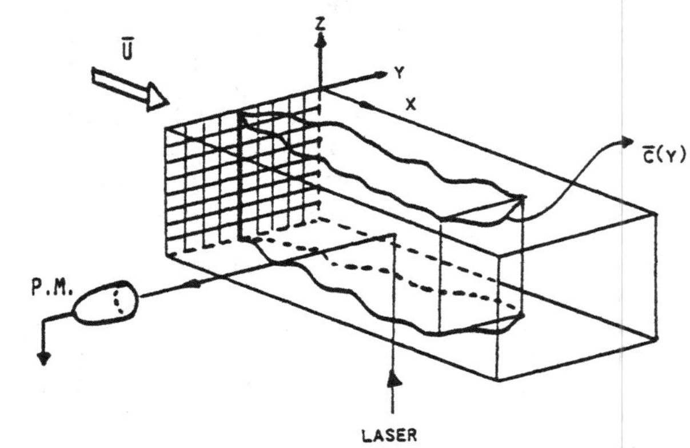

Fig. 3: Experimental set-up in the channel flow.

The measurement volume is defined by the intersection of the incident laser beam with the image of the pinhole of 50 μm. The characteristic size of this volume can be estimated at 30 μm; the focus point can be moved at will in the tank. A high-pass optical filter with cut-off frequency of 515 nm is placed between the pinhole and the photomultiplier to ensure that the excitation light, scattered or reflected by suspended particles, is not measured. The incident and analyzed beams are at an angle of 90° to reduce the intensity of scattered light and achieve maximum spatial selectivity. The measured light intensity, due only to the fluorescence of the sampled volume, is converted to a small current by a photomultiplier, which is then converted to voltage, amplified, filtered and digitized. The data are processed by a F.F.T. analyzer that yields the power spectral density S(f), the average component μ, and the variance σ^2 of the signal.

DESCRIPTION OF A TYPICAL EXPERIMENT IN THE STIRRED TANK.

Two sets of measurements are always carried out. The first with mixing of two fluids: one with a fluorescent dye, the other without. It results a fluctuating concentration field. The second experiment is carried out with one single fluid of constant concentration, equal to the mean concentration of the mixture in the first experiment, all other experimental conditions being the same. This experiment, termed the baseline, yields only the component of the spectrum and variance due to the Poisson process (for more details see Gaskey et al. 1988 and 1990).

Fig.4 shows the power spectral densities we obtained in these two different kinds of experiments.

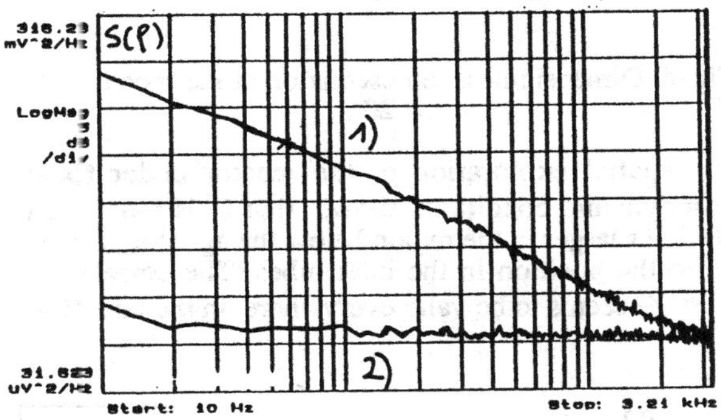

Fig.4: Power spectral density of the signal.
1) with concentration fluctuations
2) baseline

Combining the two sets of measurements we can remove the contribution due to the Poisson process to arrive at the desired values: the reduced variance β^2 and spectrum S_c of the fluctuating concentration, by using

$$S_c(f) = \frac{1}{\mu}\left[\frac{S(f)}{\mu} - \frac{S_b(f)}{\mu_b}\right] \quad (1)$$

and

$$\beta^2 = \frac{1}{\mu}\left[\frac{\sigma^2}{\mu} - \frac{\sigma_b^2}{\mu_b}\right] \quad (2)$$

The subscript b denotes the baseline measurement and the other is the measurement with a fluctuating concentration field. Time macroscales and microscales values can also be obtained:

$$L_t = \frac{\pi S_c(0)}{\beta^2} \quad (3)$$

and

$$\lambda_t^2 = \frac{\beta^2}{2\pi^2 \int_0^\infty f^2 S_c(f)\, df} \quad (4)$$

Corresponding spatial values can be obtained using Taylor's assumption, with the local average velocity U. One obtains

$$L_c = U L_t \quad (5) \qquad \lambda_c = U \lambda_t \quad (6)$$

RESULTS

We have studied the influence of hydrodynamic conditions (stirring speed N (s^{-1}) and space time τ (s)) on S_c and on the reduced variance β^2 at a fixed measurement point (point 8). Fig. 5 shows the effect of N on S_c. The spectra are shown for speeds of 5, 9, 13 s^{-1} and for $\tau = 20$ s. At low frequencies, the fluctuations decrease with increasing stirring speed and consequently β^2 decreases too. At all speeds there is a medium portion of the spectra with a slope of -1, which is an indication of a mixing progress in which viscous effects are important, but diffusive effects are not (viscous-convective subzone). This result agrees with Batchelor's theory (1967).

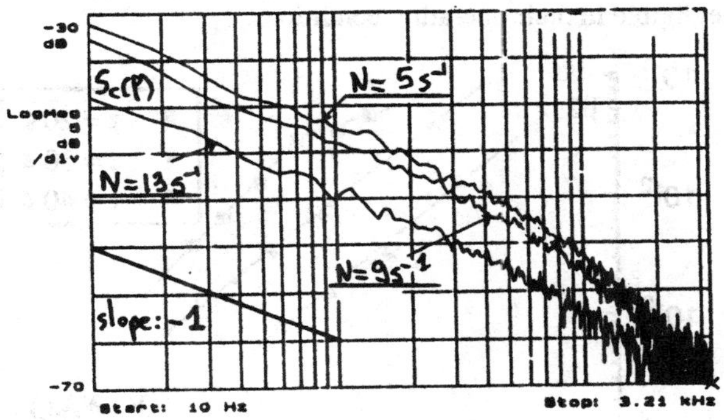

Fig.5: Influence of the stirring speed on the spectra.

Fig. 6 shows the effect of τ on S_c. The spectra are represented for space times of 20, 30, 40 s and for N = 9 s^{-1}. An increase of the space time makes the segregation decrease.

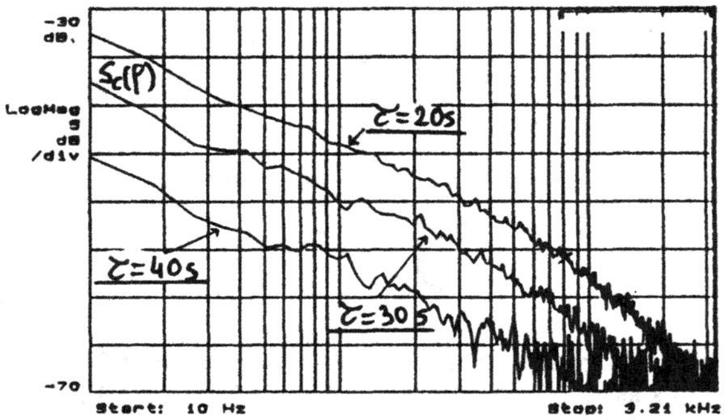

Fig. 6: Influence of the space time on the spectra.

From the values of N and τ we determine the power P (W/kg) given to the fluid per mass unit. At the opposite of Gaskey et al. (1988 and 1990) we must take into account the contribution due to the kinetic energy of the feed streams which is predominant at short space times and low stirring speeds. So, we have:

$$P = a\,N^3 + b\,\tau^{-3} \quad (7)$$

with $a = N_p D^5 / V_r$

N_p is the power number of the agitator; its value is 5 for a six bladed Rushton turbine. This leads to

$a = 3.5\ 10^{-4}\ m^2$ and $b = 2V_r^{\,2} / \pi^2 d^4 = 1002\ m^2$.

The dependence of β^2 on P gives the evolution of the quality of micromixing. We observed $\beta^2 \sim P^{-1.25}$ (fig.7) instead of $\beta^2 \sim P^{-0.67}$ for Gaskey et al. without taking into account the power of the feed streams which was negligible in their operating conditions.

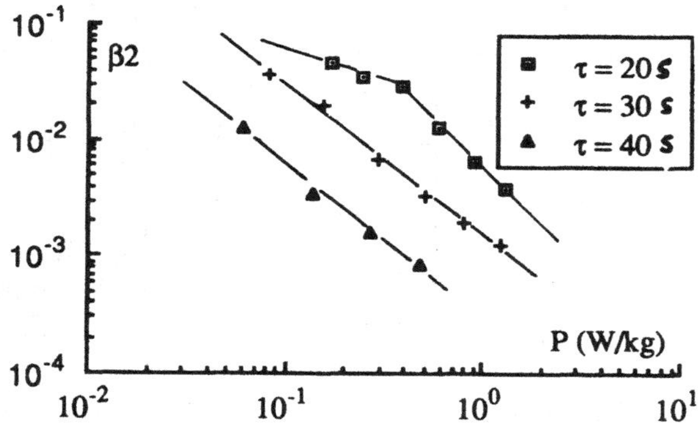

Fig. 7: Reduced variance vs. P

We also obtained an unique dimensionless representation of the spectra in the plane S_cN/β^2 vs. f/N (fig. 8).

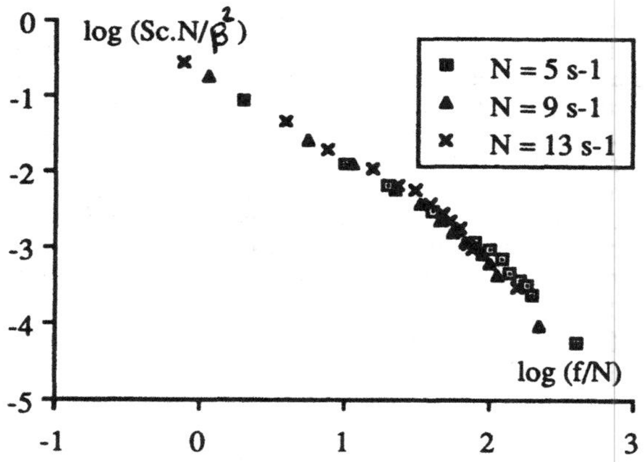

Fig. 8: Dimensionless representation of the spectra for τ = 20 s.

A spatial exploration of the reactor under fixed hydrodynamic conditions shows that β^2 is more than 100 times larger in the region below the agitator: This is due to the position in the inlet tubes. The correlation $\beta^2 \sim P^{-1.25}$ seems to be valid everywhere in the tank (fig. 9).

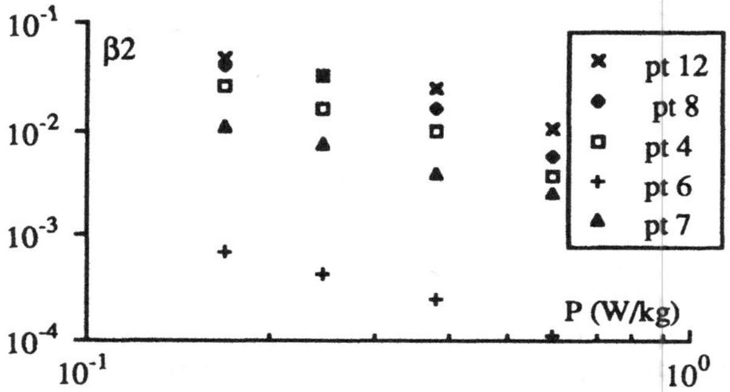

Fig.9: spatial exploration of the tank

But, when changing the measurement plane by turning simultaneously the inlet tubes and the baffles, we observed at point 8 for a space time of 20 s different power laws of β^2 against P. For a rotation of 60°, we observed $\beta^2 \sim P^{-0.48}$ and for 120°, $\beta^2 \sim P^{-0.76}$. A rotation of 180° corresponds to a simple inversion of the position of the two inlet tubes and does not change the observed power law (fig. 10).

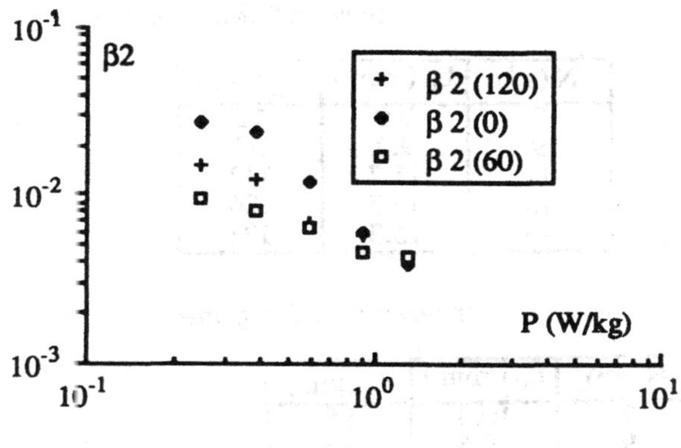

Fig. 10: Influence of the measurement plane position.

The influence of the height of the two inlet tubes was also investigated. We varied this height while keeping it the same for both tubes. We observed that the variance increases when the tubes level is deeper in the reactor. The experiments were made at point 8 with a space time of 20 s and a stirring speed ranging between 5 and 11 s^{-1}. The results are shown in fig. 11 representing the evolution of β^2; the height z of the two tubes is positive when moving the tubes upwards with the plane of agitator as origin.

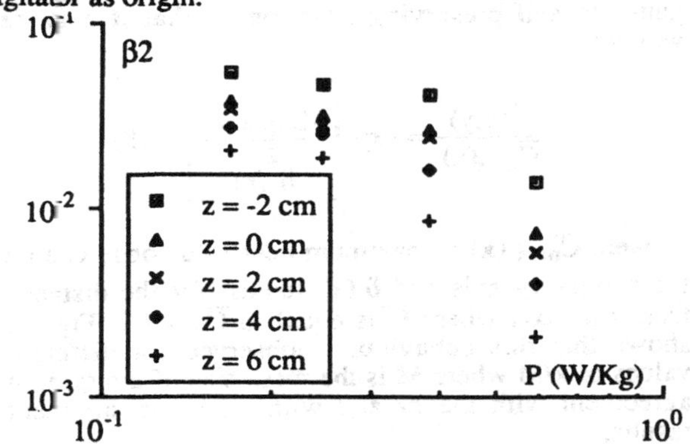

Fig. 11: reduced variance vs. P at different heights z

The ratio of the two inlet flowrates was also varied keeping constant the space time and the mixture concentration. By keeping the space time constant, we fixed the statistical aspect of the mixing process. A simple calculation shows that the kinetic power provided by the feedstreams is minimum for a given space time when the two inlet flowrates are equal. Thus, the concentration fluctuations decrease when unequal flowrates are used. The smallest fluctuations are obtained when injecting with a high flowrate the feed which contains the dye, due to lower injected concentration in that case. Fig. 12 shows the influence of the ratio of the two inlet flowrates on the segregation spectra Sc for a space time $\tau = 20$s and a stirring speed $N = 5\ s^{-1}$.

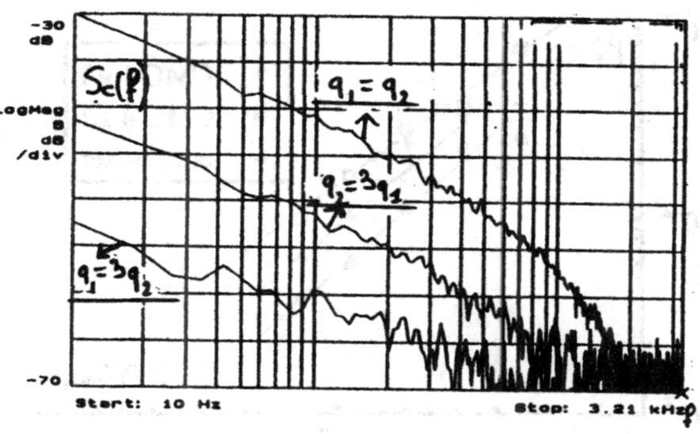

Fig. 12: Influence of the ratio of the inlet flowrates. q_1 is the inlet flowrate of the stream containing the dye and q_2 the one of the other stream.

The viscosity of the liquid was changed. Several kinds of experiments were made. We used water/glycerol mixtures with 38 % glycerol and a temperature of 25° which resulted in a newtonian fluid with a viscosity of 3 cp. Experiments were made at point 8 for $N = 5\ s^{-1}$ and $\tau = 20$ s. The comparison with the experiment in the same hydrodynamic conditions and at the same measurement point agrees with the prediction of Batchelor (1967) concerning the influence of the viscosity on the segregation spectra and the resulting reduced variance: One can note an increase of the fluctuations at low frequencies which correspond to larges structures (fig. 13) and a decrease at high frequencies.

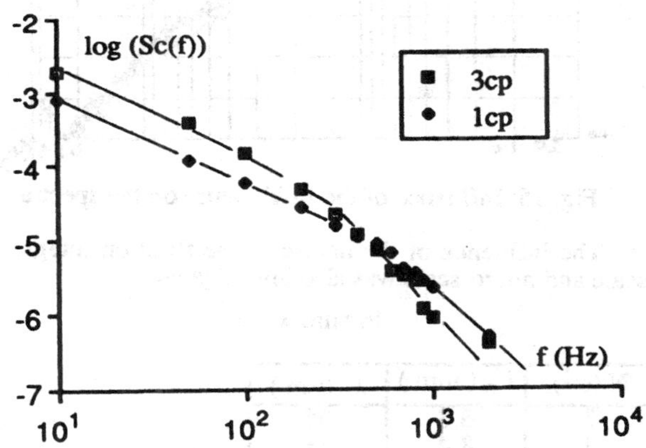

Fig. 13: Influence of the viscosity on the spectra

Other experiments were carried out with water/carboxyl-cellulose mixtures at 1 g/dm^3 and 4 g/dm^3. These mixtures are known for beeing viscoelastic. Experiments were made at the same point (point 8) and space time ($\tau = 20$ s) and with a stirring speed varying between 7 s^{-1} and 15 s^{-1}. The power law $\beta^2 \sim P^{-1.25}$ was always observed (fig. 14).

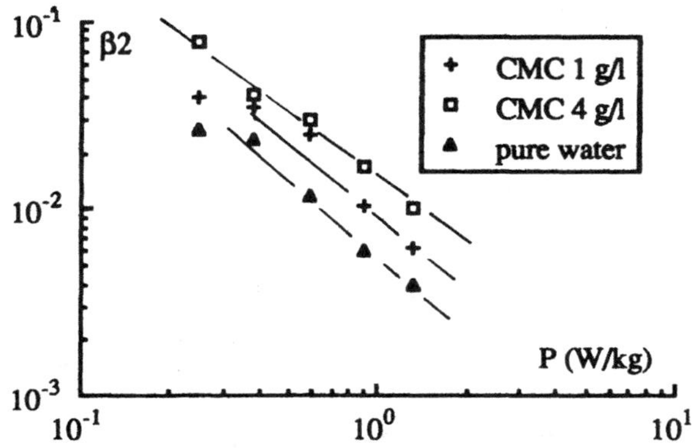

Fig.14: Influence of the fluid nature on the reduced variance.

Fig.15 shows the comparison between these experiments and the one made in pure water under identical conditions.Only the low frequency region of the spectra is influenced by CMC addition.

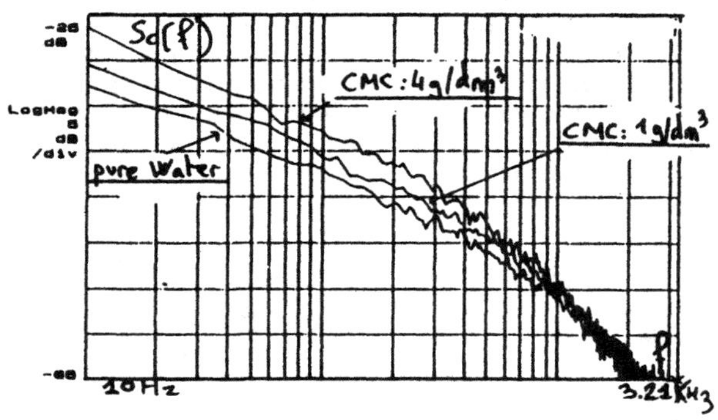

Fig. 15: Influence of the fluid nature on the spectra

The influence of the nature of the fluid on integral scale and micro-scale was also investigated.

In pure water

N (s^{-1})	L_c (mm)	λ_c (μm)
7	6.8	65
9	8.4	73
11	10.5	89
13	13.2	110
15	15.2	119

One may remark that L_c varies from about the tube diameter up to the characteristic size of the paddle when N increases. These suggest that, in general, concentration fluctuations are generated by the feed inlet tube at low stirring speeds, but at high velocities, L_c increases, indicating that the mode of creation of concentration fluctuations is no more controlled by the inlet tube.

In water/C.M.C 1 g/dm³

N (s^{-1})	L_c (mm)	λ_c (μm)
7	7.9	72
9	9.2	90
11	11.7	112
13	15.3	126
15	15.3	123

In water/C.M.C 4 g/dm³

N (s^{-1})	L_c (mm)	λ_c (μm)
7	10.3	95
9	10.5	96
11	12.4	119
13	15.3	126
15	15.3	123

The influence of stirring speed on scales seems to be smoothed out by increasing CMC concentrations.

Concerning the experiments made in the channel flow, some simple well-known theoretical considerations indicate that very near the source and far from it , the mean concentration field $\overline{C}$ (x,y) is a gaussian self-preserving function so that it may be written:

$$\frac{\overline{C}(x,y)}{\overline{C}_{max}(x)} = \exp\left[\frac{-\text{Log}\, 2\, y^2}{\delta^2(x)}\right] \quad (8)$$

were $\overline{C}_{max}$ (x) is maximum near value observed on the symmetry axis and δ (x) stands for the distance from this axis where $\overline{C}$ is equal to $\overline{C}_{max}$ / 2. Fig. 16 shows that this behaviour is observed for different values of x/M where M is the mesh size of the grid, in agreement with theory and with other measurement results.

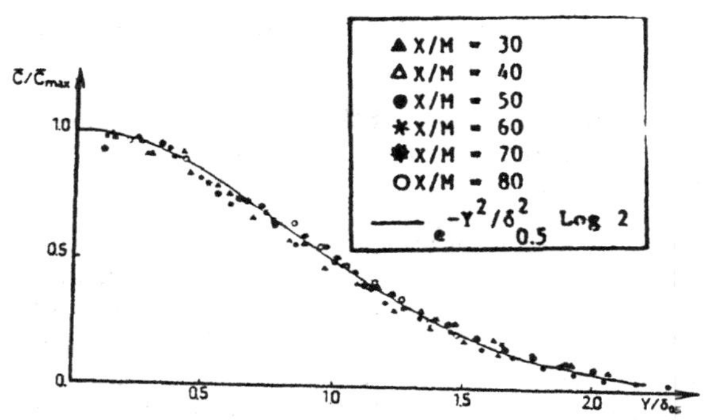

Fig. 16: Evolution of the mean concentration field.

Such a test is of course not very severe since the mean values C are influenced by low frequencies of low frequencies of the concentration fluctuations.

As the main advantage of this method is its high frequency response, the time frequency spectra S_c (f) of the concentration fluctuations was also measured on the plume axis up to 5 kHz as shown in fig. 17.

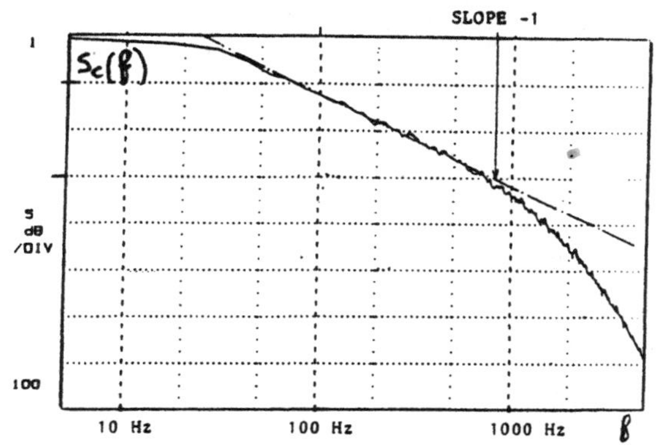

Fig. 17: Concentration fluctuations spectra in the channel flow.

It is well-known from Batchelor theory that when the Schmidt number is greater than 1, as in our situation, the spectra of concentration fluctuations has to exhibit a law of the form S_c (f) ~ f^{-1} for a well-defined range of the time frequency f. This range corresponds to frequency associated with spatial length scales flowing through the measurement volume and which are smaller than the dissipation scales of the velocity field but are large enough to remain unaffected by the molecular diffusivity. In our experiment, the lower and upper bounds of that range are respectively 100 Hz and 1000 Hz. This typical behaviour is well observed so that this measurement method is quite convenient in that frequency range. Some other experiments indicate that the method may be used up to 8 kHz so that the major part of the spectrum is observed.

CONCLUSION

These results clearly demonstrate that fluorescence spectroscopy can be a useful tool in the study of turbulent mixing. The method gives frequency resolution up to 5 kHz, which represents more than one order of magnitude improvement over previously reported methods. Similarly, the characteristic size of the measurements, 30 μm, is more than one order of magnitude smaller than previously reported measurements. If we consider that such spatial and temporal resolutions are reached without an intrusive probe, it is clear that the method offers an advance in the study of turbulent mixing.

As a demonstration of the method, we have presented some results from our study of turbulent mixing in a continuous stirred tank and in a channel flow with grid turbulence.

The experiments in the channel flow were very useful because such an experimental set-up is simple enough that it allows a easy comparison between experimental results and theoretical predictions. This test was so successful that it allowed us to use this method in the study of the turbulent mixing in a stirred tank which is a more complex experimental set-up.

In the stirred tank, measurements of the spectra up to a frequency of 800 Hz are sufficient for the calculation of the intensity of segregation β^2. Using the spectra, we also estimated the integral length scale L_c. The upper limiting frequency is large enough that we find the beginning of the viscous-diffusive subzone, which is the region where the concentration fluctuations rapidly decay. This allows the estimation of Corrsin microscale λ_c especially at low stirring speeds, Gaskey et al. (1990) were the first to determine this value in a stirred tank by using this new method.

AKNOWLEDGMENTS

The authors thank Dr.Jean-Claude ANDRE for useful advices and helpful discussion.

REFERENCES

Barthole J.P., Maisonneuve J., Gence J.N., David R., Mathieu J., Villermaux J., 1984, Measurement of mass transfer rates, velocity and concentration fluctuations in an industrial stirred tank, Chem. Eng. Fundam., 1, 17-26.

Batchelor G.K., 1967, The theory of homogeneous turbulence, Cambridge Unisersity Press.

Becker H.A., 1967, On the light scatter technique for the study of turbulence and mixing, J.F.Mech., 30, 259-284.

Gaskey S., Vacus P., David R., André J.C., Villermaux J., 1988, Investigation of concentration fluctuations in a continuous stirred tank by space resolved fluorescence spectroscopy, Proc. 6th European Conference on Mixing, Pavia, Italy, AIDIC Ed.,129-136.

Gaskey S., Vacus P., David R., André J.C., Villermaux J., 1990, A method for the study of turbulent mixing using fluorescence spectroscopy, Exp. in Fluids, 9, 137-147.

Gibson C., Schwartz W.H., 1963, Detection of conductivity fluctuations in a turbulent flow field, J.F.Mech., 16, 357-364.

Holland F.A., Chapman F.S., 1966, Liquid mixing and processing in stirred tanks, New-York, Reinhold Publ. Co., pp. 11-14.

Papanicolaou P.P., List E.J., 1988, Investigations of round vertical turbulent buoyant jets, J.F.Mech., 195, 341-391.

Patterson G.K., Bockelmann W.,Quigley J., 1982, Measurements of mixing effects on local conversion in stirred tanks, Proc. 4th European Conference on Mixing, Cranfield, U.K., BHRA Ed., J1, pp.303-313.

Rao M.A., Brodkey R.S., 1972, Continuous flow stirred tank turbulence parameters in the impeller stream, Chem. Engng. Sci., 27, 137-156.

Vacus P., Gaskey S., David R., Villermaux J., Lievre J., Bennani A., Gence J.N., 1989, A method for the study of turbulent mixing, using fluorescence spectroscopy, Euromech 253: Concentration fluctuations in turbulent diffusion, London, 30.August 1.September.

Nomenclature

$\overline{C}$	average concentration, mol.m^{-3}
$\overline{C}_{max}$	average concentration, mol.m^{-3}
D	reactor diameter, m
d	internal diameter of the inlet tubes, m
f	frequency, s^{-1}
h	height, m
L_c	spatial macroscale, m
L_t	temporal macroscale, s
M	mesh size of the grid, m
N	stirring speed, s^{-1}
N_p	power number of the agitator, (-)
P	Power delivered to the fluid per mass unit, W.kg^{-1}
q_1	inlet flowrate of the stream with the dye, m^3.s^{-1}
q_2	inlet flowrate of the other stream, m^3.s^{-1}
r	distance to stirrer axis, m
S(f)	power spectral density, s
S_b(f)	baseline power spectral density, s
S_c(f)	segregation spectrum, s
V_r	reactor volume, m ^{3}
U	average velocity, m.s^{-1}
x,y	spatial coordinates, m
z	position of the inlet tube, m

Greek symbols

β^2	reduced concentration variance, (-)
λ_c	spatial microscale, m
λ_t	temporal microscale, s
μ	signal average, s^{-1}
μ_b	baseline average, s^{-1}
σ^2	signal variance, s^{-2}
$\sigma_b{}^2$	baseline variance, s^{-2}
τ	space time, s

REYNOLDS STRESSES AND STIRRING REGIMES OF A RUSHTON TURBINE

M. MAHOUAST, LEMTA-CNRS-INPL, Nancy, F

INTRODUCTION

It is now well established that the mean velocities and the rms values of the velocity fluctuations in a tank stirred by a Rushton turbine are proportional to the rotational speed N of the agitator and to the size of the apparatus[1]. The shear rates vary in the same way. This result is of a great interest for the scaling up of reactors, but has also consequential effects in many mixing processes such as fermentation , emulsification, dispersion (gas-liquid , liquid-liquid), etc...

However, data about the Reynolds stresses are scarce[5,10]. As the turbulence strongly deviates from homogeneity and isotropy it is necessary to determine them and moreover to know the influence of the stirring speed on these quantities. The results may be useful for the modeling of turbulent mixing in stirred tanks.

In the present work, the Reynolds stress $\overline{v'_r v'_\theta}$ is measured by means of a two-component laser-Doppler-velocimeter in a 6.3 dm^3 standard tank stirred by a six-flat-blade Rushton turbine.The rotational speed N is varied up to 7 s^{-1}. Measurements are carried out in the vicinity of the blade. The results show that two stirring regimes exist according to the evolutions of the Reynolds stress versus N. And although the corresponding curves $\overline{v'_r v'_\theta}$ versus N^2 are overlying beyond a critical value N_c of the stirring speed, the two regimes can be still distinguished by the respective time microscales of the radial velocity fluctuations. But prediction about the operating regime seems impossible as it establishes itself as soon as the turbine starts to rotate.

APPARATUS AND EXPERIMENTAL TECHNIQUE

MIXING VESSEL

The standard reactor of the Holland and Chapman type[4] comprises a vessel with four baffles and a stirrer. These parts are made of a transparent material (PMMA). Their dimensions and the position of the agitator are determined from the knowledge of the tank diameter (Fig.1).
The cylindrical tank, whose base is circular and flat, is machined in a cube. The four baffles, vertical and flush with the wall, make an integral assembly which can be angularly positioned around the vessel axis, making it easier to measure. The bottom is provided by an optical window for the passage of laser beams reflected by a plane mirror fitted under

the reactor. The vessel is filled with water at a 25 °C temperature. On the free surface of the liquid, a circular optical window can be displaced through articulations and allows a good focusing of the photomultiplier optics on the measuring point.

The agitator (Rushton turbine composed of a disc with six rectangular blades) is coupled with a variable speed motor set above the tank on a frame through shock absorbers. The entire facility (tank, frame + motor) is on a 3D displacement table (precision: 0.025 mm) employed for positioning the measuring volume.

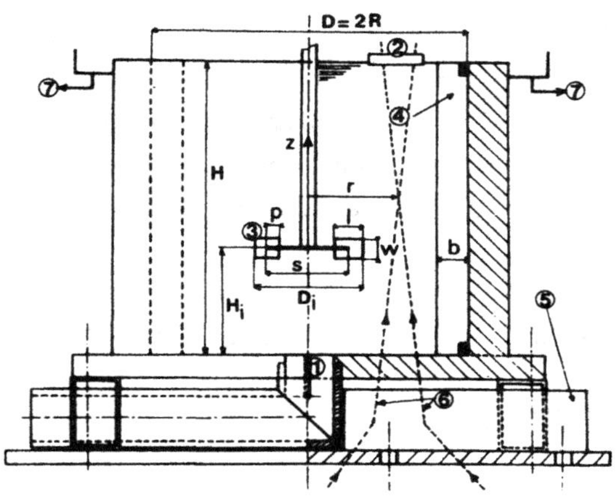

Fig.1 Mixing vessel; 1 Injector, 2 Optical window, 3 Rushton turbine, 4 Baffle, 5 Mirror, 6 Laser beams, 7 Liquid evacuation. In the present work, as the reactor runs in batch mode, the supplies of fresh water and salt water shown on the figure are not used. Only two laser beams, not three, are drawn.

Cylindrical (r,θ,z) coordinates are used to locate the measuring point. The θ angle is oriented in the direction of the rotation of the turbine, the origin corresponding to a baffle. The origin of the vertical z axis is the center of the turbine disc.

2D-LDV WITH FREQUENCY SHIFTING

The method of the laser Doppler velocimetry will not be discussed here, since it has been detailed in many books. The two-component LDV used to measure the Reynolds stresses in this work will be now shortly described.

The light source is an Argon laser. The power of the emitted beam (blue-green) can be regulated up to 5 W. This beam is splitted by a modular optical system into three parallel beams equidistant to the optical axis: a blue one B (488 nm), a green one G (515 nm), and the blue-green (BG). B and G beams are 40 MHz in frequency shifted with respect to the blue and green rays of the BG beam, by means of a Bragg cell, in order to determine measured velocity components. The plane of B and BG beams is perpendicular to the plane of G and BG beams. At the focus of a converging lens having the same optical axis as the system, these three beams give, at their intersection, two perpendicular networks of blue fringes and green fringes. Owing to the frequency shifting, each network moves at a constant velocity normal to the fringes. Receipt of forward scattered lights (Doppler signals) emitted by particles and moving through this measuring volume is made by a photomultiplier placed above the tank.

The receiving optics comprise a color separator and two interference narrow band filters (blue and green). Then, each signal is processed through a frequency mixing unit and through a frequency tracker. In the latter case, an analog output is available.

The signals are then numerically converted and processed to obtain any usual statistical function and to analyze the spectra. The frequency sampling was 2 KHz. At a given measuring point, 60 consecutive samples of 2,048 points were recorded on a disk, simultaneously for the two velocity components. Processing was most time deferred to compute the following quantities and functions as: mean values of the velocities, root-mean-square values of the fluctuations, Reynolds stresses, coefficients of auto and cross correlations, spectra and cross spectra, macro- and micro-scales of the turbulence.

Before carrying out the measurements in the vessel, the LDV has been proofed (value and sign of the velocity) in a known flow field: the solid rotation of water in a cylinder made of PMMA. So, accuracy on the velocity has been found to be lower than 1 %.

Demineralized water was used in the mixing tank to reduce the presence of natural particles, and then a progressive seeding of the liquid by silicon oil particles (1-3 µm) was made until best-quality Doppler signals were obtained.

EXPERIMENTAL RESULTS

STIRRING REGIMES AND REYNOLDS STRESSES

Preliminary measurements in the discharge flow showed that the dimensionless Reynolds stress $\overline{v'_r v'_\theta}/(\pi N D_i)^2$ is very sensitive to the stirring speed N near the blade, and much less influenced by N a little far from the blade (Fig.2).

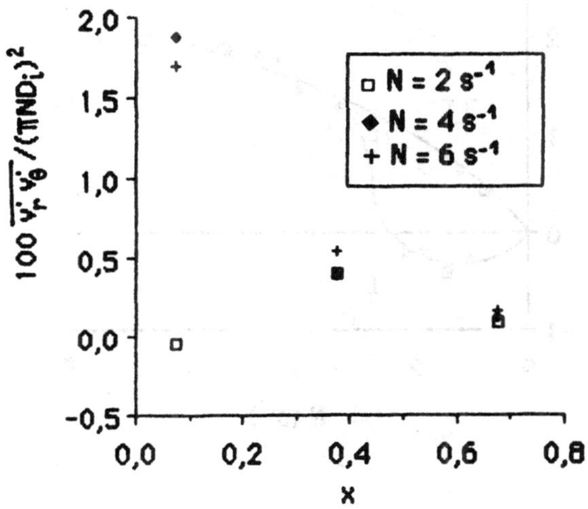

Fig.2 Influence of the stirring speed on the Reynolds stress $\overline{v'_r v'_\theta}$ along the discharge flow (z = 0, θ = 45°).

So, a systematic investigation of the behaviour of this stress near the blade was needed. Measurements have been then carried out at the location ($r-D_i/2$ = 5 mm, θ = 45° ,z = 0).

Results are plotted in Fig.3. They have been obtained from three independent runs:

- the first experiment is related as curve I,
- the second experiment corresponds to curve II,
- the third experiment has given exclusively points sometimes on curve I, sometimes on curve II.

In the first two experiments, N was progressively increased, whereas in the third one, which was done in order to insert new measurements or remake some, the values of N were not arranged in definite order.

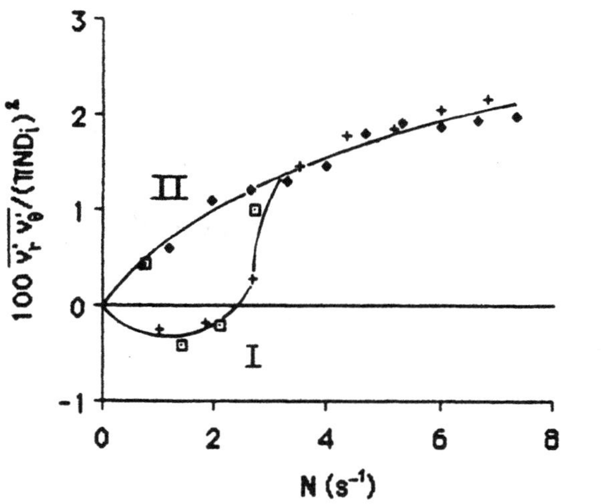

Fig.3 Evolutions of $\overline{v'_r v'_\theta}$ vs N at point ($r-D_i/2$ = 5 mm, θ = 45°, z= 0).

The results point out two possible behaviours for the Reynolds stress at stirring speeds lower than a critical value N_c = 3.3 s^{-1} , which corresponds to the Reynolds number R_{ec} = 17,000. In regime I, negative values of the stress are observed, then there is a sudden increase of this quantity towards positive large values at about N=2.5 s^{-1}. In regime II, the stress is always positive and continuously increasing. For N greater than N_c, a single curve is obtained for the two regimes, showing that $\overline{v'_r v'_\theta}/(\pi N D_i)^2$ tends to the constant value 0.02.

Notwithstanding, the regimes can be still distinguished by the respective time microscales of the radial velocity fluctuations, as it will be seen hereafter.

TIME MICROSCALES ACCORDING TO THE REGIMES

Fig.4 shows the Taylor's time microscales of the radial and tangential velocity fluctuations. Each scale is computed from the slope at the origin of the curve representing the function:

$$\sqrt{1-R_{Ei}(\tau)} = \tau/\tau_{Ei} + O[(\tau/\tau_{Ei})^3] \qquad (1)$$

where $R_{Ei}(\tau)$ is the auto-correlation coefficient of the v'_i fluctuation.

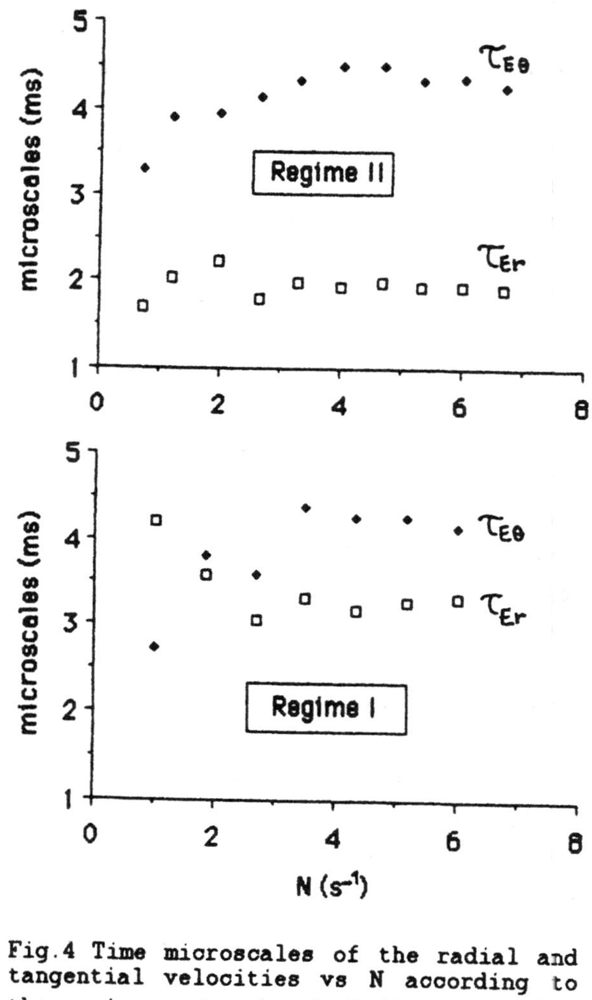

Fig.4 Time microscales of the radial and tangential velocities vs N according to the regimes, at point ($r-D_i/2$ = 5 mm, θ = 45°, z = 0).

For N > Nc, it can be seen that these scales are nearly constant. The tangential time microscale keeps the same magnitude in the two regimes, whereas the radial one is almost twice smaller in regime II than in regime I. Now, as these scales represent characteristic times of the dissipation of turbulent kinetic energy, this mechanism should not be affected by the stirring speed N. On the other hand, the time macroscales decrease when N increases, as showed by Mahouast[8]. This last result can be linked to the decrease of the mixing (or circulation time) proportionally to N^{-1} (Nagata[9]). In other words, the turbulence at large scales is affected by the rotational speed of the turbine, and does not depend on it at small scales.

DISCUSSION

Warmoeskerken and Smith[12] made a detailed study of the flow around the stirrer in a gassed system. They distinguished three stable stirring regimes according to the shapes kinds of the gas cavities behind the blades: vortex, clinging and large cavities. All these cavities owe their primary existence to the presence in the wake of the blades of trailing vortices that have been analyzed in single phase flow by several authors[3,11]. The two forementioned stirring regimes, according to the behaviour of the Reynolds stresses, have probably an influence on the formation of the different gas cavities observed in gas-liquid system. May be that in single phase flow, the two regimes correspond to different arrangements of the trailing vortices. Either regime I or regime II is established as soon as the turbine starts to rotate. Prediction about the operating regime seems impossible. In addition, when referring to the existence of periodic components (frequency of blade passage and harmonics) in the velocity fluctuations[3,6], the question of a chaotic nature of the system arises.

This work must be completed by visualizations of the trailing vortices and 2D-LDV measurements, in which N will be increased as before, but then decreased progressively. Besides, the other Reynolds stresses $\overline{v'_r v'_z}$ and $\overline{v'_\theta v'_z}$ must be also studied for a better analysis of the stirring regimes around the agitator.

ACKNOWLEDGMENT

This work was supported by the Interdisciplinary Program for Energy and Material Saving (PIRSEM) of the Centre National de la Recherche Scientifique (CNRS), France.

LIST OF SYMBOLS

D	=vessel diameter
D_i	=turbine diameter
H=D	=vessel height
$H_i=D_i$	=turbine clearance above the base
N	=stirring speed
N_c	=critical value of N
R=D/2	=vessel radius
$R_e=ND_i^2/\nu$	=Reynolds number
R_{ec}	=critical value of R_e
r	=radial coordinate
$R_{Ei}(\tau)$	=autocorrelation coefficient of the v'_i fluctuation
$x=(r-D_i/2)/D_i$	=dimensionless coordinate
z	=axial coordinate

Greek letters

ν =kinematic viscosity

θ =angular coordinate

τ =time

T_{Ei} =time macroscale of the v'_i fluctuation

Subscript

i =index for r,θ,z

Superscript

— =time averaging

REFERENCES

[1] Costes,J., and J.P. Couderc, "Pumping capacity and turbulence intensity in baffled stirred tanks; influence of the size of the pilot unit", Proc. 4th European Conference on Mixing, BHRA Fluid Engineering, Noordwijkerhout, Netherlands 1982, p.5.

[2] Goldstein, R.J., "Fluid Mechanics measurements", Hemisphere Publishing Corporation, 1983, p.155.

[3] Günkel, F.A., and M.E. Weber,"Flow phenomena in stirred tanks", AIChE J., **21**,n° 5, 1975, p.931.

[4] Holland, F.A., and M.E. Weber, "Liquid mixing and processing in stirred tanks", Reinhold Publishing Corporation, New-York, 1966, p.11.

[5] Ito, S., K. Ogawa, and N. Yoshida, "Turbulence in impeller stream in a stirred vessel", J. of Chem. Eng. of Japan, **8**,n°3, 1975.

[6] Mahouast, M., and G. Cognet, "Analysis of turbulent mixing in a CFSTR from LDV measurements", 3rd ASME International Symposium on Laser Anemometry, Boston, FED, **55**, 1987, p. 127.

[7] Mahouast, M., R. David, and G. Cognet, "Periodic phenomena generated in the discharge flow by a Rushton turbine and investigation of the Reynolds stresses", Proc. 6th European Conference on mixing, BHRA Fluid Engineering, Pavia, Italy, 1988, p.23.

[8] Mahouast, M., "Etude du mélange en cuve agitée par analyse des champs turbulent et de concentration. VLD mono et bidimensionnelle, conductimètrie", Ph.D. thesis, INPL, France, 1988.

[9] Nagata, S., "Mixing, principles and applications", Kodansha LTD Tokyo, Halsted Press New-York, 1975.

[10] Mahouast, M., R. David, and G. Cognet,"Two-component LDV measurements in a stirred tank", AIChE J., 35, 1989, p.1770-1778.

[11] Van't Riet, K., and J.M. Smith, "The hydrodynamics of Rushton turbine agitator blades", Proc. 1st European Conference on Mixing and Centrifugal separation", BHRA Fluid Engineering, Cambridge, England, 1974, paper B2, p.17.

[12] Warmoeskerken, M.M.C.G., and J.M. Smith, "Description of the powercurves of turbines stirred gas-liquid dispersions", Proc. 4th European Conference on Mixing, BHRA Fluid Engineering, Noordwijkrhout, Netherlands, 1982, p.237.

MICROMIXING BEHAVIOUR IN A BATCH VESSEL WITH SEVERAL IMPELLERS USING ALKALI-FLUORESCENT DYE SYSTEM

Y. HIROFUJI, T.INOUE, Y. NAGASE, Hiroshima University, Dept. of Chemical Engineering, J

ABSTRACT

A series of the experiments is planning to achieve a better mixing in an incomplete baffled and tall vessel with a modified impeller, which is an object of the industrial engineers. This work is the first one and is planned to establish the method of the measurement of the micromixing through examinations of well known combinations of a complete baffled vessel with the disc turbine(DT), marine propeller(PM). An incomplete baffled vessel with the inclined fan turbine(IFT) was also examined. A method of the continuous injection of the NaOH solution into the fluorescent dye solution is presented. Degree of the dispersion of the solute into the molecular level in the impeller region is very delayed and approximately depends on the power consumption per unit volume. Homogenization with the PM is performed at lower power input compared with that with the DT. The IFT is not preferable to apply for the tall vessel. However, it is found that a small baffle near liquid surface could promote the local micromixing near it.

INTRODUCTION

The standard configuration in most of the previously reported mixing researches is a disc turbine or an inclined fan turbine in a cylindrical vessel with a complete baffled condition and with liquid height equal to the vessel diameter, for example.[1,2,3,4,5] However, industrial engineers many times have to design tall vessels (H_L>1.5T) with an incomplete baffled condition and with a peculiarly modified impeller to achieve an economic mixing for a given reaction. Its configuration is quite different from the standard one and includes almost infinite number of variations.

Researches are needed for these industrial configurations parallel with basic researches for the standard one. Since there are many variations of vessel and impeller combinations, how to develop a timesaving but physically sound way of research to characterize mixing of each of these combinations. The method presented here is rather an orthodox one and its objective is a problem to give a mixing map in a given vessel through the local measurement of concentration variation at various points. Local concentration measurements are not so often applied in the mixing researches since concentration fluctuation detected at an arbitrarily chosen point in a batch vessel represents the result of repeated mixing during recirculation of the tracer material along the circulation loop, and therefore, the difference in concentration fluctuation between two arbitrarily selected points is relatively small or minor in a well mixed vessel. This fact presents a physical background for the usual mixing measurements, that is, two- point measurement at the inlet and outlet (continuous flow system[6]) or at opposite corners (mixing time concept[7])). However, characteristics of concentration fluctuations at different points become diverse in a weak mixing vessel and the difference in fluctuations represents the progress of mixing in the given vessel. Thus, local measurement of mixing is advantageous to recognize mixing in the weak mixing system. It is more advantageous of macro as well as micromixing measurements, because (it is well known that) high concentration means little mixing in the former measurement and good mixing in the molecular level in the latter. Both the measurement method are develop and measurements are made for several vessel-impeller combinations by the authors. This paper details the micromixing, whereas the detail of macromixing is given in Ref.8.

MEASUREMENT SYSTEM

Details of micromixing measurement developed by the authors are given in Ref.9. The method is briefed here for better understanding.

One of the fluorescent dyes, 8-hydroxy 1,3,6-pyrene trisulfonic acid sodium salt(HOPSA), is sensitive to hydroxide ion(OH^-) and gets excited from pH=6.5 to 9.5 with increase in fluorescence as pH increases.[10] Important properties are that HOPSA excitation spectrum range is quite different from its fluorescence spectrum range and fluorescence intensity is high. Hence, one could easily separate and detect the fluorescence with a good resolution through an optical band pass filter.

Micromixing measurement was made in the pH range of 7.7 to 8.3, which is equivalent to 5.0×10^{-7} to 2.0×10^{-6} $kmol/m^3$ of OH^- concentration, by injecting a small amount of sodium hydroxide solution, pH=12.5 (3.2×10^{-2} $kmol/m^3$), into the weak alkali solution (pH=7.7) well mixed with HOPSA. Fluorescence intensity is given in Fig.1. The reason for the selection of OH^- concentration range in this experiment will be discussed later.

The measurement system is shown in Fig.2. Excitation light from a Xe light source passes through an optical filter of band width 360nm to 400nm. Excitation light was focused to a light guide tube made of quartz, and illuminates the area around the measuring point. Fluorescence is detected by a photo-multiplier after filtering with another optical band pass filter centered at 510nm with 40nm band width. Output of photo-multiplier was amplified after current-voltage conversion, and saved to the personal computer with sampling frequency 1kHz. The number of data sampled is 160,000 per run. Measurement area is decided by a pin-hole section 2mm diameter at a measurement point.

The use of light guide tube disturbs the flow field. However, it is unavoidable since measurements have to carried out at various positions in the vessel.

CONTINUOUS INJECTION OF TRACER SOLUTION IN A BATCH VESSEL

The authors had pointed out that the instantaneous flow profile in the vessel with a propeller or a flat blade fan turbine was not axisymmetric but are of the unstable and fluctuates very much.[11,12] Its space and time scale are of the order of the size of the vessel, and second to minute, respectively. It is believed, but difficult to confirm, that this unstable flow phenomena has link with the generation of the largest turbulent eddy in the vessel. If it is true, unstable flow phenomena are expected more or less in the vessel with any type of impellers. These phenomena are confirmed later in this paper. Then, mixing experiment should be carried out continuously even in a batch vessel, since a number of repeated experiments are necessary to deduce an average behaviour of mixing in the unstable flow when the delta response method is applied. Continuous flow method which is usually applied in the mixing research should also be avoided, since the fixed position of inlet and outlet are not preferable for the purpose of local measurement of mixing.

The principle proposed here for a quasi-steady state measurement of the continuous injection of the high concentration tracer solution in a batch vessel is as follows.

Turbulent mixing in a large eddy which moves around with the flow is schematically shown in Fig.3 by a broken circle. A large blob in the eddy of which concentration is the same as that of the initial value, C_i, (Fig.3(a)) is stretched and teared out into smaller ones together with molecular diffusion from the surface of the blob (Fig.3(b)-(c)). At the final stage of turbulent mixing, there are many Batchelor scale eddies and molecules surrounding these eddies or clouds of molecules prior to uniform dispersion within molecules which are already dispersed uniformly (Fig.3(c)). After that, mixing reaches equilibrium (Fig.3(d)).

Assume that the scale of measuring space, r_C, is larger than the Batchelor scale, say circle in Fig.3. Then the average concentration, C, detected over the circle increases with the progress in micromixing until state (c) since concentrated clouds are still present. After that, C decreases to C_∞ at the completely mixed state. Higher concentration detected in the micromixing measurement corresponds to the state (c).

Let the molecular clouds change to Batchelor eddies before molecular diffusion occurs and let the total volume of these Batchelor eddies be V_i. Let the volume of measurement space be V and the concentration of molecules which are already dispersed uniformly be C_b (background concentration). Then, the increase of C until state (c) depends mainly on the turbulent nature and could be given schematically by Eq.(1), since the concentration in Batchelor eddy is equal to the initial concentration, C_i.

$$C=C_i x(V_i/V)+C_b x[(V-V_i)/V] \qquad (1)$$

Assume that C is too small compared to C_i. That is

$$C_i >> C > C_b \qquad (2)$$

Eq.(1) can be written with Eq.(2) as:

$$C-C_b=(C_i-C_b)x(V_i/V) \simeq C_i x(V_i/V) \qquad (3)$$

The term V_i/V in Eq.(3) represents schematically the turbulent nature of the given system. Eq.(3) shows that the difference between C and C_b is independent of C_b and hence it makes possible the quasi-steady state measurement. It is to be noted that the above assumption is not special but is usually encountered in most measurements.

EXPERIMENTALS

The same impeller-vessel combinations as the macromixing experiment in Ref.8 but for the flat blade fan turbine (FBT) were used for this work. Since concentration response with FBT was very curious as discussed in Ref.8.

The impeller rotation speed was adjusted so as to give the same discharge flow rate, Q_e, for all cases. Flow-follower method proposed by Porcelli and Marr[13] was applied to measure the total circulation flow rates, Q_c and discharge flow rate, Q_e.

The rate of injection of NaOH solution in Fig.2 was $5x10^{-7}m^3/s$ for the disc turbine(DT) and propeller(PM) in a complete baffled vessel and $2.5x10^{-7}$ m^3/s for the inclined fan turbine in an incomplete baffled tall vessel with round bottom(IFT) and is much smaller than Q_e. Locations of impellers and tracer injection and liquid height are shown in Fig.4, which also gives the locations of the local measurement of micromixing in matrix form.

RESULTS AND DISCUSSION

Nature of micromixing

An illustration of the experimental results is given in Fig.5, after converting light intensity into concentration. Time delay is found until point A, after NaOH solution is injected. Average of this time delay corresponds to that observed in the macromixing measurement. After point A, concentration fluctuation appears suddenly and the average increases gradually, reaches a maximum as shown by the spline fitting curve and then approaches asymptotically but very slowly toward the ideal mixing line indicated by a dash line. To show the average feature of the fluctuation, the total time interval from point A to the end of injection, which is 40s in this experiment, is divided into four equal time interval regions. The eighth-order spline function is used to fit the data in each region. Then, the fitted curves are connected for the full time interval and is shown in Fig.5. Dashed line is the average of the fitted curve and the chain lines represent maximum deviation of the fitted curve from the average line.

Degree of the approach of the average spline fitting line toward the ideal mixing line is dependent on the degree of the micromixing as shown later, but it is to be noted that the quasi-steady state measurement discussed in the foregoing section is achieved except before t_d for any cases.

The time delay, t_d, in the micromixing measurement is defined in this work at the intersection point of spline fitting curve and the lower line of the maximum deviation as indicated in Fig.5.

Results of the micromixing measurement presented us many informations about the mixing nature. These are given, for example, for the DT as follows:

1. Figure 5 shows that micromixing progresses very fast, since concentration in Fig.5 decreases 10^{-4} times to that of the injected solution. Another confirmation can be seen in the following mixing time t_M.

Let t_M be the time interval between the end of injection of the NaOH solution and the falling down of the fluctuation to zero. The average of t_M is shown at the top of Table 1. Mixing time in the micromixing is less than twice as that in the macromixing.

2. Though micromixing in the vessel progresses fast, Fig.5 shows that there still exists very sharp variations in the concentration. Let the intensity of the concentration fluctu-

ation be defined as the rms deviation from the spline fitted curve. The definition is unusual; however, after the concentration fluctuation is reduced with mixing progressed along the stream far away from the impeller, large time scale gentle waves still exist in the concentration response. If one defines the intensity of the concentration fluctuation as the rms deviation from the average line, then the high intensity can be calculated due to the large time scale wave far away from the impeller. Therefore, the spline fitting curve is introduced in this work.

The intensities, I, from t_d to the end of tracer injection are normalized by the final concentration after complete mixing, C_∞, and are listed in Table 1. Table shows that I increases from (5,2) to (5,3) which means that the micromixing has progressed after being discharged from the impeller. Such a phenomenon was not observed in the macromixing measurement. The I decreases, that is homogenization of the micromixed fluid improves from (5,3) till (8',1') along the circulation loop. Compare with Fig.6.

3. Above results emphasize that micromixing is promoted much by the DT. However, there remains very large time scale fluctuations over all the measurement points as mentioned above. Illustrations of spectrum analyses of the concentration responses at (5,3) and (8',1'), that is just after discharge and before return to the impeller, are shown in Fig.6. Spectral density at (8',1') is less over a wide frequency range than that at (5,3). However, it does not vary much for frequencies less than 0.3Hz. Experiments were repeated and confirmed the results. Absolute velocity at (5,3) is higher than that at (8',1'), and hence the spatial scale of fluctuation at (5,3) is larger than that at (8',1'). Therefore, it is sure that the large spatial scale of eddies has disintegrated while they stream from (5,3) to (8',1'). But Fig.6(b) shows that large eddies still remain in the vessel after streaming though higher frequency components have almost disappeared.

This is due to the turbulent nature of mixing. Mixing within large and energy containing eddies could be promoted to an almost uniform state, since enough power could be transferred to the fluid by the DT. However, mixing between large scale eddies would happen presumably when they can get into the impeller region at the same time. The probability of this occurring is surely very low. This is one of disadvantages of the mixing system. It is noted that, as seen in Figs.5 and 6, fluctuations with very large time scale, that is over ten seconds, exist in the vessel with the DT.

4. Other disadvantages of the DT is already given in the macromixing work[8] that the macromixing between the upper and the lower area from the impeller plane is poor.

5. Let discuss finally the local concentration distribution, $\overline{C}$, in the vessel - a measure of the homogeneity - at the quasi-steady state. One could define $\overline{C}$ as the mean concentration deviation between the spline fitting curve and the ideal mixing line during the quasi-steady state realized. In this experiment. The injected solution is at first diluted from pH=12.5 to 9.5 (constant light intensity region), and then light intensity is reduced until pH=8.3 with further progress of the micromixing (second step of the micromixing). Therefore, homogenization through the micromixing is very delayed in this experiment. That the average of the spline fitting curve approaches toward the ideal mixing line with time (Fig.5) means the micromixing in the second step progressed more or less. However, as the second step of the micromixing delays, the light intensity therefore the concentration converted from it is higher than that at the equilibrium state. As a result, the $\overline{C}$'s listed up in Table 1 give plus value in every points. When the equilibrium state is achieved during the tracer injection, the plus and minus of the $\overline{C}$ should appear to satisfy the material balance. Thus, the $\overline{C}$ listed up in Table 1 could be estimated to represent the result of the micromixing from the first step to the initial stage of the second step.

Keeping this in mind, it is concluded from the $\overline{C}$ and I in Table 1 that the distribution of the micromixed fluid is well performed quite regularly along streaming of it from point (5,3) to (8',1'). The maximum difference of t_d in Table 1 is small to be 0.2.

Micromixing with PM and IFT

It was found that micromixing with the PM and IFT were very delayed. Therefore, the $\overline{C}$ represents mostly the micromixing during the first step. Even so, negative $\overline{C}$ often appears and means that the distribution of the micromixed fluid is very weak due to the weak circulation flow.

Non-dimensionalized $\overline{C}$, t_d and I with the PM and IFT are listed up in Table 2 and 3, respectively, together with t_M. Illustrations of spectrum analyses of the concentration fluctuations are shown in Figs.7 and 8. It is noted that the injection period of the tracer solution with the IFT is 80s, while 40s with the PM and DT. So much worse the micromixing was observed with the IFT. Brief difference between the DT and PM or IFT are as follows.

PM: Mixing is degraded. The t_M is enlarged, which cannot observe in the macromixing measurement. Concentration fluctuation is not so much reduced even at the end of the circulation flow as seen in Fig.7 and Table 2. The $\overline{C}$ distribution is rather irregular in the view of the average circulation flow profile. The negative $\overline{C}$ and large t_d appear at (10,3) and (10,2). Therefore, area near the liquid surface is recognized stagnant, which can confirm through the flow visualization. Irregularity of the $\overline{C}$ comes from the unstable flow phenomenon explained in the macromixing research,[8] though the $\overline{C}$ distribution is regular in the macromixing experiment. Indeed, data scattered much in each experimental run at the same point, which may be due to the micromixing (not macromixing) variation with time at the impeller region. Considering this, it can conclude from the $\overline{C}$ and t_d in Table 2 that even the micromixing degraded in the impeller region, distribution of the micromixed fluid into the bulk in the vessel is well performed except row 10.

IFT: Reducing the rate of the tracer injection is very advantageous for the micromixing. However, the I distribution in Table 3 and Fig.8 show that the micromixing is so weak in the impeller region that the concentration fluctuation remains, that is, dispersion of the micromixed fluid is not enough, even in the lower area in the vessel. Step wise reduction of $\overline{C}$ is observed as shown in Table 3 by a dashed line. The I in the area along this line also falls down compared with those in the adjacent area. Therefore, it is concluded that the circulation flow is very weak in the area above the dashed line. Zone near the liquid surface is almost completely stagnant.

Attention was paid why the very large difference appeared between row 11 and 12. Detail examination of the concentration responses on the row 11 shows that the micromixing is progressed, of course at low level, near the baffle which locates near the liquid surface. Circulation flow goes up intermittently along the wall toward the baffle. Weak vortex, that is weak turbulence, generates near the baffle when the circulation flow turns toward the radial direction. The vortex promotes gradually an additional micromixing on its way toward the impeller. It is believed that at least the $\overline{C}$'s at (9,1) and (11,1) include this effect.

CONCLUSION

In this series of the micro- and macromixing experiment, tracer solution is introduced into the impeller region so as to promote the mixing. Degree of the micromixing performed in the impeller region and degree of the dispersion of the micromixed fluid into the bulk of fluid in the vessel could estimate with the help of the macromixing informations. Detail analysis is not yet given at present, since only one combination of the impeller and the incomplete baffled vessel is examined here. Following are a rough estimation at present.

1. Degree of the dispersion of the solute into the in molecular level in the impeller region is very delayed and, different from the macromixing result, approximately depend on the power consumption per unit volume as well known.[1,7,14]

2. However, the PM could perform homogenization in the vessel at lower power input compared with that with the DT.

3. The IFT can not apply for the tall vessel as well as known also.[15] However, it is found that a small baffle located near the liquid surface could promote the local micromixing, therefore homogenization also, at much more low power input. The IFT seems not to be preferable for this purpose.

REFERENCES

1) K.S.M.S.Raghav and J.B.Joshi: "Liquid phase mixing in mechanically agitated vessels," Chem.Eng.Sci., 74, 1988, pp.1-25

2) W.Bujalski, A.W.Nienow, S.Chatwin and M.Cooke: "The dependency on scale of power numbers of Rushton disc turbine," Chem.Eng.Sci., 42, 1987, pp.317-326

3) H.Wu and G.K.Patterson: "Laser Doppler measurements of turbulent flow parameters in a stirred mixer," Chem.Eng.Sci., 44, 1989, pp.2207-2221
4) I.Fořt: "Flow and turbulence in vessels with axial impellers." chapter 14 in V.W.Uhl and G.B.Gray(Eds.): "Mixing -theory and practice." London, Academic Press, 1986, vol.3, pp.133-197
5) J.Bertrand, J.P.Couderc and H.Angelino: "Power consumption, pumping capacity and turbulence intensity in baffled stirred tanks -Comparison between several turbines," Chem.Eng.Sci., 35, 1980, pp.2157-2163
6) E.B.Nauman: "Residence time distributions and micromixing," Chem.Eng.Commun., 8, 1981, pp.53-131
7) M.Kamiwano, K.Yamamoto, and S.Nagata: "Mixing performance in various agitator," Kagaku Kogaku, 31, 1967, pp.365-372 (In Japanese)
8) Y.Hirofuji, Y.Sasaki and Y.Nagase: "A measurement method of batch mixing and impeller mixing performance," Proc.7th European Congress on Mixing, 1991
9) Y.Hirofuji and Y.Nagase: J.Chem.Eng.Japan, to be submitted
10) O.S.Wolfbeis, E.Fürlinger, H.Kroneis and H.Marsonar: "Fluometric analysis 1.A study on fluorescent indicators for measuring near neutral ("Physiological") pH-values," Z.Anal.Chem., 314, 1983, pp.119-124
11) S.Winardi, S.Nakao and Y.Nagase: "Pattern recognition in flow visualization around a paddle impeller," J.Chem.Eng.Japan, 21, 1988, pp.503-508
12) S.Winardi and Y.Nagase: "Unstable phenomenon of flow in a mixing vessel with a marine propeller," J.Chem.Eng.Japan, to be published
13) J.V.Porcelli and G.R.Marr: "Propeller pumping and solids fluidization in stirred tanks," Ind.Eng.Chem.Fundam., 1, 1962, pp.172-179
14) J.G.van de Vusse:" Mixing by agitation of miscible fluids part I," Chem.Eng.Sci., 4, 1955, pp.178-200
15) S.Komori and Y.Murakami: "Turbulent mixing in baffled stirred tanks with vertical-blade impellers," AIChE J., 34, 1988, pp.932-937

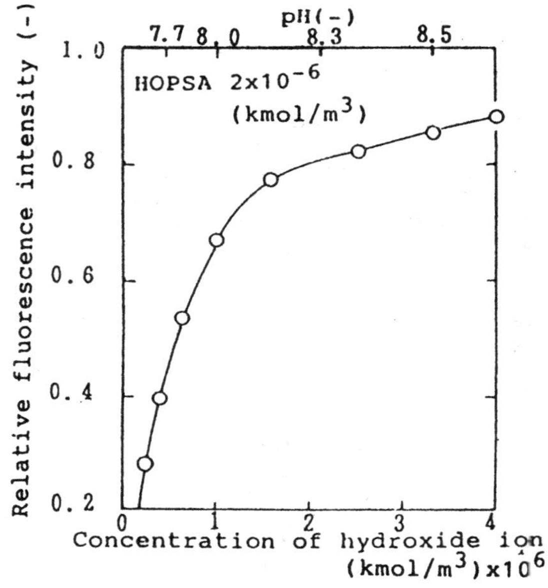

Fig.1 Fluorescence intensity of dye in sodium hydroxide solution

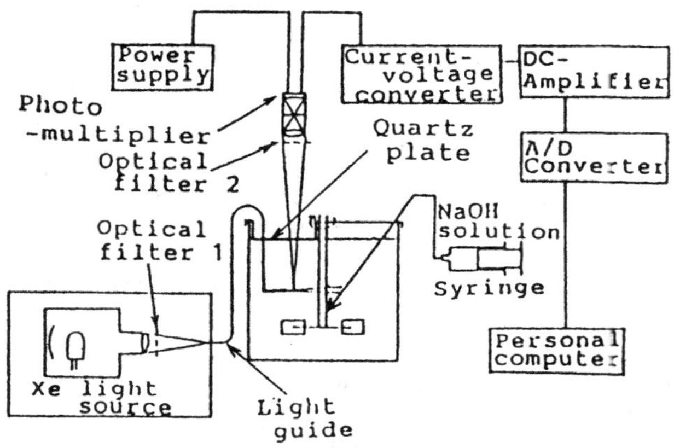

Fig.2 Measurement system

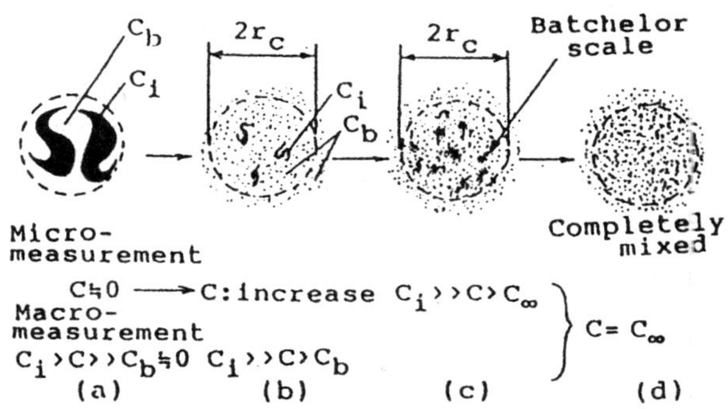

Fig.3 Schematic diagram of turbulent mixing in concentration blob

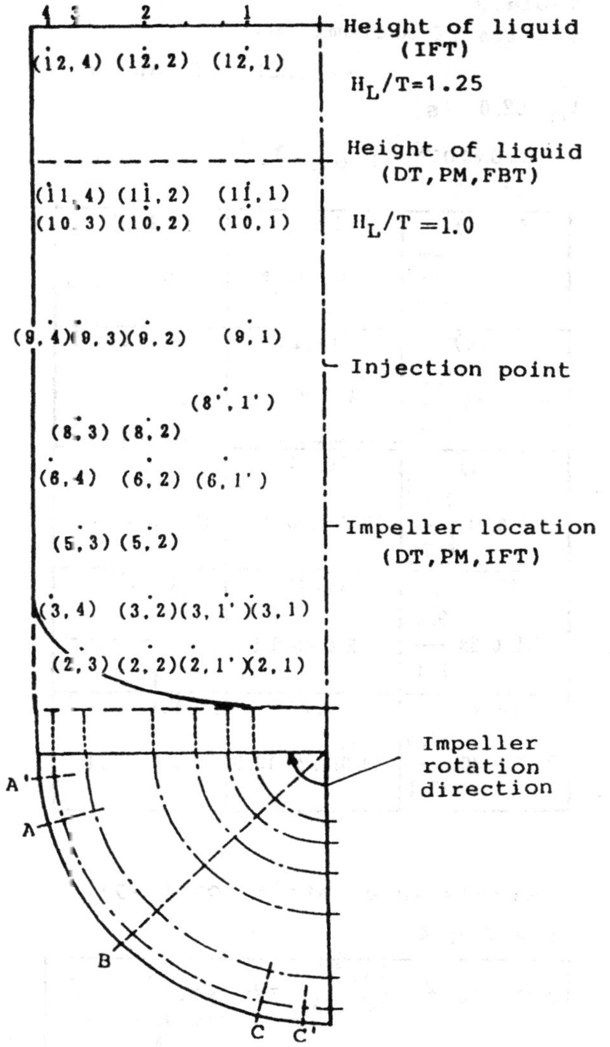

Fig.4 Locations of impellers, injection and measurements
A',A,B,C,C': Measurement points between baffles.

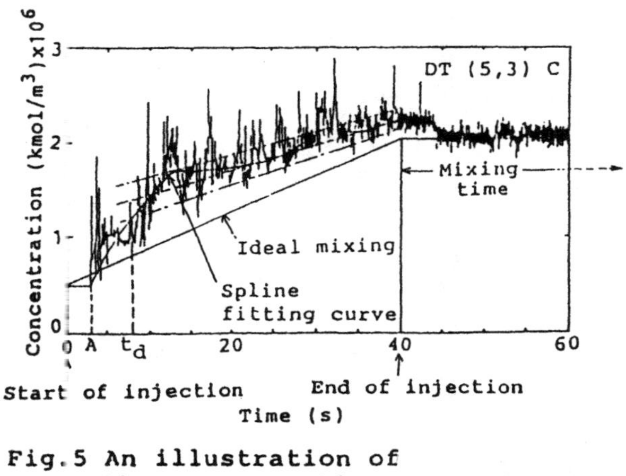

Fig.5 An illustration of the concentration response with time

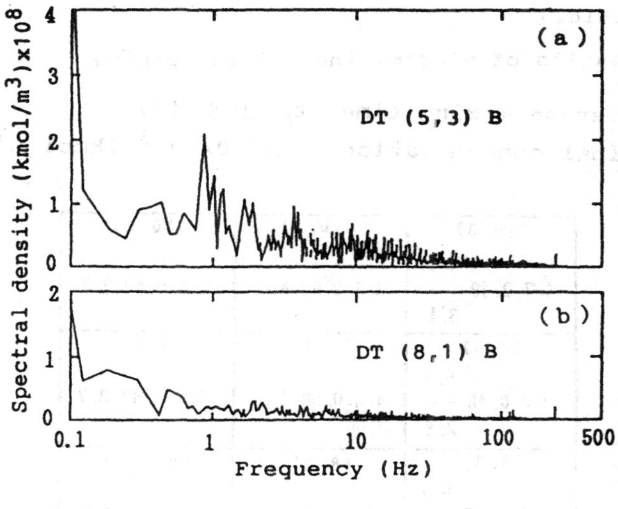

Fig.6 Examples of concentration spectra - Disc turbine

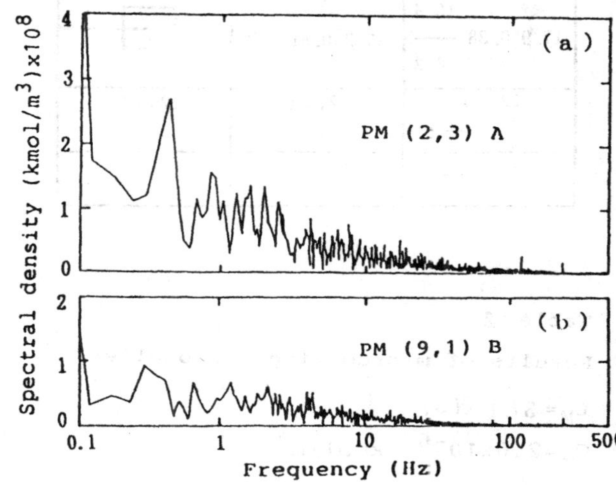

Fig.7 Examples of concentration spectra - Propeller

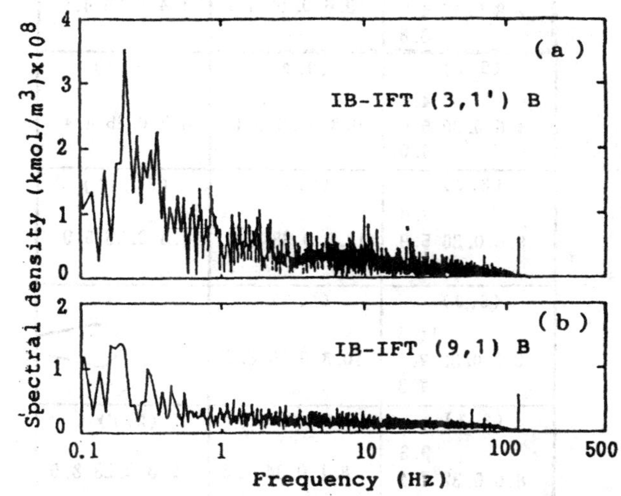

Fig. 8 Examples of concentration spectra - Inclined fan turbine

Table. 1

Results of micromixing - Disc turbine

average mixing time t_M= 27.6 (s)

final concentration $C_\infty=2.0\times10^{-6}$ (kmol/m^3)

(10,3) 4.7 4.7 0.49 — 3.1	(10,2) 3.3 0.50 4.0	(10,1) 2.3 0.49 3.2
(9,3) 5.7 7.2 0.48 — 2.8	(9,2) 4.2 0.58 3.9	(9,1) 0.9 0.47 3.7
(8,3) 8.7 10.5 0.45 — 4.0	(8,2) 6.1 0.45 3.9	(8',1') 1.1 0.45 2.6
(5,3) 10.4 17.9 0.38 — 8.2	(5,2) 17.3 0.41 7.7	
(2,3) — — — — —	(2,2) — — —	(2,1) — — —

Table. 2

Results of micromixing - Propeller

t_M= 37.1 (s)

$C_\infty=2.0\times10^{-6}$ (kmol/m^3)

(10,3) 7.2 -2.8 0.48 4.9 6.8	(10,2) -3.6 0.44 3.7	(10,1) 1.4 0.36 4.0
(9,3) 4.2 4.6 0.39 5.0 4.0	(9,2) 6.8 0.39 5.1	(9,1) 0.7 0.35 4.4
(8,3) 7.0 6.6 0.36 5.9 3.1	(8,2) 12.5 0.38 8.5	(8',1') 4.8 0.35 5.9
(5,3) 11.2 5.2 0.36 7.2 7.8	(5,2) 10.8 0.36 8.7	
(2,3) 9.3 8.0 0.32 7.4 9.5	(2,2) 8.1 0.34 7.3	(2,1') 9.3 0.23 8.0

Table. 3

Results of micromixing - Inclined fan turbine

t_M= 62.0 (s)

$C_\infty=2.0\times10^{-6}$ (kmol/m^3)

(12,4) — — — — —	(12,2) -21.5 0.52 2.6	(12,1) -21.5 0.53 3.5
(11,4) 5.6 0.3 0.44 — 3.9	(11,2) 0.4 0.45 6.1	(11,1) 4.4 0.47 3.2
(9,4) 7.7 4.8 0.41 — 6.7	(9,2) 10.3 0.48 9.1	(9,1) 12.9 0.47 6.5
(6,4) 9.0 17.5 0.38 — 8.4	(6,2) 6.2 0.40 9.5	(6,1) — — —
(3,4) 11.5 26.4 0.30 — 13.1	(3,2) 43.0 0.26 19.5	(3,1') 35.0 0.27 11.7

numerals in a matrix box (i,j)

:see Fig.4

(i,j)		
$\bar{C}/C_\infty\times10^{-2}$	t_d/t_M	$I/C_\infty\times10^{-2}$ angle A or A' B C or C'

SUPPLEMENT

For this paper : Reader is also invited to consult the Referee's comments and questions concerning the Ref.8.

NOMENCLATURE

C	=	measured concentration	($kmol/m^3$)
$\overline{C}$	=	average concentration deviation	($kmol/m^3$)
C_b	=	back ground concentration	($kmol/m^3$)
C_i	=	initial concentration	($kmol/m^3$)
C_∞	=	final concentration	($kmol/m^3$)
H_L	=	height of liquid from bottom of vessel	(m)
I	=	intensity of the concentration fluctuation	($kmol/m^3$)
Q_c	=	total circulation flow rate	(m^3/s)
Q_e	=	discharge flow rate	(m^3/s)
r_c	=	radius of measurement space	(m)
T	=	inside diameter of stirred tank	(m)
t_d	=	time delay	(s)
t_M	=	mixing time	(s)
V	=	volume of measurement space	(m^3)
V_i	=	volume of measured concentration clouds	(m^3)

REFEREES' QUESTIONS AND AUTHORS' ANSWERS

Q. What is the volume of zone where fluorescence is measured.

A. In this equipment, measuring volume is controlled by the pin-hole size of the photo-multiplier optics, and is estimated to be 2mm in diameter at measurement point.

Q. What is the level of the injection point? What are the values of 2r/T for your radial points 1, 2, 3 & 4 (r = distance of point to the system axis)?

A. They are indicated in Fig.4 in Ref.8.

The level of the injection point:
0.16m from the tank bottom

2r/T for radial points 1,2,3 & 4:
0.25, 0.61, 0.86 & 0.96

Q. You say that the degree of the dispersion of the tracer principally depends on the power consumption per unit volume P/V_L. More generally, don't you think that in a given vessel, for geometrically similar impellers, that depends on P/D^3 with D=impeller diameter?

A. It was stated in this paper "The degree of dispersion into molecular level is <u>approximately</u> dependent on the power consumption per unit volume as well known". The expression expected to consider that Batchelor scale depends on the local energy dissipation rate (liquid properties are constant in this experiment).

AN EXPERIMENTAL STUDY OF MICROMIXING USING TWO PARALLEL REACTIONS

J.R. BOURNE, S.YU, Technisch-Chemisches Lab., ETH, Zürich, CH

ABSTRACT

It is well-known that the product distribution of fast multiple reactions can be influenced by mixing. Fast, single-phase competitive-consecutive reactions of the type

$$A + B \longrightarrow R \qquad R + B \longrightarrow S$$

have been studied theoretically and experimentally as "tracer reactions" to characterise micromixing. Little attention has been given to fast parallel reactions of the type

$$A + B \longrightarrow P \qquad A + C \longrightarrow Q$$

and how their product distribution is influenced by the way in which the reactions are conducted, especially with reference to micromixing(1). This paper presents some experimental results. The two competitive reactions reported here are acid-base neutralization and alkaline hydrolysis of ethyl chloroacetate. They were carried out in a standard stirred tank reactor with a Rushton turbine. The product distributions at different feed positions, volume ratios, turbine speeds and initial concentrations were measured. Qualitative conclusions about the rate of micromixing for two parallel reactions are drawn and discussed. In addition to competitive-consecutive reactions, the two parallel reactions presented here provide another useful tracer system for micromixing studies.

INTRODUCTION

Mixing plays a significant role in determining the product distribution of fast multiple reactions. Using competitive-consecutive reactions, many studies of micromixing have been published. However few dealt with parallel reactions. For two fast parallel reactions like

$$A + B \xrightarrow{k_1} P \tag{1}$$

$$A + C \xrightarrow{k_2} Q \tag{2}$$

the degree of molecular-scale segregation can be characterised by the product distribution, Q/P, which may be generally defined by

$$Q/P = r_Q / r_P \tag{3}$$

r_Q and r_P are the specific rates of formation of Q and P, respectively. When mixing is perfect, no molecular-scale segregation exists, and the volume-averaged rates are given by the chemical kinetic expression:

$$Q/P = k_2 c_C / k_1 c_B \tag{4}$$

On the contrary, when segregation is intense, the product distribution becomes independent of the kinetics and is given by

$$Q/P = c_C / c_B \tag{5}$$

When segregation is partial, the product distribution varies with the mixing intensity, and a micromixing model is needed to relate these quantities(1).

This paper presents some experimental results of micromixing studies using parallel reactions for various experimental conditions in a baffled tank with a Rushton turbine.

THEORETICAL BASIS

Engulfment (E) model of micromixing is

$$\frac{d c_i}{d\alpha} = E (< c_i > - c_i) + r_i \qquad (6)$$

i=A, B, C. c_i and $<c_i>$ are the concentrations of substance i in the uniform reaction zone and in the surroundings, respectively. The volume of the reaction zone grows by the engulfment of the surroundings. E (engulfment rate coefficient) is given by

$$E = 0.058 (\varepsilon / \upsilon)^{1/2} \qquad (7)$$

An analogous expression, with ε_{av} written instead of ε, represents the average engulfment rate coefficient E_{av}. For a semi-batch reactor with B and C premixed in reactor, A is slowly added and mixes with the average contents of the reactor. The surrounding concentration $<c_i>$ of any part of the feed can be found by discretising the feed into σ equal parts (1). When A and B react instantaneously, A and B can not coexist at a given point. A simple transformation of the concentrations then reduces the number of ODE from 3 to 2. Details see reference (1).

The yield of Q in eqs(1) and (2) is defined by the ratio between the quantity of Q produced and the quantity of A added.

$$X_Q = \frac{V_o c_{Co} - (V_o + V_{Ao}) c_C}{V_{Ao} c_{Ao}} \qquad (8)$$

Variables determining X_Q, when equal molar quantities of A, B and C are mixed, are

$$X_Q = f(k_1/k_2, \ a, \ \overline{Da}, \ T) \qquad (9)$$

where

$$a = V_o / V_{Ao} \qquad (10)$$

$$T = \alpha E_{av} \qquad (11)$$

$$\overline{Da} = \frac{k_2 c_{Ao}}{E_{av} (1+a)} \qquad (12)$$

For an instantaneous first reaction ($k_1 \rightarrow \infty$), limits on X_Q are 0 for perfect mixing and 1 for highly segregated reagents(1).

CHARACTERISTICS OF THE SELECTED REACTIONS

REACTION KINETICS

Two parallel reactions, consisting of acid-base neutralization and alkaline hydrolysis of ethyl chloroacetate, were employed.

$$NaOH + HCl \xrightarrow{k_1} NaCl + H_2O \qquad (13)$$

$$NaOH + CH_2ClCOOC_2H_5 \xrightarrow{k_2} CH_2ClCOONa + C_2H_5OH \qquad (14)$$

where NaOH, HCl and $CH_2ClCOOC_2H_5$ correspond to A, B and C in eqs(1) and (2). The general rate expression for the hydrolysis of ethyl chloroacetate is given by(2)

$$\frac{dc_C}{dt} = - (k_o + k_{H^+} [H^+] + k_{OH^-} [OH^-]) c_C \qquad (15)$$

where $[H^+]$ and $[OH^-]$ are the concentrations of the ions H^+ and OH^-. k_o, k_{H^+} and k_{OH^-} are the rate constants for neutral, acidic and alkaline hydrolysis, respectively. In dilute aqueous solutions in which pH control is maintained, hydrolysis follows pseudo-first order kinetics, and the observed rate coefficient is given by

$$k_{obs} = k_o + k_{H^+} [H^+] + k_{OH^-} [OH^-] \qquad (16)$$

Many authors measured values of k_o , k_{H^+} and k_{OH^-} (2~5). Some scatter among these values exists. Nevertheless the orders of magnitudes of k_o , k_{H^+} and k_{OH^-} are 10^{-7} s^{-1}, 10^{-8} m^3 mol^{-1} s^{-1} and 0.03 m^3 mol^{-1} s^{-1} at 25 °C, respectively. This means that neutral and acid-catalyzed hydrolyses are very slow compared to alkaline hydrolysis. Therefore k_o and k_{H^+} can be neglected for alkaline hydrolysis, and eq(16) becomes

$$k_{obs} = k_{OH^-} [OH^-] \qquad (17)$$

Finally the reaction rate of alkaline hydrolysis of eq(14) follows

$$\frac{dc_C}{dt} = - k_2 \; c_A \; c_C \qquad (18)$$

showing that eq(14) is irreversible and second order. Nolan and Amis(5) obtained the rate constant k_2(given here in SI units)

$$k_2 = 2.0*10^5 \exp(-3.887*10^4/RT) \qquad (19)$$

Eq(19) was employed in this work.

For acid-base neutralization(eq(13)), Eigen and Meyer(6) reported $k_1 = 1.3*10^8$ m^3 mol^{-1} s^{-1} at 25 °C. The half life times of eqs(13) and (14) at 70 mol/m^3 are $1.1*10^{-10}$ and 0.47 second, respectively. It is therefore reasonable to treat neutralization as instantaneous. These times were compared with the time needed for micromixing. When 5 vortex generations are needed for the reactions to be completed, the time for micromixing is $t_m=5*12(\upsilon/\varepsilon)^{1/2}$. The stirrer speeds in the experiments range from 1.67 s^{-1} to 10 s^{-1}. Then t_m falls in the range 1.14~0.10 second. Here one can see that the half life times of reactions(eqs(13) and (14)) are much shorter than or similar to the micromixing time, so that segregation occurs at molecular scale. This implies that the reaction system of eqs(13) and (14) is suitable to be used as "test reactions" for micromixing studies.

CHEMICAL ANALYSIS

When two parallel reactions are carried out to study micromixing, chemically equivalent quantities of all reagents are usually employed(1). Here sodium hydroxide is the limiting reagent. In principle, according to the definition of product distribution(eq(8)), X_Q can simultaneously be determined by measuring (1) the concentration of $CH_2ClCOOC_2H_5$,c_C , after the reactions were completed; (2) the concentration of C_2H_5OH, c_{alc}, in the product mixture;(3) the concentration of HCl after the reactions were completed (NaOH is the limiting reagent and because the quantity of NaOH consumed by eq (13) corresponds to the measured consumption of HCl, it's quantity participating in eq (14) followed by difference). However, in practice, HCl can not be measured by back-titration, because alkaline hydrolysis of $CH_2ClCOOC_2H_5$ also occurs during the titration. Nevertheless c_C and c_{alc} can readily be determined by gas chromatography(GC) using a flame ionisation detector.

In the experiments, an Internal Standard was used for GC quantitative analysis. Equal amounts of standard solution and unknown solution were mixed. The ratio of the corresponding areas was used to prepare calibration curves and determine the unknown concentration. With this method, c_C and c_{alc} in the reacted mixture can be determined. X_Q was then obtained from either c_C with eq(20) or c_{alc} with eq(21).

$$X_Q|_{exp} = 1 - (a+1)\, c_C / c_{Ao} \qquad (20)$$

$$X_Q|_{exp} = (a+1)\, c_{alc} / c_{Ao} \qquad (21)$$

Mass balances were checked by comparing these two values. The mean relative deviations were less than 4%. Quantitative analysis of the reaction products by GC is simple and convenient for the reaction system of eqs(13) and (14).

EXPERIMENTAL

Experiments were carried out in a 2.5 liter reactor stirred by a standard Rushton turbine. Equal quantities of A, B and C were used, since product distribution is then sensitive to mixing(1). A small volume of concentrated A solution is gradually added to a large volume of less concentrated premixed B and C solution. The feed time of A solution is long enough to neglect macroscopic concentration gradients. The solubility of ethyl chloroacetate at 15 °C measured in this work is 2.25 Wt%. The maximum initial concentration of ethyl chloroacetate is 120 mol/m^3, that is 66% of the solubility. Since ethyl chloroacetate and HCl are premixed, one should make sure that ethyl chloroacetate has been completely dissolved by premixing before reactions take place. In this work more than 5 minutes premixed time were used. Fig.1 illustrates the dimensions of the tank and locations of the feed points. Experimental variables are listed in Table 1. Average energy dissipation per unit mass ε_{av} was calculated with eq(22) taking Np=4.83 for the Rushton turbine(7).

$$\varepsilon_{av} = Np\, d^5 N^3 / V \qquad (22)$$

E_{av} and $\overline{Da}$ were obtained from eqs (7) and (12),respectively. The temperature of the reacting mixture was measured, showing changes during the reactions are less than 1.5 °C. Therefore the average temperature was used to determine k_2 from eq(19). The concentrations of reagents were lower than 120 mol/m^3 , therefore , the physical properties of

Table 1 Experimental variables for different feed points (see fig.1)

Feed points	positions		experimental variables		
	h_F (m)	r_F (m)	a	c_{Co}(mol/m^3)	N(rpm)
F1	0.087	0.0065	20, 50, 80	90, 120	100~400
F2	0.097	0.0065	50	90	100~600
F3	0.108	0.0065	50	90	100~600
F4	0.120	0.0065	50	90	100~600
F5	0.135	0.0065	20, 50, 80	90, 120	100~600
F6	0.150	0.0065	50	90	100~600
F7	0.150	0.025	50	90	100~600

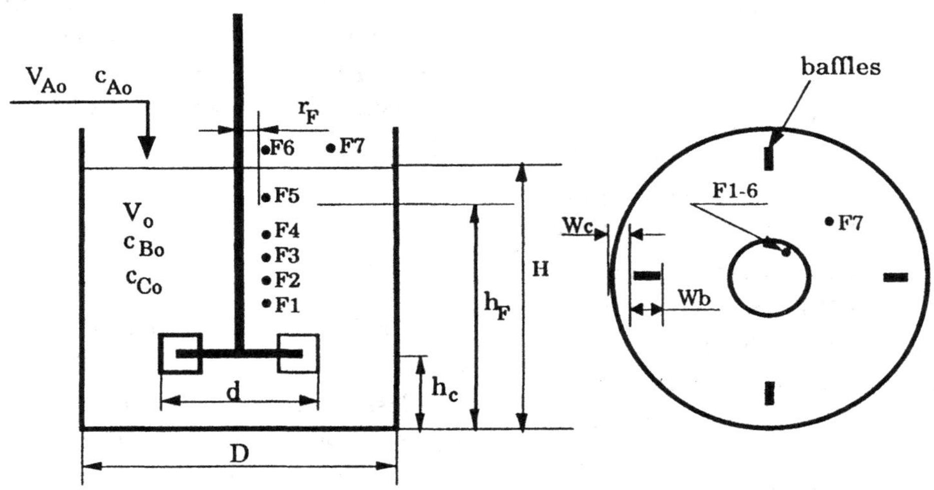

Fig.1 Dimensions of the stirred tank and locations of feed points
(D=H=3d=0.15 m , h_c =0.07 m , Wc=0.015 m , Wb=0.012 m)

water were taken as those of the reaction solution. After the reactions were completed, more sodium hydroxide was used to treat the residual ethyl chloroacetate before disposal. Finally for a certain degree of micromixing, represented by $\overline{Da}$, the corresponding X_Q can be determined by GC analysis after the reactions were completed. Calculated X_Q from the E model can then be compared with experimental results.

RESULTS AND DISCUSSION

INFLUENCE OF FEED TIME

The feed time used must exceed the critical feed time in order to make sure that experiments are carried out in the micromixing-controlled regime. Strictly speaking, t_{crit} should be determined for each new set of experimental conditions. This is however not very realistic. One solution to this problem is the determination of t_{crit} under the worst experimental conditions, for instance, smallest N and a, highest c_{Co} and feed point near surface and so on. Such t_{crit} is then the longest t_{crit} of all the experiments. If this t_{crit} is taken as the feed time, it is certain that all experiments are carried out in the micromixing-controlled regime. Fig.2 shows some experimental results, which indicate that X_Q is constant and only dependent on micromixing when the feed time is longer than 13 min. A value of 15 min. was subsequently used.

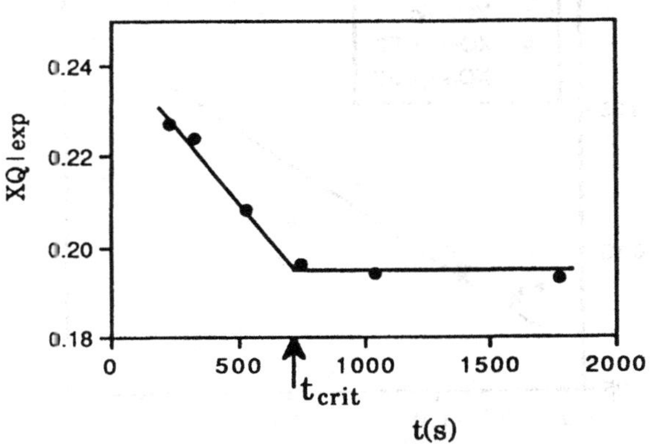

Fig.2 Influence of feed time on product distribution (a=20, N=200 rpm, c_{Co}= 90 mol/m^3)

INFLUENCE OF VOLUME RATIO AND INITIAL CONCENTRATION

For feed positions near the surface (F5) and in the impeller suction(F1), fig.3 shows the experimental results for different volume ratios, a, and initial concentrations, c_{Co}. For a=20, 50 , 80, calculated results fall within the two curves given in fig.3. The difference between these two curves is so small that it is reasonable to neglect the calculated influence of a on X_Q. The influences of c_{Co} (90 and 120 mol/m^3) and N are accounted for by $\overline{Da}$. The experimental results for the two feed points also show no significant effect of a, and indicate that the influence of c_{Co} on X_Q is well correlated by $\overline{Da}$. Thus the model described well the influences of volume ratio, turbine speed and initial concentration. The results in fig.3 show a large effect on product distribution of the feed point.

INFLUENCE OF FEED POSITION

Fig.4 shows experimental results at feed points with different heights(see fig.1). Calculated results were obtained by setting the rate of turbulent energy dissipation in the reaction zone equal to the mean value. As a feed point moves towards the stirrer, $X_Q|exp$ deviates more from the calculated results and becomes smaller. This is owing to the increase of local energy dissipation rate towards the turbine. For a given feed point, for instance F5, the difference between calculated and experimental results increases as $\overline{Da}$ increases or N decreases. Micromixing time is inversely proportional to engulfment rate, that is , $t_m \sim E^{-1} \sim \varepsilon^{-1/2} \sim N^{-3/2}$. Distance travelled by reacting fluid ~ vt_m, where v is local fluid velocity. Hence distance ~ $(Nd)N^{-3/2}$~$N^{-1/2}$, so with decreasing N, reacting fluid spreads further and ε rises towards the turbine. Consequently X_Q | exp falls. Success in predicting results from the E model relies on good estimates of local energy dissipation rates.

Fig.5 presents results for feed points near the surface at different radial distances. X_Q | exp first exceeds and then becomes smaller than X_Q | calc as $\overline{Da}$ increases. The reason is as for fig.4 . The radial position of the feed point has only a slight influence on the product distribution and hence also on the energy dissipation near the surface.

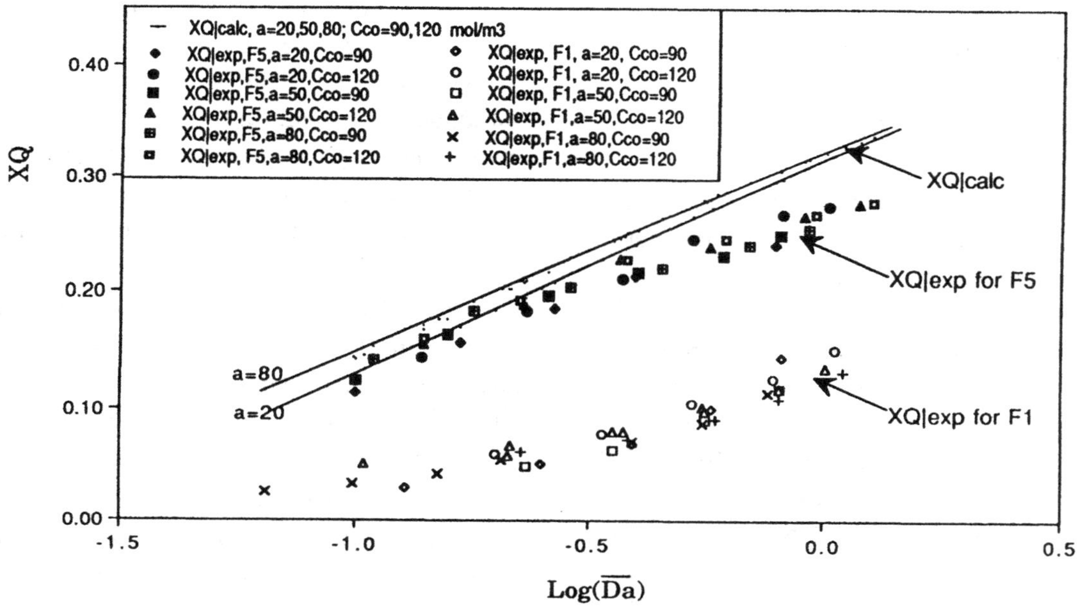

Fig.3 Influence of volume ratio, stirrer speed and initial concentration

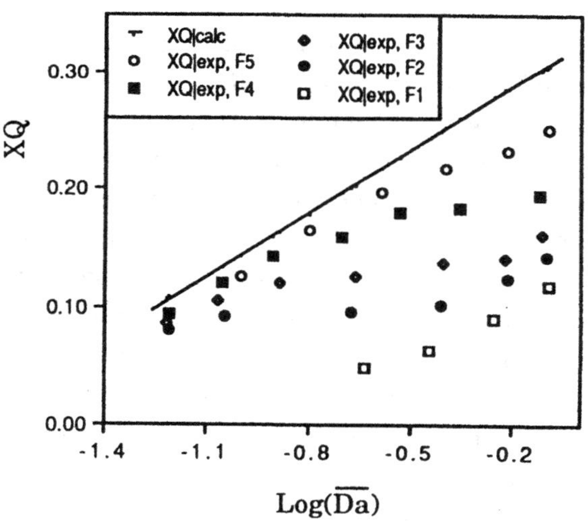

Fig.4 Influence of height of feed points (a=50, c_{Co}= 90 mol/m^3)

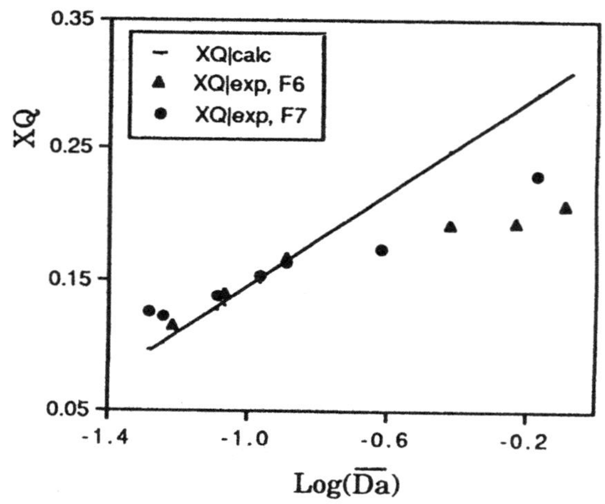

Fig.5 Influence of radial position of feed points (a=50, c_{Co}= 90 mol/m^3)

CONCLUSIONS

(1) The kinetics of the two parallel reactions are known. The neutralisation of HCl(B) and hydrolysis of ethyl chloroacetate(C) are second order irreversible reactions. (2)The half lives of C and B are similar to or much shorter than the micromixing time, so that segregation occurs at molecular scale. (3)The product distribution , which can readily be obtained by simple GC analysis of the residual ester and the ethanol formed, characterises the degree of micromixing. The two parallel reactions are suitable for studying micromixing.

Experimental results for various stirrer speeds, initial concentrations, volume ratios and feed positions were obtained. The effects of the first three variables on the product distribution are well described by the engulfment model of micromixing. The influence of the feed point requires reasonable estimates of the trajectory of the reaction zone and the distribution of the energy dissipation rate.

NOTATIONS

A,B	Reagents NaOH and HCl ,respectively
C	Reagent ,$CH_2ClCOOC_2H_5$
a	Volume ratio of reagent solutions
c_i	Concentration of i in reaction zone
$< c_i >$	Surrounding concentration of i
d	Impeller diameter
D	Tank diameter
$E, E_{\varepsilon v}$	Engulfment rate coefficient, defined by eq(7)
Fi	Feed point i
H	Liquid height
h_F	Height of feed point
k_i	Rate constant of reaction i
N	Impeller speed
P, Q	Reaction products
r_F	Radius of feed point
r_i	Rate of formation of i
T	Dimensionless time or temperature
t	Time
t_{crit}	Critical feed time
t_m	Micromixing time
V_o	Initial volume of solution of B and C
V_{Ao}	Volume of A solution added to reactor
X_Q	Yield of Q
$\overline{Da}$	Damköhler number defined by eq(12)
Np	Power number

GREEK

α	Age of reaction zone
$\varepsilon, \varepsilon_{av}$	Local and average turbulent energy dissipation rate per unit mass
υ	Kinematic viscosity
σ	Number of parts of feed

REFERENCES

(1) J.Baldyga and J.R.Bourne, "The effect of micromixing on parallel reactions", Chem. Eng. Sci., 45, 1990, 907

(2) A.J.Kirby, "Comprehensive chemical kinetics, Chapter 2: Hydrolysis and formation of esters of organic acids", ed. C.H.Bamford and C.F.H.Tipper, Elsevier Publishing Company, 1972, 10

(3) J.Barthel,G.Bäder and G.Schmeer, "Untersuchungen zum Substituenten- und Lösungsmitteleinfluss auf Solvolysereaktionen. I.Der Einfluss polarer α-Substituenten auf die Geschwindigkeit der alkalischen Esterhydrolyse in Wasser bei verschiedenen Temperaturen", Z. Phys. Chem. Neue Folge, 62 , 1968 , 63

(4) E.K.Euranto and A.L.Moisio, "Kinetics of alkaline hydrolysis of alkyl halogenoacetates", Suomen Kemistilehti , B37 , 1964 , 92

(5) G.J.Nolan and E.S.Amis, "The rates of the alkaline hydrolysis of ethyl α-haloacetates in pure and mixed solvents", J. Phys. Chem. 65 ,1961, 1556

(6) M.Eigen and L.Maeyer, "Untersuchungen über die Kinetik der Neutralisation I", Z. Elektro chem. 59, 1955 , 986

(7) W.Bujalski, A.W.Nienow, S.Chatwin and M.Cooke, "The dependency on scale of power numbers of Rushton disc turbines", Chem. Eng. Sci., 42, 1987, 317

Referees' Questions and Authors' Corresponding Answers

General comments

Your paper is very, very interesting. It's a pity that you don't specify in your abstract, in your introduction and in your conclusion that you consider and work in a feed-batch reactor.

Answer to general comment

A fed-batch reactor was used.

1. Question:

Is $(V_0 + V_{A0})$ constant ?

Answer:

V_0 is constant. V_{A0} depends upon volume ratio $a=V_0/V_{A0}$. For a given volume ratio, $(V_0 + V_{A0})$ is constant.

2. Question:

Why do you use X_Q instead of P/Q ?

Answer:

Use of X_Q or Q/P is a matter of definition/convention. In fact, if one is known, the other can be calculated.

3. Question:

What is the feed point corresponding to your Fig.2 ?

Answer:

Fig.2 corresponds to feed point F5 as shown in Fig.1.

4. Question:

Bruxelmane (BRUXELMANE M., p. XA23, XA38, XB3, XD18, Proc. Int. Symp. on Mixing (M. Bruxelmane and S. Lefebvre, Ed.), Faculte Polytechnique de Mons, Belgium, 1978) has proposed that is was necessary to keep τ/t_c = cst to obtain the same residence time distribution curves (and thus the same relative degrees of conversion and the same yields) in two continuous stirred tank reactors at a given scale (same impeller type) or at two different scales (same impeller type). τ is the mean residence time of the solution in the reactor. The same author has even asserted that a continuous reactor perfectly if $\tau/t_c = Q_c/V_0^* > 100$ (Q_c being the circulation flow rate and V_0^* the feed flow rate).

For your Fig.2, $Q_c/V_0^* = 440$! However, the agitation Reynolds number is equal to ±8500 and the initial dispersion of the feed has surely an important effect. Nevertheless, in my opinion, in order, the feed time of a solution to be long enough to neglect macroscopic concentration gradients, the characteristic variable is Q_c/V_0^* whatever the size and the geometry of the stirred tank reactor may be liquid (same impeller type and same relative location of the feed point P_F and the impeller center point P_I that is $P_F P_I/D$=cst).

What do you think about those affirmations ?

Answer:

It is true that a continuous reactor is perfectly mixed if $\tau/t_c = Q_c/V_0^* > 100$. However the criterion of $\tau/t_c = Q_c/V_0^* > 100$ does not apply to mixing and fast reaction in a semi-batch reactor (SBR). For this case, whether or not macroscopic concentration gradients exist is determined by feed time relative to bulk blending time, which is a small multiple of t_c. This can be characterised by the curve of product distribution vs feed time. If feed time is not long enough relative to blending time ($t<t_{crit}$ in Fig.2), complete bulk blending is not ensured because the fresh A-solution is being fed to the feed point faster than B and C arrive there by circulation from other regions in the tank. This incomplete macromixing will interact with incomplete micromixing (inhomogeneity on molecular scale) and both influence the product distribution X_Q. On the contrary, if feed time is long enough relative to blending time ($t>t_{crit}$ in Fig.2), only micromixing is incomplete, the feed mixes with the average contents of the tank and X_Q is uniquely determined by micromixing. If the degree of micromixing is constant, i.e., constant feed position, stirrer speed and ε, X_Q is then independent of feed time (as shown in Fig.2). Finally it can be concluded that it is better to determine t_{crit} by the curve of X_Q vs feed time which represents the interaction between macromixing and micromixing not by Q_c/V_o^* which represents the macromixing only.

5. Question:

Don't you think that your engulfment rate coefficient E is a function of P/d^3 rather than of ε_{av} ? In my opinion, E also depends on the mean pumping time t_p or better on the mean circulation time t_c ? To this end, it would interesting to investigate the effect of the impeller diameter upon P/Q or X_Q. What scale up rule(s) do you advise for semi-batch stirred reactors with complex reactors ? Personally, I recommend the use of the same impeller type, the same relative location of the feed point and the impeller center, $Q_c/V_o^* = cst$ as well as $P/d^3 =$ cst.

Answer:

(1) Engulfment rate coefficient E is a function of ε_{av} , or P/d^3 if the density is constant. $\varepsilon_{av} = P/\rho V$, $V \sim d^3$. If $\rho = cst$, $\varepsilon_{av} \sim P/d^3$.

(2) t_c or t_p is important for macromixing, i.e., flow or circulation rate in the tank. However ε or E is important for micromixing because,according to Kolmogroff 's isotropic theory of turbulence, fine-scale structure of turbulence depends upon ε and υ.

(3) We have already investigated the influence of impeller diameter on X_Q. As d/D varies, flow rate Q_c certainly changes, but more important for micromixng, the ε distribution changes.

(4) We are investigating the scaleup rule for micromixng and have got results at two scales (2.5 liter and 19 liter). We will get the results in a 60 liter reactor soon. Conventional scaleup rule of constant ε will face a challenge in the case of SBR with fast complex reactions, because the trajectory of reaction zone will, in some case will significantly, vary on scaleup.

LDA MEASUREMENTS OF FLOW IN STIRRED GAS-LIQUID REACTORS

V.P. MISHRA, J.B. JOSHI, Dept. of Chemical Technology,
University of Bombay, IND

ABSTRACT

Mean flow, turbulence intensities and fractional gas hold-up have been measured using laser Doppler anemometer Fast Fourier transform analyser (FFT) was used for filtering the noise in the presence of gas. Fully baffled, flat bottomed cylindrical vessel of 500 mm i.d. was provided with 168 mm diameter pitched blade down flow turbine. Air and water were used as gas and liquid phases. The maximum gas hold-up was 30 per cent which was measured using the FFT analyser. A procedure was developed for the measurement of local gas hold-up using the velocity time data. It has been shown that simultaneous measurement of liquid and bubble velocities is possible using LDA.

The energy balance was established around the impeller at all the gas velocities. The values of hydraulic efficiency have been estimated. The mathematical model developed by Ranade and Joshi was extended to the case of gas-liquid system. A comparison has been made between the model predictions and the experimental observations.

INTRODUCTION

Stirred vessels are very commonly used in industry for a variety of reactions. Inspite of great utility of this equipment, the existing design procedures are not completely reliable. This status is basically because of the complex fluid mechanics in stirred vessels. Advances in experimental techniques and useful theoretical concepts of turbulence make it possible to analyse the fluid mechanics of these vessels in much more detailed manner. Ranade and Joshi (1990a, 1990b) have critically reviewed the literature on the experimental and modelling aspects of the flow generated by disc turbines. They have reported exhaustive flow information and have shown a good agreement between model predictions and experimental observations for the case of single phase liquid flow.

In the presence of gas, the two phase flow pattern is extremely complex. The impeller pumping capacity strongly depends upon the interactions between the impeller and the gas. The interactions result into a variety of regions of impeller operations (Mann (1983)). These include the regions of impeller loading, impeller flooding, cavity formation, cavity break-up and the recirculation of gas-liquid mixture in the impeller zone. The flow pattern in the vessel depends upon the region in which the impeller operates.

For the measurement of flow, laser Doppler anemometer is widely used. The presence of gas bubbles poses some problems in the LDA measurements. Since the refractive index of gas is different from that of liquid, the laser beams change their paths whenever a gas bubble comes in their way. As a result, the measurement point frequently goes out of focus. Secondly the bubbles provide light scattering centres and contribute to noise in a photomultiplier signal. Because of these difficulties there is practically no information in the published literature on the gas-liquid flows in stirred vessels.

The aims of the present paper are : (i) development of a technique for the measurements of flow in the presence of gas (ii) development of a method for segregation of bubble velocity from the liquid velocity (iii) development of a technique for the measurement of local values of fractional gas hold-up (iv) to present new data on the flow in gas-liquid stirred vessels.

EXPERIMENTAL

AGITATION EQUIPMENT AND INSTRUMENTATION

The experimental system was composed of : agitator assembly, cylindrical vessel, a DANTEC 55X modular series laser Doppler Anemometer (LDA) and its associated electronic instrumentation; a fast Fourier Transform analyser and a personal computer (80386 + 80387).

The basic components of the agitation set-up are shown schematically in fig.1. The agitator assembly comprised a one HP motor with pulley arrangement to allow eight different speeds. The diameter of the flat bottomed cylindrical vessel was 500 mm with four baffles of T/10 width (with no gap between the baffles and the wall). To minimise the effect of curvature on the intersecting beams, the cylindrical vessel was placed inside a square vessel. The vessel was constructed from transparent Plexiglass. The outer container was filled with same liquid as the vessel, in order to reduce the errors in locating the probe volume. The impeller was pitched blade turbine (PBT) of 168 mm diameter and having six blades. The blade width was 50 mm and the blades were attached at 45° to the hub. Water and air formed the liquid and gas phases, respectively. The gas was sparged using a ring sparger having four holes of 1.5 mm diameter. The ring had outer and inner diameters of 107 and 135 mm respectively. The gas flow rate was varied in the range of 0-150 ml/s. The impeller speed was 3.5 r/s in all the experiments. The entire agitation assembly with vessel was firmly mounted on a two dimensional (x,z) traversing mechanism. The traversing mechanism could displace the entire assembly through a span of ± 250 mm with 1 mm increments along either of the two coordinates. The movement of the traversing mechanism was controlled manually.

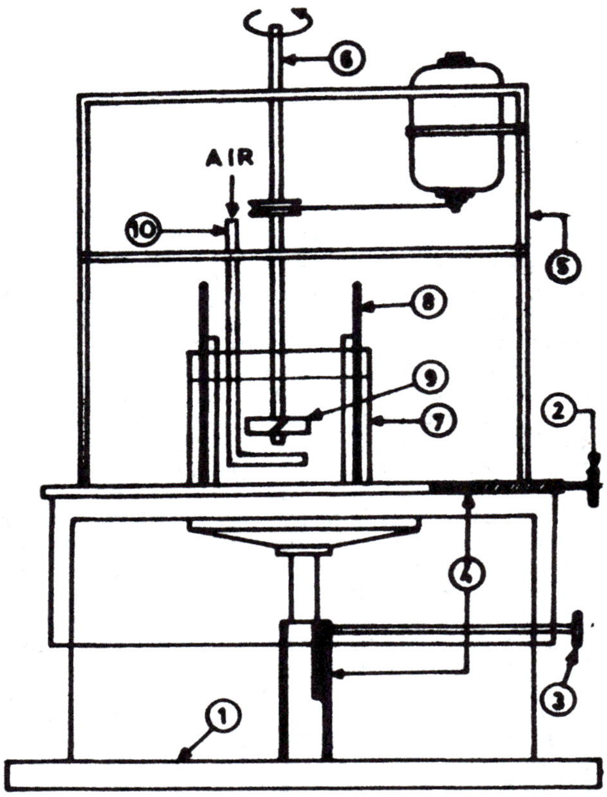

FIGURE 1

① Screw jack ② Wheel for horizontal movement ③ Wheel for vertical movement ④ Scale ⑤ Agitator mount assembly with motor ⑥ Agitator shaft with pulleys ⑦ Square tank ⑧ Cylindrical vessal ⑨ Impeller – PBT ⑩ Gas sparger

Measurements of the mean flow and rms fluctuating velocities were made using LDA in forward scatter mode. A 5W, Spectra Physics Ar-ion laser was used as the light source. The laser and the other optics were mounted on a bench which had a unidirectional traversing mechanism. The traversing mechanism could displace the front lens by ± 200 mm in 5 μm increments by means of a linearly encoded stepping motor monitored by the personal computer. To identify the flow reversals correctly, a frequency shift was given to one of the beams by means of a Bragg Cell. Data validation and signal processing were carried out with the help of a DANTEC 57N Burst Spectrum Analyser (BSA) (fast Fourier transform) analyser. The personal computer functioned as the central data acquisition and reduction controller via an interface (National Instrument PC2A card). Additional details pertaining to the equipment have been given by Ranade and Joshi (1990a).

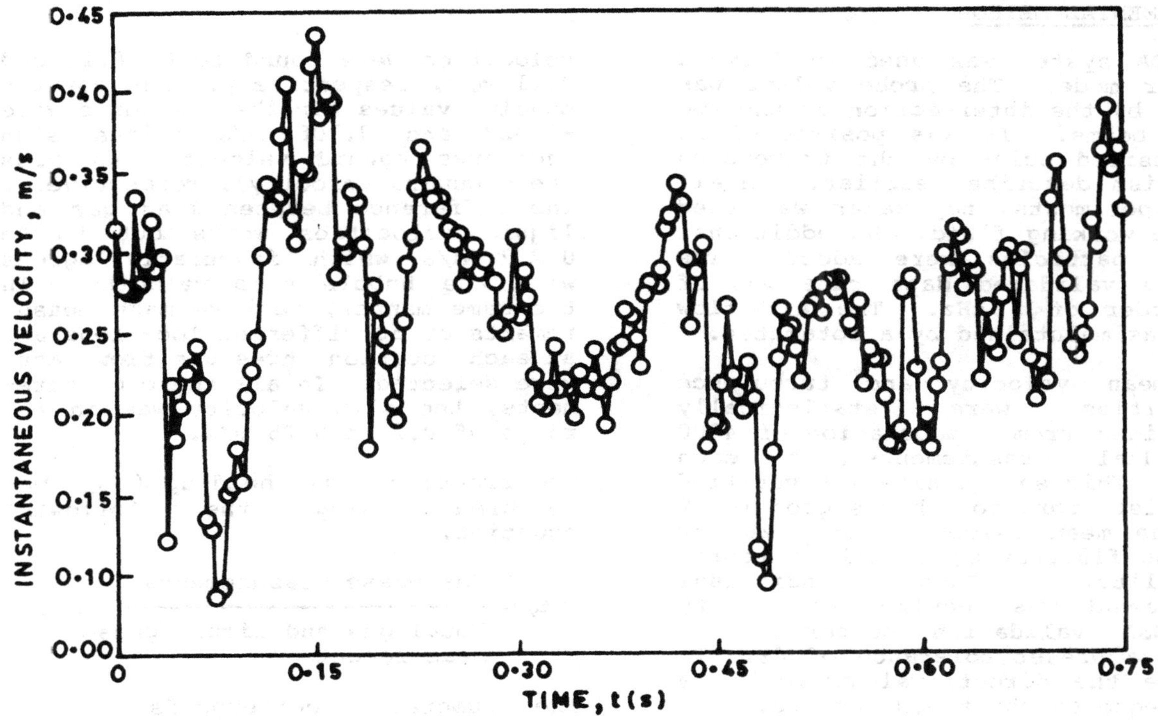

FIGURE 2A

Velocity-time data in the absence of gas

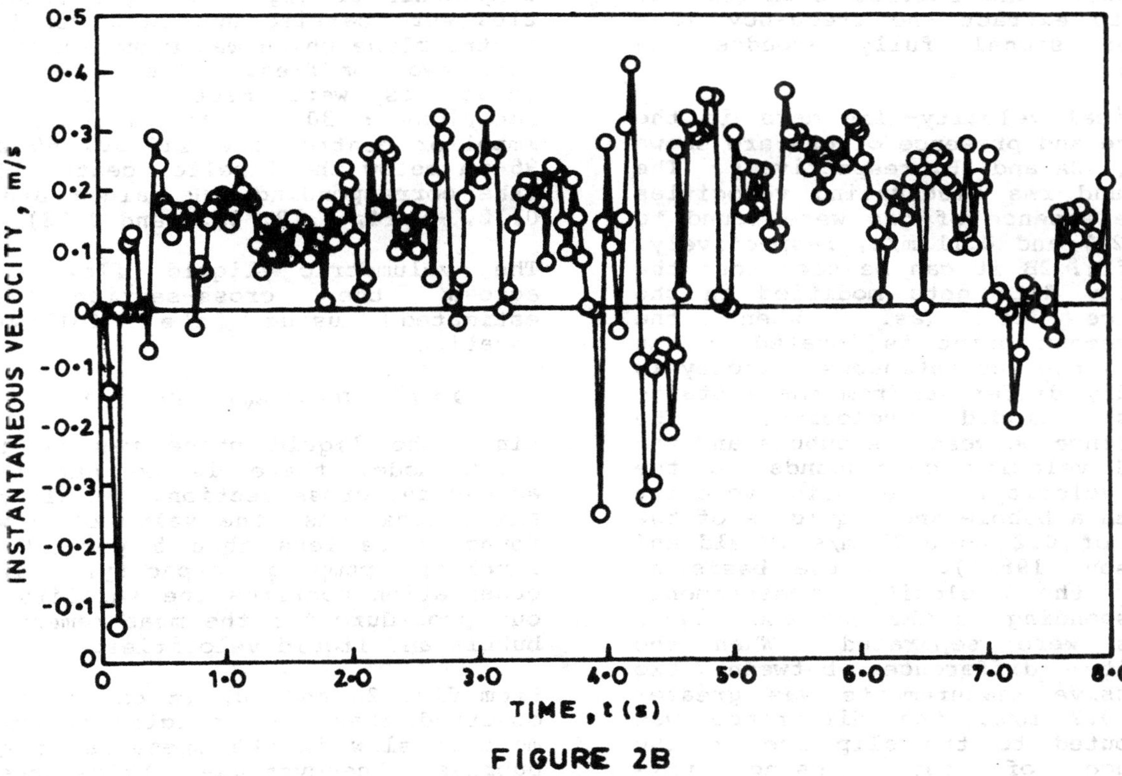

FIGURE 2B

Velocity-time data in the presence of gas

EXPERIMENTAL METHOD

The LDA system was used in forward scatter mode. The probe volume was formed by the intersection of the two laser beams. It was positioned at the desired point by the traversing mechanism described earlier. In all the experiments, tap water was used as the working fluid. No additional tracer particles were added. The average validated data rate was of the order of 5 kHz. The air flow rate was maintained by a rotameter.

The mean velocity and turbulence intensities were statistically determined from a population of 4000 individual measurements in each batch. This sample size was verified as satisfactory to achieve good ($\pm$ 1% for the mean velocity and $\pm$ 2% for the rms fluctuating velocity) reproducibility. Each individual measurement was required to satisfy the BSA validation comparison to within a preset tolerance of 6%. To resolve the directional ambiguity a net frequency shift was imposed.

The LDA measurements in a gas-liquid mixture result in a very noisy signal. The noise was filtered using a fast Fourier transform analysis incorporated in the Burst Spectrum Analyser. The Fourier transform is able to extract the frequency of a Doppler signal fully embedded in noise.

A typical velocity-time data in the absence and presence of gas are shown in fig. 2A and 2B, respectively. The mean and rms fluctuating velocities in the absence of gas were found to be 0.237 and 0.11 m/s, respectively. From fig. 2B it can be seen that the velocity data gets modified in the presence of gas. When the measurement point is located in the bubble, the instantaneous velocity is markedly different from the instantaneous liquid velocity. The difference between the bubble and the liquid velocity corresponds to the slip velocity. The slip velocity between a bubble and liquid is of the order of 0.2 to 0.25 m/s (Field and Davidson (1980)). On the basis of this, the velocity measurements corresponding to the gas and liquid phases were separated. When the velocity difference between two successive measurements was greater than 0.2 m/s, the difference was attributed to the slip due to the presence of gas. Using this procedure, the mean and rms liquid velocities were found to be 0.21 and 0.11 m/s, respectively. The corresponding values for the gas phase were -0.027 and 0.108 m/s (minus sign indicates upward velocity and plus the downward velocity), respectively. The difference between mean gas and liquid velocities works out to be 0.237 m/s which favourably agrees with the bubble slip velocity. In the same manner, we have made measurements at 60 different locations and at each location three gas flow rates were selected. In all these measurements, the slip velocity was in the range of 0.2 to 0.25 m/s.

The fractional gas hold-up (ϵ_v) was measured using the following equation.

$$\epsilon_v = \frac{\text{Gas phase measurements}}{\text{Total gas and liquid phase measurements}} \quad \ldots\ldots 1$$

The numerator corresponds to the number of measurements corresponding to bubbles only. The reproducibility of hold-up measurement was within 3 percent.

The measurements of radial and axial components of liquid and gas velocities were carried out in a vertical centre plane which was midway between the two baffles. The scans of velocities were made at 5 axial locations : 30 and 60 mm above the impeller centre line and 30, 60 and 95 mm below the impeller centre line (the corresponding z/R values are : -0.12, -0.24, 0.12, 0.24 and 0.38).

The volumetric liquid flow rate across the cross-section was estimated using the following equation :

$$Q = \int_0^R 2\pi r (1-\epsilon_v) v_z \, dr \quad \ldots 2$$

Since the liquid phase was in the batch mode, there is no net flow across any cross-section. At all the axial locations, the value of Q was found to be less than 5 percent of impeller pumping capacity. This observation confirms the validity of our procedure for the measurement of bubble and liquid velocities.

From fig. 2A and 2B, it can also be observed that the velocity measurement is slow in the presence of gas bubbles. Whenever gas bubbles appear in the beam path, the measurement point gets shifted and goes out of focus. Therefore, the rate of valid measurements decreases with an increase in the fractional gas hold-up.

RESULTS AND DISCUSSION

A radial gas hold-up profile has been shown in fig. 3 at three axial locations. It can be seen that the value of increases from the centre and attains a maximum and further decreases as the vessel wall is approached.

The values of mean liquid velocity in the absence and presence of gas are shown in fig. 4. It can be seen that the mean liquid velocities are mostly lower when the gas phase is present. At some locations, the true liquid velocity (V) in the presence of gas is high. However, when compared on the superficial basis, the local liquid velocities in the presence of gas were found to be less than those in the absence of gas.

From fig.4 it can be seen that the PBT generates downflow (as per design) in the central region (plus sign for downflow and minus sign for upflow velocities). The flow reversal was found to occur at r/R value of about 0.7. Thus, the downflow and upflow regions occupy almost equal cross-sections. Ranade and Joshi (1989a) have discussed the details pertaining to the flow generated by PBT in liquid alone.

Fig.5 shows the axial rms velocities at various axial and radial locations. When these values are reworked on the superficial basis, the rms velocities in the presence of gas were found to be lower than those in the absence of gas.

Fig. 6 shows the comparison between the experimental values of liquid velocity in the presence of gas and the predicted values by the model of

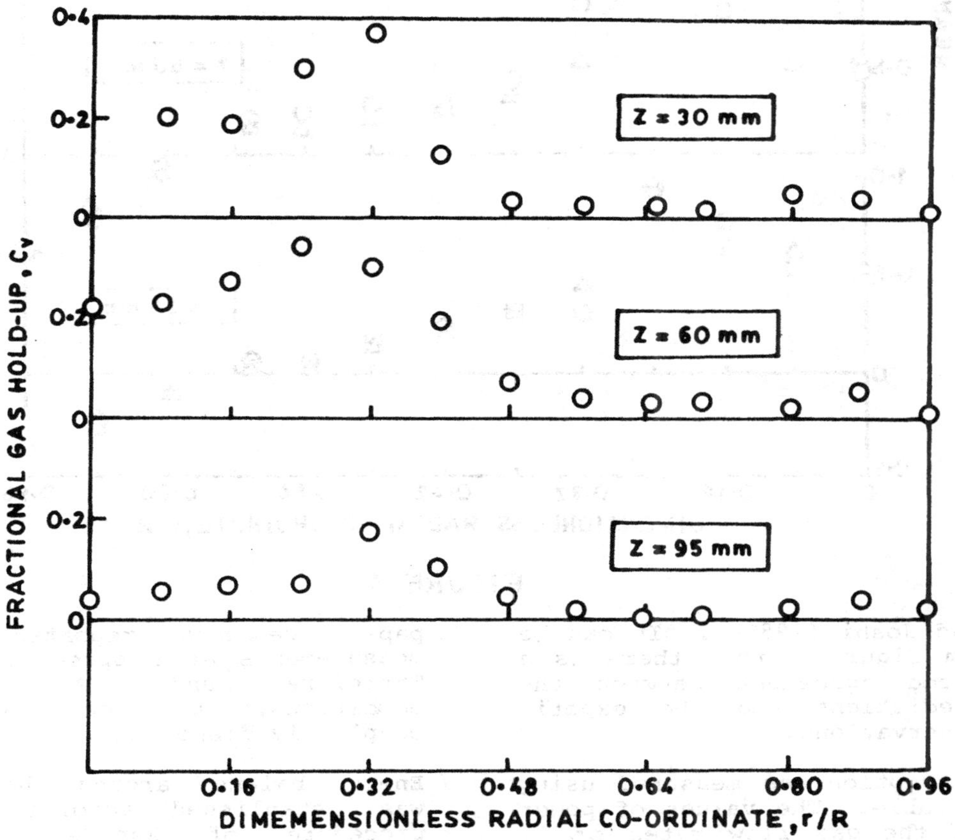

FIGURE 3

Radial and axial variation of fractional gas hold-up

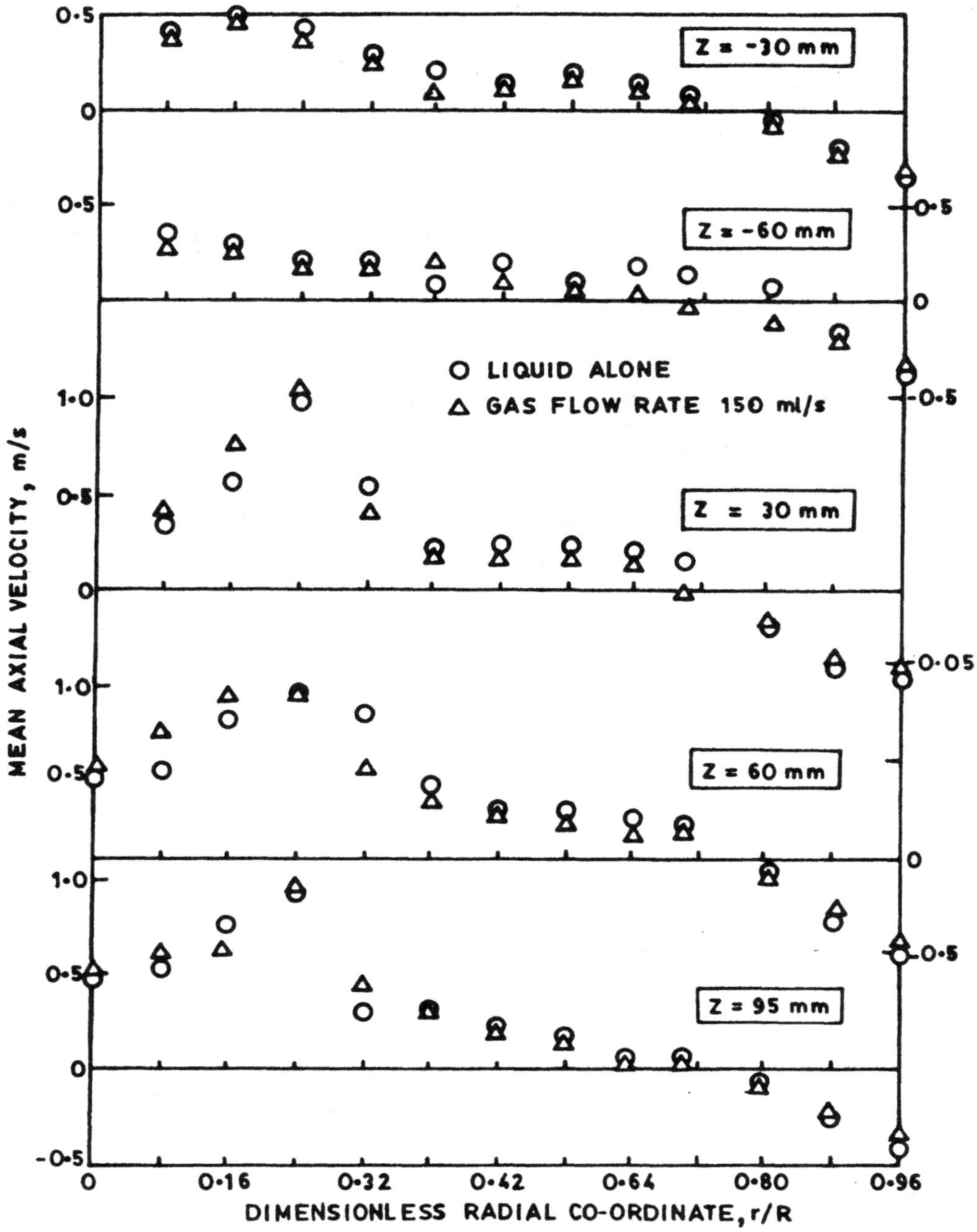

FIGURE 4

Ranade and Joshi (1989b). It can be seen from Figure 6 that there is a fairly good agreement between the model predictions and the experimental observations.

Power consumption was measured using a torque table. The values of power number of the gas flow rates of 0, 50, 100 and 150 ml/s were found to be 1.73, 1.64, 1.61 and 1.58 respectively. Further, the critical impeller speed for gas dispersion at the gas flow rates of 50, 100 and 150 ml/s were found to be 1.88, 2.37 and 2.7 r/s, respectively. In this paper, we have reported all the measurements at a speed of 3.5 r/s. Therefore, under the operating conditions, the gas phase was completely dispersed.

Energy balance around the impeller was established according to the procedure of Ranade and Joshi (1989a). They have reported that 32.6% of the input energy is dissipated in the impeller region when the gas is absent. At the gas flow rates of 50, 100 and 150 ml/s, the percent energy dissipated in the impeller region was found to be 32,

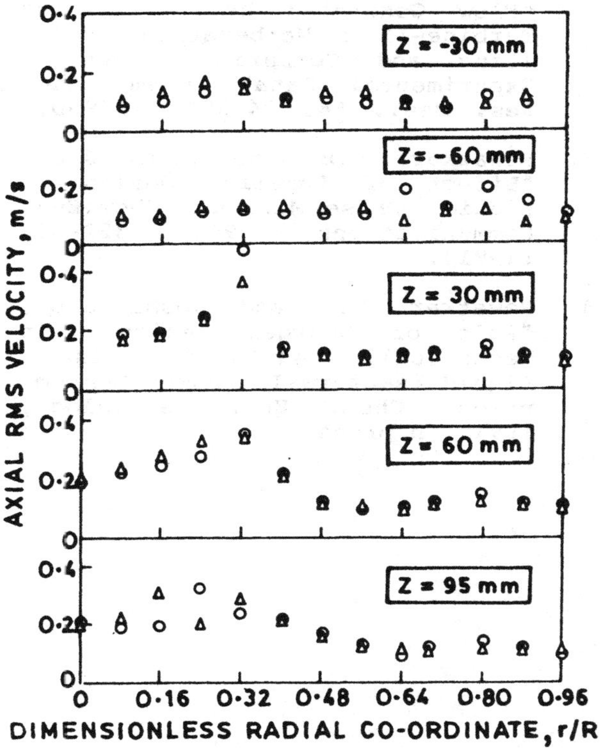

FIGURE 5

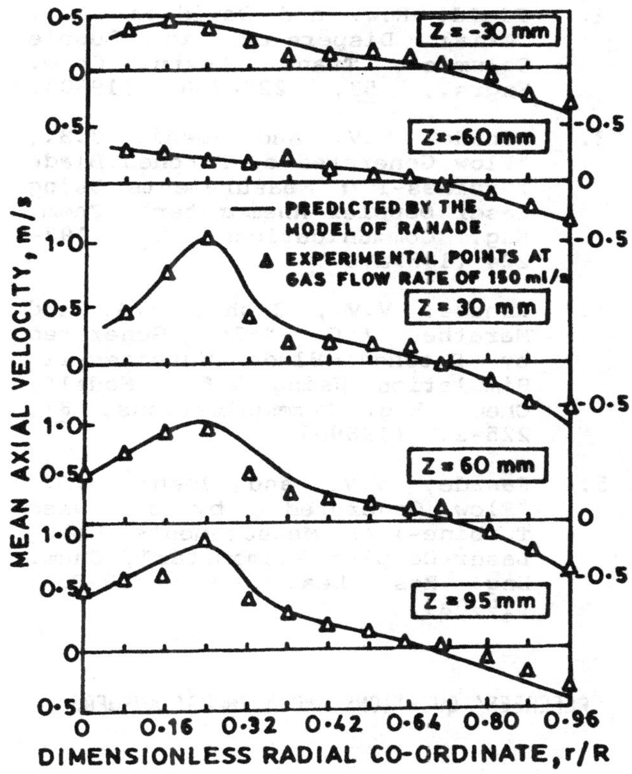

FIGURE 6

Comparison between model predictions and experimental observation

30.6 and 29.1, respectively. It can be seen that the energy dissipation pattern changes very slightly. This is probably because the operating impeller speed was well beyond the critical impeller speed for gas dispersion.

CONCLUSIONS

A technique has been developed for the simultaneous measurement of gas and liquid velocities and the turbulence characteristics. A procedure has also been developed for the measurement of local gas hold-up. Such measurements are expected to be useful for understanding the different regimes of operation such as impeller loading, impeller flooding, cavity formation, cavity break-up and recirculation of gas phase to the impeller. A systematic work is needed to cover a wide range of impeller speed, gas flow rate, impeller design and the liquid phase physical properties. Such a work has been planned in our laboratory.

ACKNOWLEDGEMENT

One of us (VPM) wishes to thank the University Grants Commission for the award of fellowship. The experimental work was supported by a grant from the Department of Science and Technology, Government of India [No. III, 4(77)/88-ET].

NOMENCLATURE

D	Impeller diameter (m)
R	Impeller radius (m)
r	Radial co-ordinate (m)
Q	Volumetric liquid flow rate (m^3s^{-1})
T	Vessel diameter (m)
v_z	Axial liquid velocity (ms^{-1})
z	Axial co-ordinate (m)
ϵ_v	Fraction gas hold-up (-) or void fraction

REFERENCES

1. Mann, R. "Gas-Liquid Contacting in Mixing Vessels", Inst. Chem. Engrs., Rugby (1983)

2. Field, R.W. and Davidson, J.F. "Axial Dispersion in Bubble Columns", Trans. Instn. Chem. Engrs., 58, 228-236 (1980).

3. Ranade, V.V. and Joshi, J.B., "Flow Generated by Pitched Blade Turbines-I : Measurements Using Laser Doppler Anemometer", Chem. Eng. Communications, 81, 197-224 (1989a).

4. Ranade, V.V., Joshi, J.B. and Marathe, A.G. "Flow Generated by Pitched Blade Turbines-II: Simulation Using k-ε Model", Chem. Eng. Communications, 81, 225-248 (1989b).

5. Ranade, V.V. and Joshi J.B., "Flow Generated by a Disc Turbine-I : Measurements Using Laser-Doppler Anemometer", Chem. Eng. Res. Des., 68, 19-33 (1990a).

6. Ranade, V.V. and Joshi J.B., "Flow Generated by a Disc Turbine-II : Mathematical Modelling and Comparison with Experimental Data", Chem. Eng. Res. Des., 68, 34-50", (1990b).

7. Rewatkar, V.B. and Joshi J.B., "Effect of Impeller Design on Liquid Phase Mixing", Chem.Eng. Communications, 91, 322-353 (1991).

8. Rewatkar V.B. and Joshi J.B., "Role of Sparger Design in Mechanically Agitated Gas Liquid Reactors-I : Power Consumption" Chem. Eng. Technology, 1991, In press.

REFEREES' QUESTIONS AND AUTHORS' ANSWERS

Q.1 Why do you use a non-standard pitched-blade turbine (blade width/diameter=W/D=0.3)?

A.1 We have selected W/D= 0.3 from energy efficiency point of view, Ranade and Joshi (1989a) have measured flow pattern generated by Pitched Blade Turbines of different blade widths and blade angles. They have shown that the pumping capacity per unit power consumption is maximum when W/D=0.3, when compared with other W/D ratios such as 0.2 and 0.4. Rewatkar and Joshi (1991) have measured liquid phase mixing time using pitched blade turbines. They have shown that, at a given power consumption, the mixing time at W/D=0.3 is minimum when compared with other W/D ratios.

Q.2 What is the value of h, the distance of agitator centre and Cs, the distance of sparger centre line from bottom of vessel ?

A.2 h = 168 mm.
Cs= 84 mm.

Q.3 In your fig. 3 & 4 is z the axial co-ordinate from agitator centre line ?

A.3 Yes, z is the axial co-ordinate from agitator centre line. Z is positive below the impeller and negative above the impeller.

Q.4 For your Fig. 3 to 5, what are the values of the gas flow rate Qg, $N_{\theta g}/N_{\theta o}$ and N_{qpg}/N_{qpo} ? (N_θ and N_{qp} being the power and (dimensionless) pumping number in aerated (index g) and non- aerated (index o) conditions)

A.4 For Fig. 3 to 5, Q_G=150 ml/S.
$N_{\theta g}/N_{\theta o}$ = 0.93 (Rewatkar and Joshi (1991))
N_{qpg}/N_{qpo}=0.74

A STUDY OF VENTILATED GAS CAVITIES ON DISC-TURBINE BLADES

S.A. MANNING, G.J. JAMESON, Dept. of Chemical Engineering,
University of Newcastle, NSW, AUS

ABSTRACT

The presence of ventilated gas cavities on the rear of impeller blades is an important characteristic of gas-liquid stirred systems. Cavities blanket the rear of the blades, reducing the impeller power consumption, the interfacial area and the rate of interphase mass transfer. Few fundamental studies involving measurements in and around cavities appear in the literature.

In this study, air was supplied through a hollow shaft to one blade of a two-blade standard disc-turbine impeller. The size of ventilated cavities produced at various impeller speed/gas flowrate combinations was measured. A specially designed bearing mounted on the impeller shaft allowed accurate measurements of the gas pressure inside the cavities to be made.

For constant gas flowrates, the impeller speed was increased from zero. At very low impeller speeds, no cavities were formed. At higher speeds cavities formed which grew in length as the speed increased. The cavities grew to a maximum size, which increased with gas flowrate. Once the maximum size was reached, further increases in the impeller speed caused the tails of the cavities to be eroded, resulting in shorter cavities. Cavity pressure was almost constant during the cavity growth region, dropping only slightly. At higher impeller speeds substantial reductions in cavity pressure were observed which mirrored the reduction in cavity size. The critical speed for the transition between cavity growth and dispersion regions was found to increase with increasing gas flowrate.

Models are presented that relate the growth region (Zone 1) and dispersion region (Zone 2) of cavity size with the cavity internal pressure for various impeller operating conditons. The implications of these results to the power consumption and gas dispersion quality in impeller stirred gas-liquid reactors are discussed.

INTRODUCTION

Effective mixing of gases and liquids is essential to the success of many industrial processes, including aerobic biological processes, recovery of minerals by flotation, treatment of sewage and fermentation. Of the many devices in use for dispersing gases in liquids, the most common is the disc- (or Rushton-) turbine agitated vessel. Their ease of manufacture and well defined operating characteristics have made disc-turbines a useful standard for gas dispersion investigations.

When gas is added to a turbine-impeller agitated liquid it is drawn to the low pressure regions at the rear of the blades, forming ventilated gas cavities. Investigations by Bruijn et al. (1974), van't Riet et al. (1975, 1976), Warmoeskerken et al. (1981), Nienow et al. (1977) and Ismail et al. (1984) clearly indicate that gas-filled cavities have a major influence on the hydrodynamics and power draw of an impeller.

Rennie and Valentin (1968) were the first to provide evidence of gas collection and cavity formation in the low pressure regions at the rear of turbine blades. More recent investigations have identified a number of cavity forms which exist at various impeller rotational speed and gas flowrate combinations. Van't Riet (1974) identified three forms and labeled them "vortex", "clinging" and "large". These cavity forms arose consecutively as the gas flowrate was increased at a fixed impeller rotational speed.

Warmoeskerken and Smith (1982) found that if a small amount of air is introduced below the impeller, six identical vortex cavities are produced. At higher gas flowrates, large cavities begin to form. One or two form at first, which hunt from blade to blade in an unstable manner. This condition exists only for a very narrow range of gas flowrates. At slightly higher gas flowrates, an alternating large and clinging cavity arrangement appears with the large cavities remaining on particular blades. This arrangement has become known as the 3-3 configuration and once formed, is stable over a wide range of impeller speed and gas flowrate combinations.

A number of investigators have suggested that cavity combinations such as the 3-3 configuration arise from uneven gas distribution to the blades (Warmoeskerken and Smith, 1982; Smith and Verbeek, 1988). The strongest argument in favour of this view is presented by Smith and Verbeek (1988) who observed natural cavity formation behind disc-turbine blades in near-boiling liquids. They found that large cavities tended to be of equal size, and not in the 3-3 configuration typical of ventilated cavities. They concluded that a 3-3 structure must occur due to differential gas sparging to different blades. Only under extreme conditions close to the liquid boiling point could they produce different sized cavities. No mechanism for the uneven ditribution of gas under the impeller disc, leading to different cavity sizes on different blades, has been found.

The complex relationships between impeller speed, gas flowrate and cavity size have been the focus of several studies. Warmoeskerken and Smith (1982) observed that the cavity size increased with increasing gas rate and impeller speed. At all speeds examined, for a given gas flowrate, the cavities have some maximum size and further increase in gas flowrate was required to flood the impeller. They conclude that at higher impeller rotational speeds, both the initial size and maximum size of large cavities is greater.

Nienow et al. (1983) found that cavity size increases with increasing impeller speed under constant gas flowrate conditions, a result valid for fluids with a range of rheological properties. This cavity growth was visually observed, and corresponded to a reduction in the impeller power number. The reduction in power number was thought to be due to the combined effects of streamlining, the reduced pumping capacity of the impeller and increased pressure in the gas filled cavities. Further increases in impeller speed resulted in an increase in power number, believed to be accompanied by reductions in the cavity size. The reduction in cavity size could not be verified visually, because at these high impeller speeds many small bubbles were ejected radially by the impeller, imparing visibility.

Investigations into cavity formation are typically carried out with multiple-blade impellers, with gas supplied to all the blades from a single source located beneath the impeller. Under these conditions, there is neither control nor metering of the flowrate of gas to a given blade. The exact flow and pressure conditions responsible for the formation of a particular cavity form is at present unknown.

The aim of this investigation is to gain knowledge of the mechanism of cavity formation. Experiments have been performed under controlled conditions, with the size and internal pressure of a single cavity being measured as a function of impeller speed and gas supply rate.

EXPERIMENTAL

A square, flat bottomed perspex vessel 0.58 m diameter, equipped with four wall baffles of width 0.058 m was used as the mixing tank.

A standard disc turbine (Rushton et al., 1950) of diameter 0.185 m was located 0.185 m from the base of the vessel. The working fluids were filtered tap water and air. The liquid depth was maintained at 0.58 m.

The impeller was driven by a 1.5 kW 3-phase induction motor mounted above the tank on a low friction bearing. A 0.5 N load cell connected to the motor by a lever arm provided accurate torque measurements.

Both two-blade and six-blade turbines were examined. Air was supplied to the rear of one of the blades through a copper tube, 3 mm in diameter, attached to the shaft at a point just above the disc. Air entered the top of the hollow shaft through a specially-made low-friction bearing, leaving through the copper tube whose open end was positioned at the centre rear of one of the blades. Air flowrate was measured using a rotameter, enabling the gas demand for a given cavity type to be accurately determined. By sparging gas to only one of the blades, a relatively small number of bubbles were produced which enhanced visual observation of the cavities under strobe lighting. A general arrangement of the apparatus is given in fig.1.

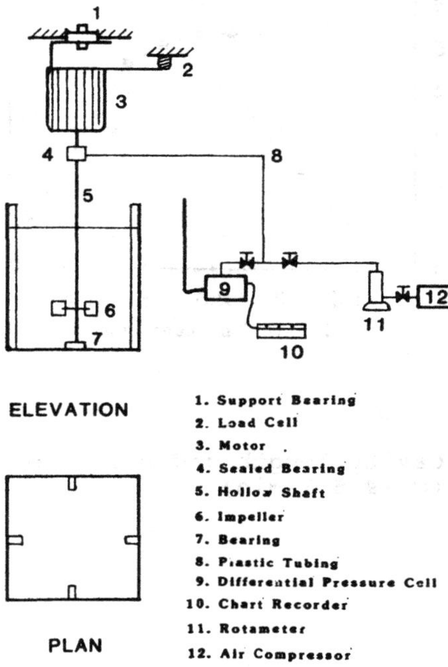

Fig.1 Experimental apparatus

By inspecting the gassed impeller under strobe lighting, with the strobe speed matched to the impeller speed, static images of the cavities were produced. The distance between the cavity tails and the blades were visually measured from calibrated markings on the side of the tank. Cavity size is reported as the dimensionless cavity length "L", defined as the distance between the cavity tail and the blade, divided by the blade height.

The gas line supplying air to the impeller shaft was joined by a line connected to a differential pressure (d.p.) cell. Simultaneous air injection and pressure measurements were made. The calibration between air-line pressure and cavity internal pressure was achieved as follows. The line pressure was monitored over a range of air outlet pressures. Pressure at the air outlet, at the rear of the blade, was varied by adjusting the immersion depth of the impeller. The calibration was conducted over a wide range of gas flowrates and air outlet pressures. Additional tests with no water in the tank indicated that the impeller rotational speed (causing the air-bearing to rotate) had no influence on the pressure calibration.

The static pressure in the liquid at the pressure tapping was used as the reference pressure.

RESULTS AND DISCUSSION

The relationship between the dimensionless cavity length, impeller tip speed and gas supply rate are shown in fig.2. Results were obtained by varying the impeller rotational speed for several fixed gas flowrates.

When the impeller tip speed is increased from zero the ventilated cavity behind the blade initially increases in length (Zone 1). At higher speeds the cavity begins to disperse and cavity length decreases (Zone 2). Both the speed required to reach the transition between Zones, and the cavity size at this speed, increase with increasing gas supply rate.

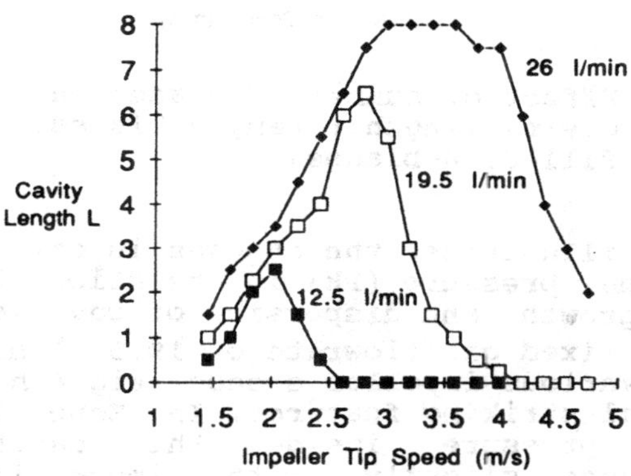

Fig.2 Cavity length for different impeller speeds and gas flowrates

Under high gas flowrate conditions (26.5 l/min), the curve of cavity size versus increasing impeller speed reaches a maximum at L = 8 (fig.2). This maximum occurs because the tails of the long cavities interact with the radial liquid flow from the opposing (trailing) blade. To examine this interaction further, experiments were repeated when four blades were added to the disc to produce a standard six-blade disc-turbine. The effect of the extra blades on the growth and dispersion of cavities is shown in fig.3 for a gas flowrate to the blade of 19.5 l/min. The addition of the extra blades reduced the distance between successive blades, resulting in a reduction in the maximum cavity size. The cavity size was restricted to a length approximately equal to the blade spacing. The tails of the cavities stopped just short of the leading face of the trailing blades. The data of fig.2 also indicate that the onset of Zone 2 and the size of Zone 2 cavities is unaffected by the proximity of other blades.

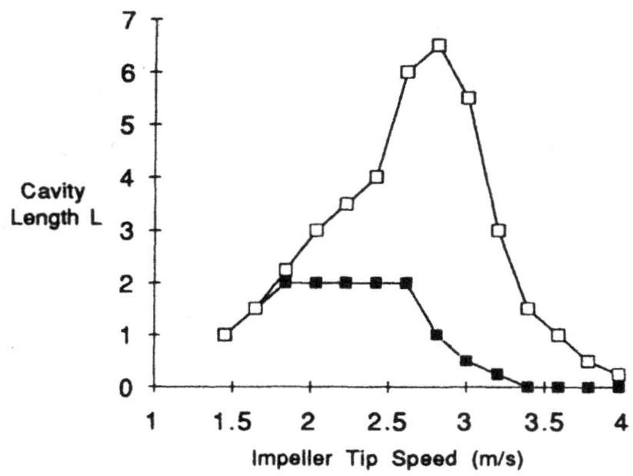

Fig.3 Effect of number of blades on cavity length (open, 2-blades; filled, 6-blades)

Fig.4 illustrates the changes in cavity internal pressure (Pk) in relation to the growth and dispersion of cavities for a fixed gas flowrate of 19.5 l/min and various impeller speeds. Fig.4 has several striking features. In Zone 1, the pressure inside the cavity decreases slightly with increasing speed up to the transition speed. At higher speeds (Zone 2) the cavity pressure decreases sharply, corresponding to the decrease in cavity size. These trends were also observed for tests conducted with gas flowrates of 12 and 26.5 l/min.

Fundamental differences between Zone 1 and Zone 2 cavities were observed during this study. Zone 1 cavities were tapered, and the largest cavities had tails which reduced to a point. Bubble breakage occurred from the tips of the cavity tails. Once formed, bubbles rose almost vertically out of the impeller stream with very little radial movement. In contrast, Zone 2 cavities had very blunt tails, and bubbles which formed at the rear of the cavities were pumped radially outward by the impeller. A sketch of the typical cavity shapes is shown in fig.5.

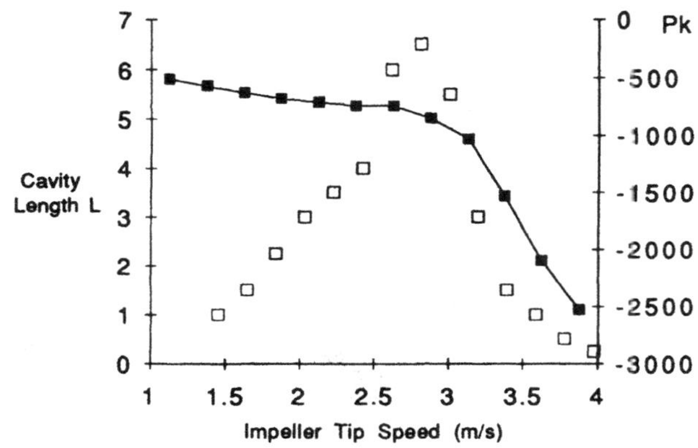

Fig.4 Cavity length and pressure (Q=19.5 l/min)

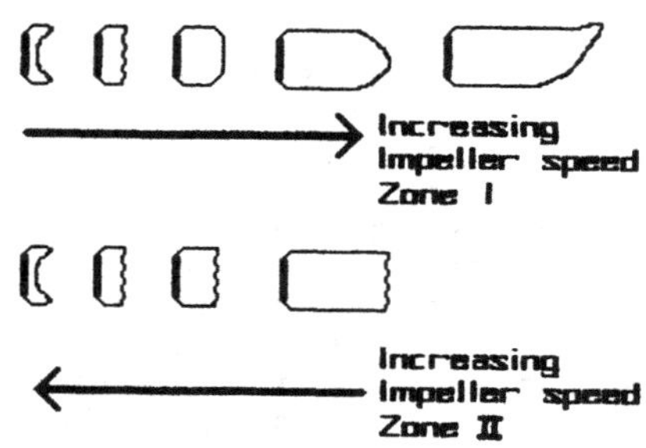

Fig.5 Typical cavity shapes

To explain the processes involved in the formation of cavities of a particular size and shape, the prevailing fluid flow conditions must be well specified. The models outlined below are based on simple concepts of fluid mechanics.

MODEL FOR ZONE 1

Consider a blade moving through a liquid at a relative velocity vL. Gas entering the impeller region (at a fixed flowrate) will be drawn to the low pressure region at the rear of the blade to form a ventilated cavity.

Liquid flowing around the cavity will result in a drag force in much the same way as air flow around a solid body such as an aerofoil. The liquid pressure at the centre rear of a bluff body, in this case the tail of a gas cavity, is termed the base pressure. The drag force on the cavity, and thus the base pressure, varies with the shape of the cavity.

Being a gas/liquid interface, the walls of a cavity are free to move under the influence of prevailing pressure gradients. The formation of a particular cavity shape results from an equilibrium condition where the gas pressure inside the cavity is the same as the liquid pressure at all points on the cavity surface.

Let us now consider what should happen to the cavity if the relative velocity were increased from vL to vL+dvL. By Bernoulli's equation (conservation of energy), the increase in relative velocity will result in a decrease in liquid pressure around the cavity. This pressure reduction will be greatest near the rear of the cavity, where the fluid streamlines on the cavity surface possess the highest degree of curvature. The gas/liquid interface at the tail of the cavity will move away from the blade in the direction of lowest pressure, in this case forming a longer cavity. As the cavity becomes longer, it assumes a more tapered shape. During this lengthening process, the liquid streamlines adjacent to the rear of the cavity contain progressively less curvature, resulting in a gradual increase in the liquid pressure adjacent to the rear of the cavity. The tail of the cavity will continue to extend until a new equilibrium state is attained.

The arguments presented here to explain the process of cavity growth in Zone 1 do not include the gas flowrate as a parameter. Inspection of fig.2 reveals that the size of Zone 1 cavities is indeed almost independant of gas flowrate for a specified tip speed. The maximum cavity size achievable in Zone 1 is however influenced by gas flowrate, but no explanation for this process has yet been found.

Whether the cavity is short or long, it has much the same cross sectional area. Within the cavity there is a net flow of gas towards the trailing edge, replacing the gas being entrained into the wake. Therefore, for a fixed gas flowrate to the cavity, the superficial gas velocity inside the cavity will be independant of the cavity length, and the cavity pressure can be expected to vary little with the imposed velocity change. The results shown in fig.4 indicate that the cavity internal pressure is indeed almost constant for Zone 1 cavities.

MODEL FOR ZONE 2

The process of increasing cavity length with increasing blade velocity at a fixed gas flowrate (Zone 1) is reversed in Zone 2 i.e. the cavities become shorter as the velocity is increased. The cavities in Zone 2 are blunt, lacking the elongated tails characteristic of Zone 1 cavities. Entrainment of bubbles from the rear of the cavities by the radial component of the fluid flow was examined as a possible explanation for these differences. Controlled tests revealed that the radially-flowing liquid affects the cavity shape only when the tail of a cavity comes very close (within 1 to 2 blade-widths) to the leading face of the trailing blade. Inspection of fig.2 reveals that for relatively low gas flowrates the maximum cavity size, corresponding to the transition between Zones, is quite small. The tails of these small cavities are many blade-widths away from the trailing blade, yet still undergo a transition into Zone 2. It appears unlikely then that the liquid flow induced by the trailing blade has much influence on the tails of the cavities in Zone 2.

Another possible explanation for the cavity behaviour in Zone 2 involves examination of the fluid flow fields around the cavities. The fluid flowing over the impeller blade will separate at the blade's leading edge. Separation is accompanied by the formation of velocity gradients in the liquid. Through the action of relaxation processes, the velocity gradients will become less intense as the liquid travels away from the blade towards the cavity tail. Unfortunately, the exact characteristics of the fluid flow conditions around cavities, especially at the rear where bubbles are formed, is extremely complex and poorly understood.

The nature of liquid flow over a gas stream differs significantly from that over a solid surface. For example, when liquid flows over a solid flat surface the fluid elements close to the surface are slowed, and consequently a boundary layer develops in the fluid which contains strong velocity gradients. In contrast, there is little slowing of liquid molecules adjacent to a gas-liquid interface. Very little shear is generated in the liquid and no boundary layer is formed. Despite these differences, there appears some interesting parallels between the behaviour of Zone 2 cavities and the transition to turbulence of fluid in the boundary layer on a solid flat surface.

When liquid flows over an infinite flat surface, a boundary layer develops, growing in thickness at points downstream of the leading edge. This boundary layer, which is initially laminar, will become turbulent at a distance Xt behind the leading edge. A turbulent boundary layer contains substantial components of the resultant fluctuating velocities normal to the direction of mean flow. Boundary layers separate soon after they become turbulent, resulting in intense backflow and fluctuating velocities [Simpson, 1989]. If a turbulent liquid boundary layer were to form on the surface of a cavity, the turbulent conditions associated with boundary layer separation would lead to dispersion of the cavity walls downstream of the separation point. The distance from the leading edge to the transition/separation point on the surface of a cavity would, by this analogy, dictate the length of Zone 2 cavities.

An expression for Xt, valid for fluid flow over an infinite flat plate, is given by Schlichting (1966) as:

$$Xt = (5*105n)/vL\rho \quad . \qquad (1)$$

This equation predicts that for water flowing over an infinite flat plate at a velocity of 4 m/s (typical of the liquid velocities in the present study), the transition to turbulence will occur at 0.125m behind the leading edge. This value is of the same magnitude as the cavity lengths measured in the present study for impeller tip speeds of 4 m/s.

Equation (1) also indicates that for fixed liquid properties, the distance to transition (Xt) is inversely proportional to the liquid velocity. This trend is clearly evident in the data of fig.2 where it is seen that the rate of cavity shortening is also inversely proportional to the liquid velocity for each of the gas flowrates examined.

To take this analogy even further, in the case of liquid flow over a gas cavity (air/water interface) the appropriate value for vL is the relative velocity between the liquid flowing around the cavity and the gas velocity inside the cavity i.e. vL-vG. Substitution into equation (1) yields:

$$Xt = (5*105n)/([vL-vG]\rho) \qquad (2)$$

Equation (2) suggests that for a constant superficial gas velocity, the transition from laminar to turbulent flow conditions on the surface of the cavity will occur closer to the leading edge if the liquid velocity (impeller speed) is higher. Thus, for the cavities of Zone 2, higher impeller speeds should result in shorter cavities. This prediction is consistent with the experimental results. Equation (2) also predicts that for a particular liquid velocity around a gas cavity, higher superficial gas velocities should generate longer Zone 2 cavities. This prediction is also consistent with the experimental results.

Futhermore, the smooth walls and tapered tails of the Zone 1 cavities suggest that the liquid flow is laminar over the whole cavity surface. If this is the case, the theoretical transition to turbulence must occur some distance behind the tails of the cavities. As the impeller speed increases through Zone 1, the cavity length increases and the theoretical value for Xt (which is greater than the cavity length) decreases. At some point, dependant on the gas velocity, cavity length and Xt become equal in value and the cavity has reached it's maximum size. Further increases in impeller speed reduce the value of Xt and the cavity tail is eroded by liquid in the turbulent boundary layer on the cavity surface, forming a shorter cavity with a blunt tail. Visual observations reproduced in fig.5 confirm that the Zone 2 cavities are much blunter than the Zone 1 cavities.

In Zone 1, the cavities become longer and more tapered as the impeller speed is increased. Under these conditions the power drawn by the impeller, relative to the ungassed condition, should decrease. In Zone 2 however, the power consumption should increase with increasing impeller speed because the base pressure increases when the cavities become shorter and blunter. These predictions are consistent with the power measurements made by Nienow et al. (1983).

The experimental results indicate that equation (2), derived for fluid flow on a solid surface is capable of predicting many of the trends observed in the relationships between cavity size, impeller speed and gas flowrate.

CONCLUSIONS

The size and shape of ventilated gas cavities on the rear of disc-turbine blades has been examined for a range of gas flowrate and impeller speed combinations.

By holding the gas flowrate constant and varying the impeller speed, two Zones of cavity characteristics were identified. Zone 1 corresponded to smooth cavities which grew in length and became more tapered as the impeller speed was increased. Bubbles formed at the rear of the cavities rose vertically with very little radial motion. Zone 2 was characterised by blunt cavities which reduced in length as the impeller speed was increased. Bubbles left the rear of these cavities with a strong radial velocity component.

Surprisingly, many trends in the properties of ventilated gas cavities appear to follow equations for the transition from laminar to turbulent flow in the boundary layer on the surface of a flat solid surface.

Unfortunately, the extremely complex fluid flow conditions present around ventilated gas cavities prevent full specification of the prevailing flow conditions. For this reason, the models presented here attempt to provide insight into the overall trends, not quantitative predictions.

NOMENCLATURE

L	Cavity length	(-)
n	Liquid dynamic viscosity	(Pas)
Pk	Cavity pressure	(Pa)
Q	Volumetric gas flowrate	(l/min)
vL	Liquid velocity	(m/s)
vG	Superficial gas velocity	(m/s)
dv	Incremental velocity	(m/s)
Xt	Cavity length (Zone 2)	(m)
ρ	Liquid density	(kg/m3)

REFERENCES

Bruijn, W., van't Riet, K.R. and Smith, J.M. : "Power Consumption with Aerated Rushton Turbines", Chem.Eng.Sci., V.18, p.355-375 (1974).

Ismail, A.F., Nagase, Y. and Imon, J. : "Power Characteristics and Cavity Formation in Aerated Agitations", A.I.Ch.E.J., V.30, p.487 (1984).

Nienow, A.W., Wisdom, D.J. and Middleton, J.C. : "The Effect of Scale and Geometry on Flooding, Recirculation and Power in Gassed Stirred Vessels", Second European Conference on Mixing, (BHRA, U.K.), p.F1-1 (1977).

Nienow, A.W., Wisdom, D.J., Solomon, J., Machon, V. and Vleck, J. : Chem.Eng.Commun., V.19, p.273 (1983c).

Rennie, J. and Valentin, F.H.H. : "Gas Dispersions in Agitated Tanks", Chem.Eng.Sci., V.23, p.663 (1968).

Rushton, J.H., Costich, E.W. and Everett, H.J. : Chem.Eng.Progr., V.46, p.395 and 467 (1950).

Smith, J.M. and Verbeek, D.G.F. : "Impeller Cavity Development in Near Boiling Liquids", Chem.Eng.Res.Dev., V.66, p.39 (1988).

van't Riet, K. and Smith, J.M. : "The Trailing Vortex System Produced by Rushton Turbine Agitators", Chem.Eng.Sci., V.30, p.1093 (1975).

van't Riet, K., Boom, J.M. and Smith, J.M. : "Power Consumption, Impeller Coalescence and Recirculation in Aerated Vessels", Trans.I.Chem.E., V.54, p.124 (1976).

Warmoeskerken, M.M.C.G., Feijen, J. and Smith, J.M. : "Hydrodynamics and Power Consumption in Stirred Gas-Liquid Dispersions", Fluid Mixing, I. Chem. E. Symposium Series, No.64, paper J1 (1981).

Warmoeskerken, M.M.C.C.G. and Smith, J.M. : "Description of the Powercurves of Turbine Stirred Gas-Liquid Dispersions", Fourth European Conference on Mixing, (BHRA, U.K.), p.237 (1982).

NUMERICAL PARTICLE TRACKING IN A TURBINE AGITATED VESSEL

K.A. KUSTERS, J.G. WIJERS, D. THOENES, Dept. Chemical Engineering,
Eindhoven University of Technology, NL

SUMMARY

Modelling orthokinetic turbulent coagulation in a stirred tank requires knowledge of parameters describing size dependent growth and breakup rates. The residence time weighted strain rate is used for estimating the initial coagulation rate. As the aggregates grow larger, breakup occurs due to the shear stresses exerted upon the aggregates in a limited region near the impeller. In this region further growth is excluded resulting in a reduced coagulation rate. The reduced coagulation rate as well as the breakup frequency as a function of the aggregate size can be obtained from numerical particle tracking. The particle tracking program predicts the movement of the aggregates in the turbulently agitated vessel. The program requires knowledge of the flow pattern and the shear field in the turbine agitated vessel, which were determined from laser doppler velocimeter measurements.

INTRODUCTION

In turbulent coagulation processes performed in stirred tanks, the final size of the aggregates is the result of a dynamic equilibrium between growth of the aggregates in the quiescent circulating bulk flow (low shear) and disruption in the impeller discharge zone (high shear). Important parameters in the coagulation process are the local strain rate s, the hydrodynamic shear stress τ acting on the aggregates and the residence times of the aggregates in the zones of different strain rate and shear stress. The strain rate s controls the orthokinetic coagulation rate and is given by the following equation[1]:

$$s = \sqrt{2\,\varepsilon / (15\,\pi\,\nu)} \qquad (1)$$

where ε is the local energy dissipation rate and ν is the kinematic viscosity of the fluid. The shear stress determines the local maximum size of the aggregates, and in the viscous sub range of turbulence the local root-mean-square (rms) shear stress is equal to:

$$\tau = \mu\gamma = \mu\,\sqrt{2\,\varepsilon / (15\,\nu)} \qquad (2)$$

where μ is the dynamic fluid viscosity. According to Sonntag and Russel[2], the maximum aggregate size is related to the shear stress through the relation:

$$d_{max} \sim (\tau / S_0)^{1/(n\,(D - 3))} \qquad (3)$$

where n = exponent in power law dependence of shear modulus on volume fraction solids
D = fractal dimensionality
S_0 = strength near centre of floc

The residence time in the zones of low shear stress represents the time during which the aggregates are able to grow beyond sizes which correspond to the zones of high shear stress. Because coagulation of larger aggregates is only possible in the zones of low enough shear stress (eq. 3), where these large aggregates are not being disrupted, their coagulation rate is reduced relative to the coagulation

rate of the smaller aggregates which do not break up. If f_t denotes the fraction of time the aggregates spend in the low shear zones and f_s represents the ratio of the strain rate the aggregates encounter in the low shear zones and the average strain rate the aggregates encounter throughout the vessel, the coagulation rate is reduced with a size dependent factor $f_r = f_t \cdot f_s$.
The residence times and strain rates in zones of different shear stress required to calculate the reduction factor f_r, are computed from a particle dynamics model that simulates the movement of a dispersed phase particle in the agitated vessel. The particle tracking enables us also to compute the breakup frequencies for the various aggregate sizes. For this purpose the simulation program keeps up a record of the number of times the aggregates cross the zones with high shear stresses which cause the aggregates to break up. If disruption is assumed to occur almost instantaneous when the aggregates enter the high shear zones, the breakup frequency is given as the number of crossings per unit time.
The simulation program requires exact knowledge of the average and rms fluctuating velocities and integral length scale of the turbulent eddies in every point of the vessel. To obtain the required information laser doppler velocimeter measurements have been performed in a baffled, Rushton turbine agitated cylindrical tank, similar to the one in which the coagulation experiments were carried out. The local energy dissipation rate was obtained from the inviscid estimate[3], which states that the energy dissipation rate is proportional to the rms fluctuating velocities raised to the third power, divided by the integral length scale. Interpolation between the measured points is done with mathematical models based on mass and momentum balances and experimental correlations, which describe the flow pattern and shear field in the stirred vessel. This static fluid flow model is combined with the concepts of turbulent dynamics in order to predict the reduction factor and breakup frequency as a function of aggregate size.

LASER DOPPLER VELOCIMETER

The laser doppler velocimeter (LDV) used is a one dimensional DISA 55L90a counter processor, operated in the forward scatter differential Doppler mode. A Bragg cell was used to allow an optical frequency shift of 40 MHz between the two beams to enable separation of pedestal and Doppler spectra. The photo multiplier output was electronically mixed with a variable frequency generator. The counter processor was set to the 'single measurement per burst' mode of operation. The digital output of the counter processor is connected to a personal computer. The data available are the electronic frequency shift f_S, frequency of the doppler burst f_D and the time between successive measurements (inter arrival time). The measured velocity is given by $(f_D - f_S) \cdot \delta$ with δ denoting the distance between the interference fringes at the crossing of the laser beams[4].

SIGNAL PROCESSING

Local energy dissipation rate ε values were obtained from the inviscid estimate[3]:

$$\varepsilon = C\, k^{3/2} / L \tag{4}$$

k = kinetic energy = $1/2\,(u_t^2 + v_t^2 + w_t^2)$
L = turbulent integral length scale
u_t, v_t, w_t = rms turbulent fluctuations of the 3 velocity components
C = constant

The integral length scale is obtained by applying the Taylor hypothesis:

$$L = U\,(T_r + T_z + T_\theta) \tag{5}$$

where T_r, T_z, T_θ = Eulerian integral time scales
$U = \sqrt{(U^2 + V^2 + W^2)}$
U, V, W = radial, axial and tangential mean velocities

The integral time scales are obtained by integration of the autocorrelation functions or from extrapolation of the power spectra to zero frequency.
To obtain reliable values for the local energy dissipation rate, non-biased time averaged values of mean velocity and especially of the rms fluctuating velocity are required. Possible velocity bias[5] errors were corrected for by weighting the velocity data with their corresponding inter arrival times. The measurements were performed at a high data rate (i.e. large amount of velocity measurements per second) in order to ensure an effective correction[6]. The autocorrelation functions or power spectra were computed using the so-called Sample-Hold technique, which gives non-biased results at high data rates[6].
Corrections were made for the periodic non-turbulent velocity pulsations produced by the turbine blades near the impeller tip.
Errors in measured mean and fluctuating velocities were less than 5%. The errors in the determined energy dissipation rate amount up to 20%.

MIXING VESSEL

The geometry of the perspex vessel of .20 m is shown schematically in fig. 1. The vessel was completely filled with tap water. The dimensions of the six bladed Rushton disc turbine impeller are summarized in table 1. The impeller distance from the bottom was equal to half the vessel diameter. The cylindrical vessel was placed inside a transparent glass container and the gap between the container and the cylinder was filled with water so that the refraction of the laser light beams from air to water did not occur over the curved surface of the cylinder, but over the flat surface of the container. The tangential and radial velocity were measured in two rotationally symmetrical points. The axial component was measured in both points and could be used to check the rotational symmetry.

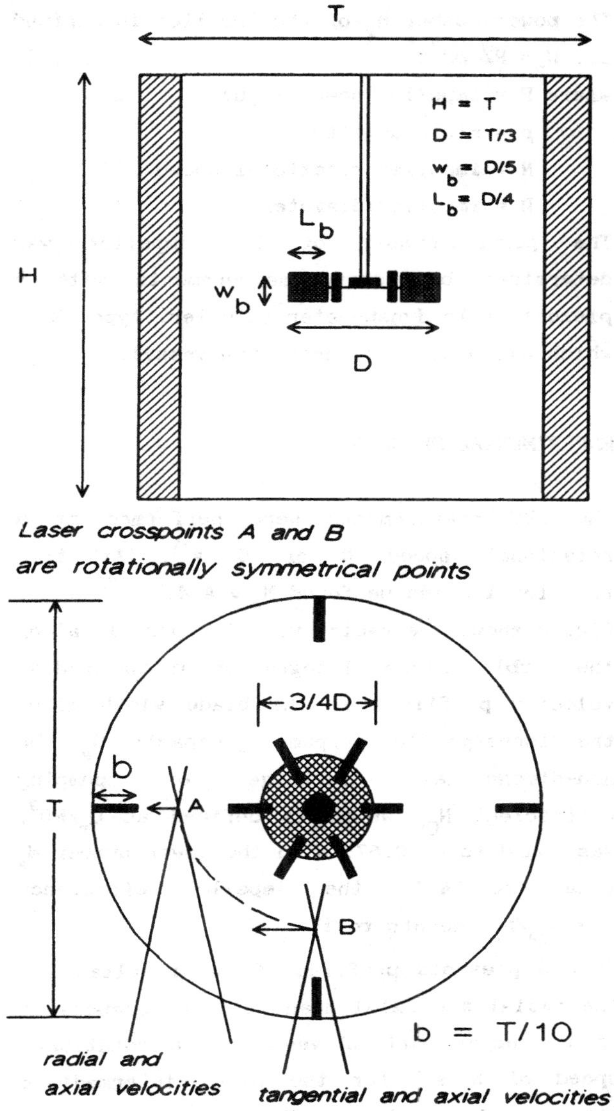

Fig. 1: Front and top view of basic geometry of the vessel. In the top view graph the measurement configurations for the radial and tangential velocities are indicated.

Table 1: Dimensions of turbine impeller in mm.

Impeller diameter	67.0
Blade width	13.4
Blade length	16.7
blade thickness	1.1
disc diameter	50.2
disc thickness	3.1
hub diameter	18.2
hub height	11.9
shaft diameter	8.1

The power number N_p of the impeller is defined as: $N_p = P/\rho N^3 D^5$ (6)

where P = impeller power input

ρ = fluid density

N = impeller rotational speed

D = impeller diameter

The power number of the impeller was determined by torque measurements with a piezoelectric dynamometer (Kistler, type 9275) which was positioned under the vessel.

EXPERIMENTAL RESULTS

The LDV measurements were performed at a rotational speed N of 5 s^{-1}. At this rotational speed we found N_p = 4.4.

Fig. 2 shows the radial velocity profile along the turbine blade. Integration of the radial velocity profile over the blade width gives the discharge flow or pumping capacity Q_p. The non-dimensional discharge or pumping coefficient N_{QP} which is defined as Q_p/ND^3, was found to be 0.67. With the power number N_p equal to 4.4, the impeller efficiency $\eta = N_{QP}/N_p$ amounts to 15%.

Fig. 3 presents profiles of the resultant of the radial and axial mean velocity components in the upper part of vessel at a rotational speed of 5 s^{-1} for the plane intermediate between the baffles. The centre of the circulation loop is located approximately at the non-dimensional radial coordinate $r_0^* = 2\ r_0/\ D = 2.4$ and the non-dimensional axial coordinate $z_0^* = 2\ z_0/\ D = 1.5$ (the z-origin being at the midheight of the liquid). The flow pattern in the lower part of the vessel proved to be the mirror image of the flow pattern in the upper part with $z^* = 0$ as the symmetry plane.

The tangential component proved to be only fairly strong near the turbine impeller. Compared to the resultant of the radial and axial movements, the tangential component remains less important in the rest of the tank (except for the centre of the circulation loop.)

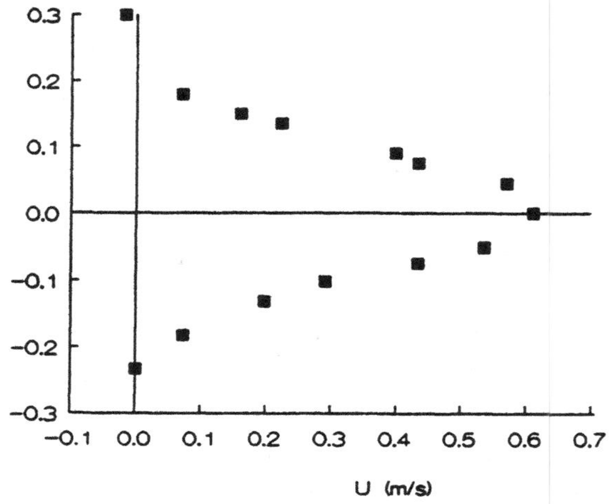

Fig. 2: Mean radial velocity profile at impeller tip.

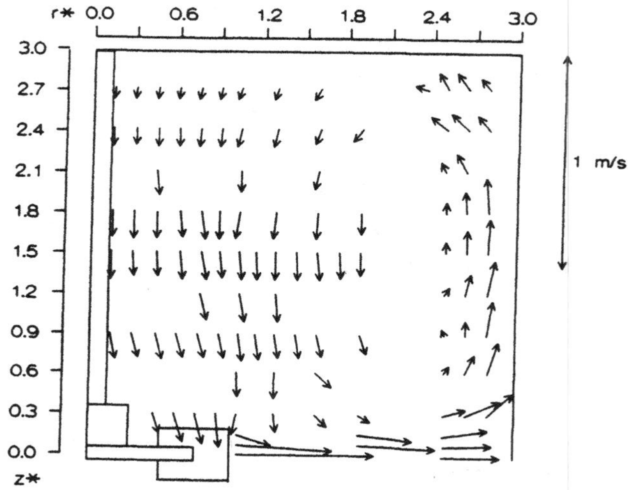

Fig. 3: Profiles of resultant of the radial and axial mean velocity components in the plane just between the baffles (θ = 45°). Stirrer speed = 5 s^{-1}.

Laser Velocimeter measurements were also carried out in the baffle plane (θ = 0°). In agreement with findings of Costes[4] we observed that the discharge flow is independent of θ except for the region close to the baffle. The circulation loops, however, change significantly. The circulation centre is closer to the centerline of the impeller stream, z_0^* = 0.8 and also nearer to the impeller axis at a radial position of r_0^*= 2.

The circulation flow capacity in the upper part of the tank at $\theta = 45^\circ$ was determined by integration of the axial mean velocity component from the impeller axis to the radial coordinate of the circulation centre at $z^* = 1.5$. The circulation flow number N_{Qc} which is defined as $2Q_c/ND^3$, is equal to 3.4 ± 0.4. The ratio of the circulation and discharge flow number is therefore $N_{Qc}/N_{QP} = 4.8$.

In the baffle plane Q_c is considerably smaller resulting in a circulation flow number of 1.7 ± 0.2 and $N_{Qc}/N_{QP} = 2.5$.

The integral length scale in the impeller discharge stream is approximately equal to the blade width. In the bulk the integral length scale varied from 2 to 4 times the blade width.

The value of the constant C in eq.(4) was evaluated through an integral power balance. For this purpose the values of $k^{3/2}/L$ were spatially integrated over the volume of the vessel and after multiplication with the fluid density compared to the total power input. The value of the constant C in eq.(4) was found to be 1 ± 0.2.

In fig. 4 the lines of equal $\varepsilon / \bar{\varepsilon}$ are plotted, with $\bar{\varepsilon}$ denoting the power input per unit mass averaged over the total vessel volume. The maximum value of $\varepsilon / \bar{\varepsilon}$ near the impeller tip is approximately 37.

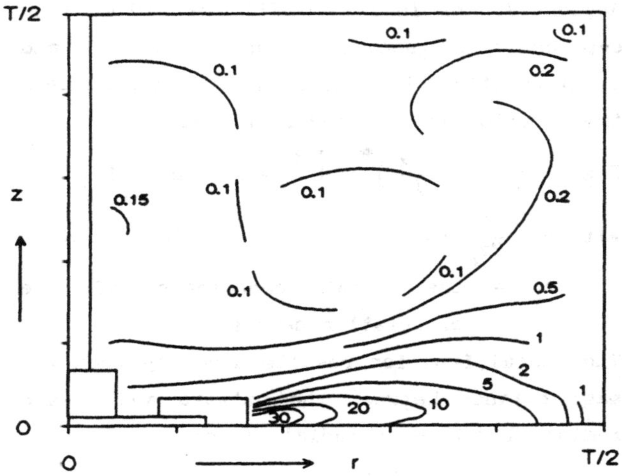

Fig. 4: Lines of equal energy dissipation rate in upper section of vessel.

MODEL FOR STATIC FLUID FLOW

The static fluid flow model provides estimates of the mean velocities, the rms fluctuating velocities, the integral length scale and the energy dissipation rate in every point of the vessel. This interpolation between the measured points was facilitated by analytic functions for the various parameters fitted to the experimental data. In the following sections the analytic expressions which have been used are discussed.

Mean velocity field

The model proposed here to describe the mean flow patterns in a fully baffled turbine agitated vessel is two dimensional. From the data of Reed[5], the influence of the baffle upon the mean velocity profiles seems to be mainly confined to a restricted area around the baffle ($-5^\circ \leq \theta \leq 5^\circ$), which suggests that the two-dimensional flow pattern in the 45° plane gives a good first approximation of the turbulent flow in a stirred tank. Furthermore with measurements conducted only in the 0° and 45° plane, the three dimensional picture is far from complete to justify three dimensional modelling.

The top section of the stirred vessel is divided into 6 subregions as shown in figure 5, for which different analytic expressions give a description of the radial velocity profiles. These subregions are:

- I & II. The flow from the impeller
- III. Jet stream impinging on vessel wall
- IV. Circulating flow
- V. Axial flow returning back to impeller
- VI. Flow of fluid in the region swept by the revolving impeller

The local axial mean velocity gradient is computed from the local radial mean velocity gradient through a mass balance:

$$\frac{\partial(rU)}{\partial r} + \frac{\partial(rV)}{\partial z} = 0 \tag{7}$$

And:

$$V(r,z) = 1/r \int_0^z \frac{\partial(rU(r,z'))}{\delta r} \partial z' \quad (8)$$

Eq.(8) is based on rotational symmetry, i.e. $\partial(rW)/\partial\theta = 0$. Outside region I this is only approximately true, and therefore only one of the two velocity components (radial or axial) can be made to fit the experimental data. We matched the axial velocity. The analytic expressions for the radial mean velocity profiles obtained from literature were adapted in order to establish the match between measured and calculated (through eq.(8)) axial mean velocities.

The following analytic expressions concern the mean radial velocity in the 6 subregions of the upper part of the tank. The radial velocity in the lower part of the tank can be computed by substituting -z instead of z into these equations:

Region I. The tangential jet stream model as proposed by Desouza[8] has been employed to describe the radial mean velocity:

$$U_I(r,z) = \frac{A\,\sigma^{1/2}}{2\,r^{3/2}}(\,r^2 - a^2\,)^{1/4}\left[1 - \tanh^2\left(\frac{\sigma\,z}{2\,r}\right)\right] \quad (9)$$

with $A = N\,D^2\,N_P^{1/2}\,(\,0.3355 - 15.91\,z\,)$

$a = .022251 + 1.9182\,z$

$\sigma = 19\,\exp(\,-0.108\,N_P)$

Region II. We modified the tangential jet model for this region:

$$U_{II}(r,z) = \frac{A\,\sigma^{1/2}}{2\,r^{3/2}}(\,r^2 - a^2\,)^{1/4}\left[1 - \tanh^2\left(\frac{\sigma\,w_b}{r}\right)\right] \quad (10)$$

with $A = N\,D^2\,N_P^{1/2}\,(.3355 - 15.91\,w_b/\,2\,)$

w_b = blade width

Region III. The basis of the analytic expression for the radial velocity in this region is the separation of the radial flow dependence into a radial and axial function[9], i.e.: $U(r,z) = F(z)\,G(r)$ (11)

On reaching the tank wall the radial velocity decreases rapidly to zero. The flow near the wall is similar to stagnation flow, for which the radial function of the radial mean velocity can be described by[9]:

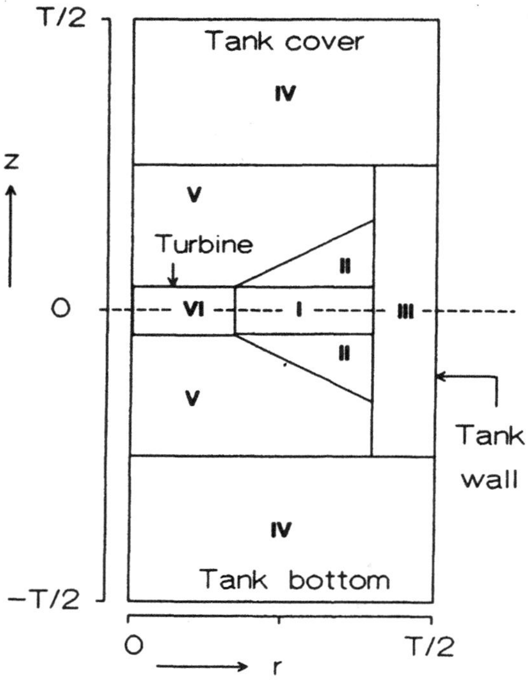

Fig. 5: Subregions used in the static fluid flow model.

$$G(r) = \frac{r}{4}\left[\left(\frac{2\,r}{T}\right)^4 - 1\right] \quad (12)$$

with T = tank diameter.

Because of continuity,

$$F(z) = U_I(r_0,z)\,/\,G(r_0) \quad (13)$$

where r_0 = the radial coordinate of the circulation centre

Region IV. As in region III the radial flow dependence is separated into a radial and axial function. For the axial function we used the equation as proposed by Platzer[9]:

$$F(z) = \left(1 - \frac{2\,z}{T}\right)^{m-1}\left[\frac{2\,z}{T}(1 + m) - 1\right]$$

$$\text{with } m = \frac{T}{2z_0} - 1 \quad (14)$$

z_0 = the axial coordinate of the circulation centre

The radial function results directly from the mass balance between the outward and inward radial flux in each radial plane:

$$G(r) = \int_0^{z_0} U\,\partial z \,/\, \int_{z_0}^{T/2} F(z)\,\partial z \quad (15)$$

Region V. The radial mean velocity U in this region is set to zero.

Region VI. This is the region swept by the impeller. No experimental data are available for this region. The radial velocity was assumed to vary linearly with the distance from the impeller axis:

$$U_{IV} = 2\ U_I(\ D/2,\ z)\ \frac{r}{D} \tag{16}$$

The calculated axial velocities are in close agreement with the experimental data except for the region just above the impeller as shown in figure 6. The differences in the measured and calculated axial velocities presented in fig. 6 are due to the somewhat simple models which are used to describe the flow in region V and VI.

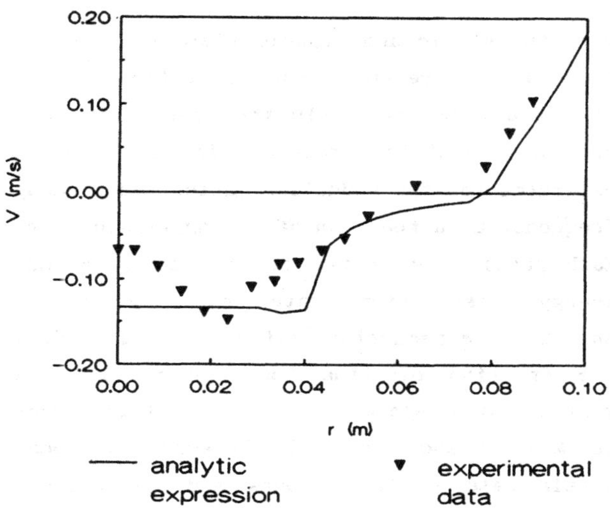

Fig. 6: Comparison of measured and calculated mean axial velocities just above the impeller ($z^* = 0.3$).

Fluctuating velocity field

The local rms fluctuating velocity is obtained by multiplying the average flow velocity **U** with the turbulent intensity I. The turbulent intensity is defined as:

$$I = \sqrt{\left(\frac{u_t^2 + v_t^2 + w_t^2}{U^2 + V^2 + W^2}\right)} = \sqrt{\left(\frac{2\ k}{U}\right)} \tag{17}$$

In addition to radial and axial mean velocities, estimates of the tangential mean velocity are required for computation of the average flow velocity. In the subregions I, II and III the following expression for the tangential mean velocity was applied:

$$W = \frac{a}{(r^2 - a^2)^{0.5}}\ U \tag{18}$$

which results from the tangential jet model[8]. If eq.(18) yields smaller values for the tangential velocity than 0.05 times the tip impeller speed[9] (= πND), the tangential mean velocity is set equal to the latter value. The tangential velocity is also set equal to 0.05 πND in the other subregions of the tank. Empirical analytic expressions to describe the variation of the turbulent intensity with the radial and axial coordinate were determined from the experimental data and implemented in the static fluid flow model for interpolation between the experimental data.

Isotropy is assumed, i.e. the rms fluctuating velocity is assumed to be equal in all three directions of space:

$$u_t = v_t = w_t = I \bullet U \tag{19}$$

Local integral length scale

The integral length scale is set equal to the blade width in regions I, II, III and VI, and equal to 2.5 times the blade width in the other regions.

Energy dissipation rate distribution

The local energy dissipation rate is calculated using eq.(4) with C equal to 1. The integration of the computed values of energy dissipation rate over the entire tank returned the measured total power input within 10%.

TURBULENT DYNAMIC FLOW

In the previous section, a model producing time averaged quantities like the mean velocity profile was developed. Extended with the turbulent dynamic flow model the instantaneous turbulent velocity field as experienced by a moving particle can be calculated. The turbulent dynamic model is therefore essential in predicting the movement of the particle which is being tracked throughout the vessel.

It is assumed that the moving particle follows the fluid flow accurately. This assumption seems reasonable considering the density of the primary particles (polystyrene) used in the coagulation experiments being almost equal to the density of the suspending medium (0.5 M NaCl solution) and the size of the aggregates being small.

The particle movement is started at randomly generated coordinates in the vessel. At this point the average and rms velocities, local integral length scale and energy dissipation rate are calculated. The instantaneous fluctuating velocity is randomly generated, with the absolute value set equal to the rms velocity. Summation of average and instantaneous fluctuating velocities give the instantaneous velocities in this point. Next the particle is moved by these instantaneous velocities in the corresponding two-dimensional flow direction during a certain time step $\Delta\tau$ much smaller than the Lagrangian integral time scale. The Lagrangian time scale is related to the local integral length scale by:

$$T_L = \frac{\beta L}{2 \sqrt{2/3\ k}} \tag{20}$$

According to Hinze[10]: $\beta \cong 0.4$

In the next points the same procedure is repeated except for the instantaneous fluctuating velocities which are now made time correlated. This is done by using an auto-regressive process[11] which provides a time correlated velocity component with a Gaussian amplitude distribution:

$$u^i = \rho(\Delta\tau)\, u^{i-1} + c^i \sqrt{1 - \rho^2(\Delta\tau)} \tag{21}$$

where u^i is the i-th velocity sample, $\rho(\Delta\tau)$ is the autocorrelation function and c^i is the i-th normally distributed random contribution. Each new velocity sample is thus generated from the previous one and the present value of the random contribution c^i. The values of c^i have a mean value of zero and are mutually uncorrelated. Their root-mean-square (rms) value is equal to the rms value of the velocity fluctuations. The autocorrelation function is an exponential time correlation: $\rho(\Delta\tau) = \exp(-\Delta\tau / T_L)$, where $\Delta\tau$ is the time step used in the simulation. Eq. (21) is valid in a homogeneous field, i.e. equal rms velocity in space. In order to incorporate the inhomogeneity of the turbulent field the instantaneous values of u^{i-1} are continually multiplied by the ratio of the rms values of u^i and u^{i-1}.

Precautions have been taken in order to avoid that the particle sticks to the wall, the impeller axis or the impeller disc.

PARTICLE TRACKING

With the static and dynamic fluid flow models combined, we are able to simulate the movement of the aggregates within the vessel. The aim of the particle tracking is to compute estimates for the reduction factor and breakup frequency as a function of the aggregate size. Each particle size is related to a maximum energy dissipation rate ε_{max} by eq.(2) and (3). The reduction factor is computed from the fraction of time the aggregates of a certain size spend in the low shear zones ($\varepsilon \le \varepsilon_{max}$) and the ratio between the local strain rate in these zones and the overall time averaged strain rate. The latter is calculated by:

$$\bar{s}_T = \Sigma\, s_i\, \Delta\tau_i \,/\, \Sigma\, \Delta\tau_i \tag{22}$$

where s_k is the strain rate in time step i of the particle's track. The strain rate encountered by the aggregates in the low shear zones was calculated by applying eq. (22) to the time steps in the zones with $\varepsilon \le \varepsilon_{max}$.

During the particle's track the residence time in zones of different energy dissipation rate is recorded, resulting in a residence time distribution as a function of the energy dissipation rate. Integration of this residence time distribution for $\varepsilon \le \varepsilon_{max}$

yields the fraction of time f_t the aggregates spend in the low shear zones.

The breakup frequency is computed from the number of times the aggregates enter the zones with $\varepsilon > \varepsilon_{max}$ from zones with $\varepsilon \le \varepsilon_{max}$, divided by the total duration of the particle's track. In the particle tracking program the particle's track is truncated at a hundred crossings of the ε_{max} boundary and the total time elapsed during the particle's track is computed. The size dependency of the reduction factor f_r and breakup frequency f_b was determined by performing the particle tracking for various values of ε_{max}.

RESULTS OF PARTICLE TRACKING

If the fluctuating velocities are set to zero, the particle follows the streamlines as shown in fig. 7. This figure is obtained by starting the particle's track at $z^* = 1.5$ and successive radial distances of 10, 20, ..., 70 mm from the impeller axis. The streamlines are not completely closed. The displacement of the end point relative to the starting point is dependent on the size of the time step $\Delta\tau$ which has been applied in the simulation and the size of the grid elements, which were used to solve eq.(8) numerically. The time step used to compute fig. 7 was $0.05 \cdot T_L$ and the grid was 250 by 400 (radial by axial coordinates). With fluctuations imposed upon the mean path, the particle is able to leave the streamlines. Plotting multiple circulations results in fig. 8.

The residence time distribution of the particle in zones of different energy dissipation rate is shown in fig. 9. The residence time distribution does not deviate much from the volumetric distribution of the energy dissipation rate, which is also presented in fig. 9, suggesting an equal flow rate through the various zones of different energy dissipation rate. The residence time averaged strain rate $\bar{s}_T$ is therefore almost equal to the volumetric averaged strain rate $\bar{s}_V$: $\bar{s}_V / \bar{s}_T = .98$. The residence time weighted strain rate does however differ much from the

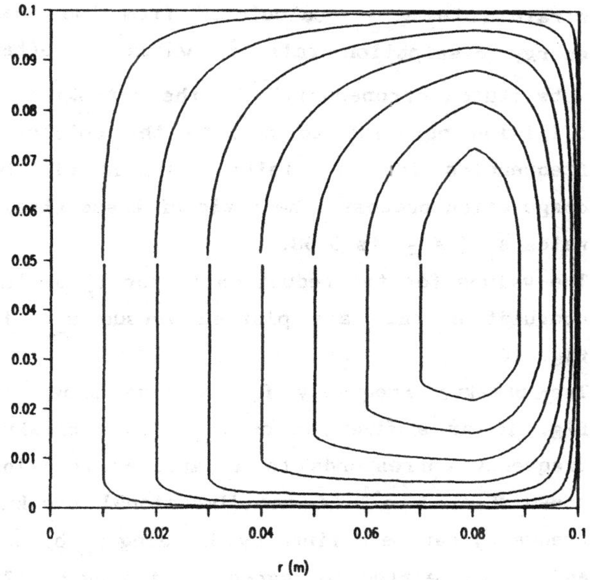

Fig. 7: Streamlines in upper section of vessel

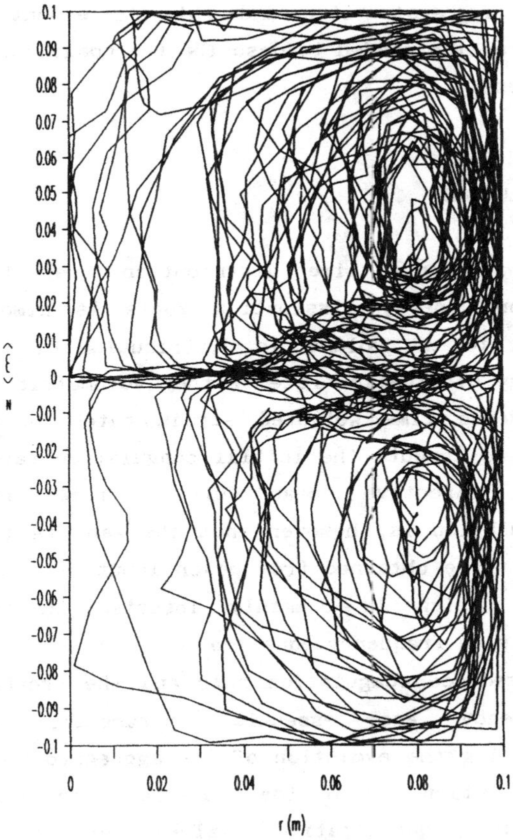

Fig. 8: The particle's track in the vessel

strain rate $s_{\bar{\varepsilon}}$ calculated from the mean energy dissipation rate $\bar{\varepsilon}$, which is often substituted erroneously[12] in the orthokinetic collision equation to compute the collision frequencies for the initial stages of the coagulation process. The ratio of these strain rates $\bar{s}_T / s_{\bar{\varepsilon}}$ is 0.68.

The values for the reduction factor f_r of the coagulation rate are plotted versus ε_{max} in fig. 10.

The breakup frequency f_b is also shown in fig. 10 as a function of ε_{max}. This breakup frequency corresponds to a impeller rotating speed N of 5 s^{-1}. A non-dimensional breakup frequency can be defined by dividing f_b by N. As ε_{max} is related to aggregate size by eq.(2) and (3) fig. 11 gives in fact the variation of the reduction factor and breakup frequency with aggregate size and this is essential information needed to describe the coagulation process.

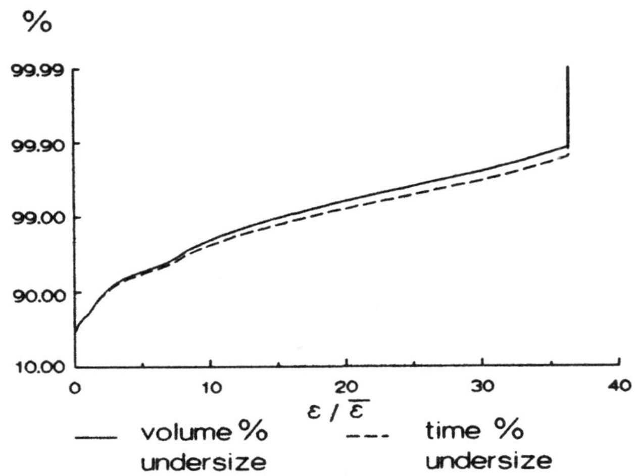

Fig. 9: Cumulative residence time distribution of the moving particle in the zones of different energy dissipation rate versus the cumulative volume distribution of the energy dissipation rate in the stirred vessel.

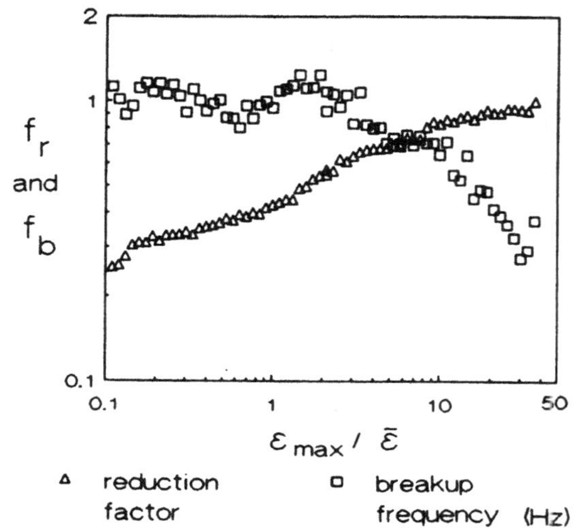

Fig. 10: reduction factor and breakup frequency as a function of ε_{max}.

CONCLUSIONS

The residence time distribution over the various energy dissipation zones is almost equal to the volumetric distribution of the energy dissipation rate. Either volumetric or residence time weighted strain rate can be used to compute the initial coagulation rate. This effective mean strain rate for coagulation is, however, not the same as the mean value obtained from power input per unit mass, which is commonly inserted in the collision frequency equation.

The reduced coagulation rate and the breakup frequency are essential parameters in modelling the evolution of the aggregate size distribution. With the numerical particle tracking quantitative values of these parameters were obtained for the various aggregate sizes.

REFERENCES

1. Saffman, P.G. and Turner, J.S., 'On the collision of drops in turbulent clouds', J. Fluid Mech. 1, 1956, 16-30
2. Sonntag, R.C. and Russel, W.B., 'Structure and breakup of flocs subjected to fluid stresses II: theory', J. Coll. Int. Sci. 115, 1987, 378-389

3. Tennekes, H. and Lumley, J.L., 'A first course in turbulence', MIT Press, 1972
4. Costes, J. and Couderc, J.P., 'Study by laser doppler anemometry of the turbulent flow induced by a Rushton turbine in a stirred tank: influence of the size of the units-I. Mean flow and turbulence', J. Chem. Eng. Sci. 43, 1988, 2751-2764
5. Edwards, R.V., 'Report of the special panel on statistical particle bias problems in laser anemometry, J. Fluids Eng. 109, 1989, 89-93
6. Kusters, K.A, Van Strien, C.J.G., Wijers, J.G. and Thoenes, D., 'Effect of velocity bias on integral time scale', Proceedings of 3rd International Conference on Laser Anemometry - Advances and Applications, Ed.: Turner, John T., BHRA/Springer Verlag, 1990, 557-566
7. Reed, X.B., Princz, M. and Hartland, S., 'Laser doppler measurements of turbulence in a standard stirred tank', Proceedings of 2nd European Conference on Mixing, 1977, paper B1
8. Desouza, A. and Pike, R.W., 'Fluid Dynamics and Flow patterns in Stirred Tanks with a Turbine Impeller', Can. J. Chem. Eng. 50, 1972, 15-23
9. Platzer, B. and Noll, G., 'Modelling of the local distributions of velocity components and turbulence parameters in agitated vessels - method and results', J. Chem. Eng. Process., 23, 1988, 13-31
10. Hinze, J.O., 'Turbulence', McGraw-Hill, New York, 1975, p. 426
11. Buchhave, P., Benzon, H.v. and Rasmussen, C.N., 'LDA Bias: Comparison of measurement errors from simulated and measured data, Proceedings of 5th Int. Symp. Appl. Laser Techn. Fluid Mech., 1990, paper 29.3
12. Koh, P.T.L., Andrews, J.R.G., Uhlherr, P.H.T., 'Flocculation in stirred tanks, Chem. Eng. Sci. 39, 1984, 975-985

NOMENCLATURE

A, a	tangential jet stream parameters
C	empirical constant in eq. (4)
c	random contribution to velocity fluctuation ($m\ s^{-1}$)
D	impeller diameter (m)
D	fractal dimensionality
d_{max}	maximum aggregate size (m)
F(z)	variation of U with z
f_b	break-up frequency (s^{-1})
f_r	coagulation reduction factor
f_s	ratio of coagulation strain rate and the average strain rate
f_D	frequency of Doppler burst (s^{-1})
f_S	electronic frequency shift (s^{-1})
f_t	fraction of time to coagulate
G(r)	variation of U with r ($m\ s^{-1}$)
I	turbulent intensity
i	iteration number
k	turbulent kinetic energy (m^2s^{-2})
L	integral length scale (m)
N	impeller speed (s^{-1})
N_P	power number
N_{QC}	circulation flow number
N_{QP}	impeller discharge coefficient
m	wall jet velocity parameter
n	exponent in power law dependence of shear modulus on volume fraction solids[2]
Q_C	circulation flow capacity (m^3s^{-1})
Q_P	discharge flow capacity (m^3s^{-1})
P	impeller power input (W)
r	radial coordinate (m)
r^*	nondimensional r, 2r/D
r_o	radial coordinate of circulation centre (m)
S_o	strength of floc matrix near centre of floc (Pa)
s	strain rate (s^{-1})
$s_{\bar{\varepsilon}}$	s calculated by substituting $\bar{\varepsilon}$ in eq. (1) (s^{-1})
$\bar{s}_T$	residence time averaged s (s^{-1})
$\bar{s}_V$	volumetric averaged s (s^{-1})
T	Tank diameter (m)
T_L	Lagrangian time scale (s)

T_r, T_z, T_θ	Eulerian integral time scales of radial, axial and tangential velocity fluctuations (s)
U	average flow velocity (m s^{-1})
U, V, W	radial, axial and tangential mean velocities (m s^{-1})
V_t	liquid volume (m^3)
u_t, v_t, w_t	radial, axial and tangential root mean square velocities (m s^{-1})
w_b	blade width (m)
z	axial coordinate (m)
r^*	nondimensional z, 2z/D
z_0	axial coordinate of circulation centre (m)

Greek symbols

β	constant in eq.(20)
δ	distance between interference fringes in LDV probe volume (m)
$\Delta\tau$	time step in particle tracking program (s)
γ	shear rate (s^{-1})
ε	energy dissipation rate (m^2 s^{-3})
$\bar{\varepsilon}$	mean ε, $P/(\rho\ V_t)$ (m^2 s^{-3})
η	impeller efficiency
θ	tangential coordinate (°)
μ	dynamic viscosity (kg $m^{-1}s^{-1}$)
ν	kinematic viscosity (m^2s^{-1})
ρ	liquid density (kg m^{-3})
$\rho(\Delta\tau)$	autocorrelation function
σ	tangential jet stream parameter
τ	shear stress (Pa)

DISCUSSION

1. What do you call the fractal dimensionality?

Answer:

The fractal dimensionality *D* as used in eq.(3) characterizes the interior structure of the floc. For a uniform internal porosity independent of floc size, *D* would be 3, the dimension of space. A smaller exponent is evidence that the density of particles within the floc decreases away from its centre. The dependency of the total number of particles in the floc *N* on the floc size R is given by:

$$N \sim R^D \tag{22}$$

and the volume fraction φ within the floc varies with radius *r* according to:

$$\varphi \sim (r)^{D-3} \tag{23}$$

Eq. (23) applies to $0 < r \le R$.

Taking into account the spatial variation of the internal volume fraction according to eq.(23), Sonntag[2] developed a breakup criterium for an isolated floc in simple shear flow. Assumptions of linear elasticity and the Mises criterion for yield under shear, with moduli and critical stress varying with volume fraction as φ^n, produce a relationship between the maximum nondimensional aggregate size $R' = R/\sqrt{(k)}$ and the shear rate γ of the form:

$$R' \sim (\mu\ \gamma/\ S_0)^{(D-1)/2n(D-3)} \tag{25}$$

With permeability $k \sim R^{(3-D)}$ eq.(25) reduces to:

$$R \sim (\mu\ \gamma/S_0)^{1/(n(D-3))} \tag{26}$$

Eq.(26) is equivalent to eq.(3) as $d_{max} = 2R$ and $\tau = \mu\gamma$, provided eq.(26) can be extended to turbulent flow. As long as the aggregates are smaller than the Kolmogorov microscale of turbulence this latter assumption does not seem to be a serious drawback.

2. Why is N_{QP} so low?

Answer:

Our value for the pumping coefficient, $N_{QP} = 0.67$, is smaller than the values reported by Revill[13] in table 2 of his article. These values lie in the range of 0.85 ± 0.10. Most of the data used by Revill to derive these N_{QP} values were based on velocity measurements performed with flow obstructive techniques such as hot wire/film anemometers and pitot probes. As pointed out by Costes[14], hot wires and pitot tubes cannot detect reversal of flow, i.e. a change of sign in the radial velocity. This results in a larger mean radial velocity value than would be obtained with laser doppler anemometry or streak photography. Costes reported N_{QP} values obtained by hot film and laser doppler anemometry of 0.93 and 0.73 respectively. This latter value is in close agreement with the value (0.75) reported by Revill which was based upon the streak photography measurements performed by Cutter[15].

Our value for the pumping capacity is only slightly lower than the value reported by Costes. This minor difference is to be attributed to the larger disc thickness of our impeller with respect to the impellers used by Costes. A larger disc thickness lowers the power number[16] of the impeller and its corresponding pumping coefficient.

REFERENCES CONTINUED

13. Revill, B.K., 'Pumping capacity of disc turbine agitators - A literature review'. 4th European Conference on Mixing, BHRA, 1982, 11-24

14. Costes, J., Alran, C. and Couderc, J.P., 'Characteristics of the discharge flow from a Rushton turbine. Measurements by thermal and laser-Doppler anemometry'. Int. Chem. Eng. 31, 1991, 55-65

15. Cutter, L.A., 'Flow and turbulence in a stirred tank', AIChE J. 12, 1966, 35-45

16. Bujalski, W., Nienow, A.W., Chatwin, S. and Cooke, M., 'The dependency on scale of power numbers of Rushton disc turbines', Chem. Eng. Sci. 42, 1987, 317-326

Karl A. Kusters

RESIDENCE TIME DISTRIBUTION OF SOLID AND LIQUID PHASE IN A STIRRED TANK REACTOR

L. GUO, S. YANG, Institute of Chemical Metallurgy,
The Chinese Academy of Sciences, Beijing, China

ABSTRACT

The residence time distribution (RTD) of solid and liquid phases in a stirred tank reactor has been studied in this article. Comparison with RTD of solid and liquid phases was also made. In a continuous stirred tank reactor (CSTR), factors influencing RTD are stirring speed, solid particle concentration, geometrical parameters of reactor as well as the location of the inlet and outlet of the fluid. Sea sand and water are used as solid particle and liquid phase. Limestone with the same particle size of sand was used as tracer of solid particle and 10% KCl solution was used as tracer of liquid phase. Two regions model was used to describe the flow pattern of the fluid in the tank, there is only one parameter on the model has to be evaluated based on the experimental data.

INTRODUCTION

Solid-liquid stirred tank reactor has been widely used in many areas, such as chemical, metallurgical and biological industries. As a multiphase reactor, it behaves as a device with better mixing, mass and heat transfer between solid and liquid phases. Therefore, mixing and flow pattern of fluid are important for evaluating the performance of the reactor. Up to now, RTD of fluid is a still better method to describe the mixing of fluid in a CSTR.

By means of RTD, the flow pattern and mixing performance of solid and liquid phases in a CSTR can be described reasonably and satisfactorily. Solid particles and liquid phase are in turbulently moving in the tank during agitation. The main factors influencing RTD of the two phases are stirring speed; solid particle concentration as well as geometrical parameters of the stirred tank, the location of inlet and outlet of fluid also influence RTD of the fluid. The aim of this paper is to study these parameters influencing RTD of the solid particles and liquid phase and using two regions model to describe the mixing and flow pattern of the fluid in the reactor.

Zweitering(1) studied the factors influencing mixing of solid particles in a stirred tank, and gave definitions of "complete suspension" and "homogeneous suspension", empirical expressions were established for the critical stirring speed to approach these above two states. Porcelli and Marr(2) studied the suspension of a single solid particle in a stirred tank, he found the rising velocity of liquid is the main factor of suspension of solid particle. Narayanan(3) thought that suspension of solid particles is due to drug force, from the view of force balance, he established an empirical expression of critical stirring speed. Riger(4) made an analysis from momentum balance. Bohnet(5) applied a laser senser to measure the concentration of solid particles, studied the distribution of solid concentration in the reactor. Tojo(6) used the sedimentation-dispersion model to calculate the distribution of concentration of solid particles with micro-computer. From the view of turbulent theory, Frantisek Penaz(7) established a turbulent equation of solid particles and calculated a distribution of concentration, but the result of calculation differs from those of experiment slightly. Bowling(8) used a magnetic tracer to study RTD of solid particles on a fluidized-bed. Krishnaiah(9) used a coating tracer to study RTD of solid particles in a fluidized-bed. Huihua Mao(10) found that size of solid particles also influence their RTD. Up to now, there is no paper to deal with the effect of locations of entrance and exit of the fluid.

MODELS OF FLOW PATTERN

Generally, mixing of solid particles in a CSTR is often considered as perfect mixing, but there is a large deviation from that condition. The flow of fluid in the stirred tank can be divided into two regions, upper region and lower region as shown in Fig. 1.

Therefore the two-regions model (each region can be assumed as a perfect mixing space.) separated by the turbine blade can be used to simulate the flow patterns of solid particles and liquid phase in the reactor. For a CSTR the RTD of solid particles and liquid phase , also varies with the location of entrance and exit of fluid. Therefore different models were used to simulate the flow patterns with different locations of entrance and exit of fluid.

Model 1: inlet is in upper section, outlet is in lower section.
Material balance equation:

$$RdC_1/d\theta = C_0 + kC_2 - (1+k)C_1 \qquad (1)$$

$$(1-R)dC_2/d\theta = (1+k)(C_1 - C_2) \qquad (2)$$

Initial conditions:

$$C_1 = 0, C_2 = 0, \quad When \quad t = 0 \qquad (3)$$

The solution of eq. (1)(2)(3) is below:

$$F(\theta) = C_2(\theta)/C_0 = 1 + (\lambda_1 e^{\lambda_2\theta} - \lambda_2 e^{\lambda_1\theta})/(\lambda_2 - \lambda_1) \qquad (4)$$

where

$$\lambda_{2,1} = \{-(1+R) \pm [(1+k)^2 - 4R(1-R)(1+k)]^{1/2}\}/\{2R(1-R)\} \qquad (5)$$

It can be seen , that more complete mixing of fluid can be obtained with greater value of flow rate coefficient k between two regions in the model. If k=0, this model will become tow tanks in series model.

model 2: inlet is in upper section, outlet is also in upper section. Material balance equation:

$$RdC_1/d\theta = C_0 + kC_2 - (1+k)C_1 \quad (6)$$

$$(1-R)dC_2/d\theta = k(C_1 - C_2) \quad (7)$$

Initial conditions:

$$C_1 = 0, C_2 = 0, \quad when \quad t = 0. \quad (8)$$

The solution of eq. (6), (7), (8) is below

$$F(\theta) = C_1(\theta)/C_0 = 1 + [(\lambda_2 + 1/R)e^{\lambda_1\theta} - (\lambda_1 + 1/R)e^{\lambda_2\theta}]/(\lambda_1 - \lambda_2) \quad (9)$$

$$\lambda_{1,2} = \{-(1+k-R) \pm [(1+k-R)^2 - 4kR(1-R)]^{1/2}\}/[(2R(1-R)] \quad (10)$$

From above expression, the more complete suspension of fluid will be approached with greater value of k. If k=0, this model will be an ideal mixing region with a dead space.

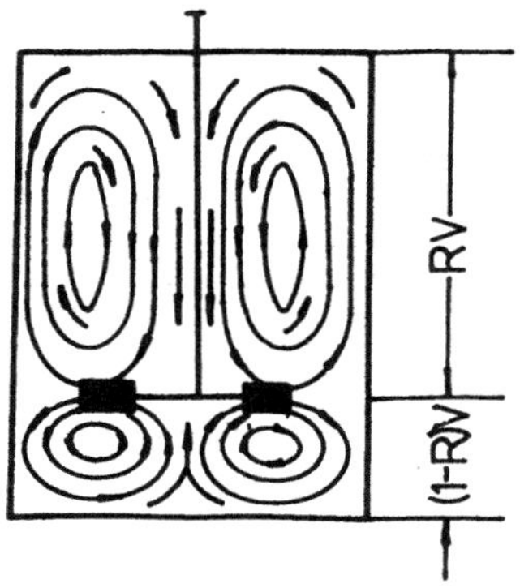

Fig.1 Flow pattern of the fluid in the tank

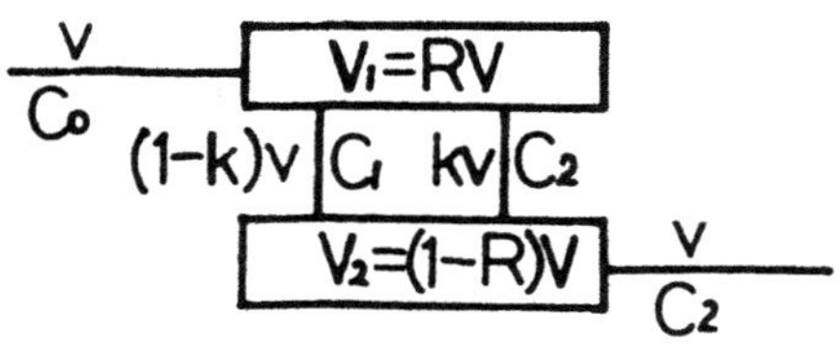

Fig.2a Model 1

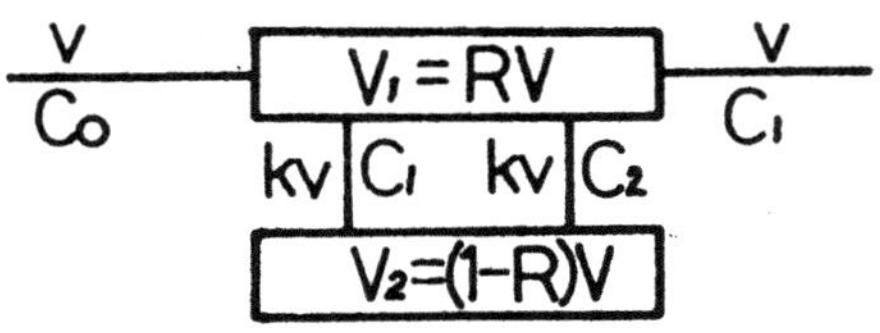

Fig.2b Model 2

EXPERIMENTAL

The experiment was carried out in a standard stirred tank made by plexiglass with 0.12m in diameter, there are four baffles on the wall, a Rushton turbine with six flate blades which are shown in Fig.3. The positions of entrance and exit of fluid are located on the opposite side of the wall, the height of which from bottom are 0.13m and 0.03m respectively.

Sea sand is used as solid particle, its density is 2670 kg/m^3, its mean diameter is 0.195 x $10^{-3}m$ (0.15 to 0.30x$10^{-3}m$), its shape is nearly spherical. Tap water is used as liquid phase. Concentration of solid particles in the reactor is from 3% to 7%(wt), stirring speed is from 1200 to 1600 rpm , the level of slurry in the tank is 0.14 to 0.18m from the bottom.

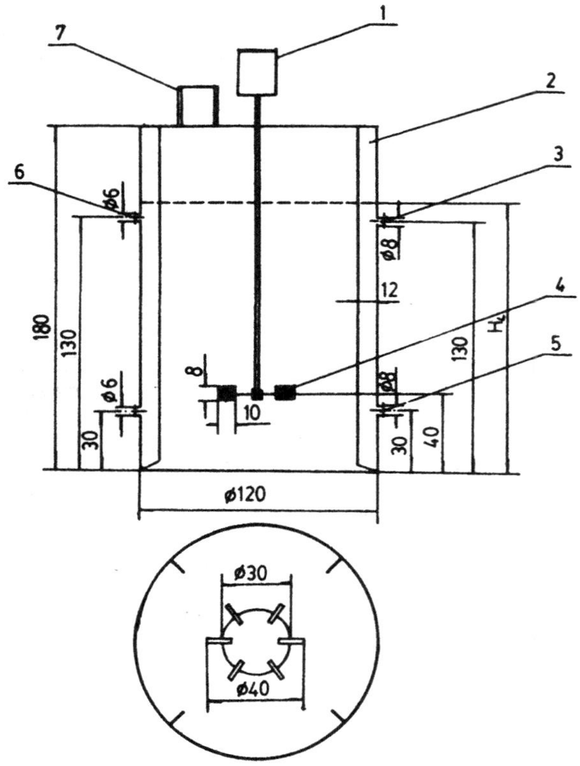

Fig.3 Experimental stirred tank reactor
1.d.c. motor 2.baffle 3.upper outlet 4.turbine blade 5.lower outlet 6.inlet of liquid 7.tracer inlet

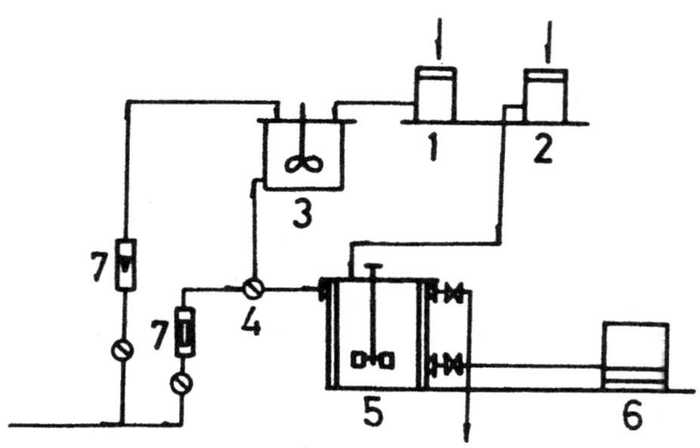

Fig.4 Experimental flow sheet
1,2.particle feeder 3.premixing tank 4.three way valve 5.experimental tank 6.signal adaptor 7.rotary flow meter

Limestone is used as tracer of solid particles whose density is 2700 kg/m^3, the diameter of limestone equals to that of sand particle. Tracer of liquid phase is 10% KCl solution. The tracer of solid particle is introduced with step-responce method after a steady state approached. In the beginning of the experiment, sand particles are transported by particle feeder 1 into pre-mixing tank 3, whose volume is about half of the experimental tank. After agitation with water homogeneously, the slurry flows down to the experimental tank 5 via threeway valve 4; by which the flow of slurry will be replaced by water after the steady state approached, at the sametime another flow of sand particle containing tracer particle with the same flow rate from particle feeder 2 is introduced into the reactor, while the steadystate can be maintained. The tracer of liquid phase is introduced by the impules-response method, 50 ml of liquid tracer is introduced during less than 3 seconds into the tank. It does not influence the steadystate, since the amount of the liquid tracer is small. Liquid and solid particles are collected as samples at the exit of the fluid, then the tracer of solid particles is detected with the chemical analysis. The tracer concentration of liquid is detected by a chloride ion-selective electrode. After A-D converter, the signal of the electrode is recorded by a micro-computer.

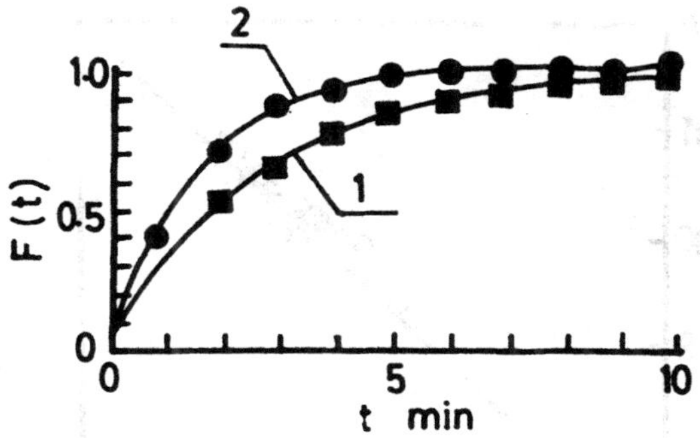

Fig.5 RTD of solid particles at the outlet of reactor
1. C_s=3%, H_L/D=5.33, R=0.75, Re=19000, k=7.42, τ=2.73min.
2. C_s=7%, H_L/D=5.33, R=0.75, Re=18000, k=4.95, τ=1.53min.
outlet in upper section

EXPERIMENTAL RESULTS AND DISCUSSION

According to the experimental results, the function of residence time distribution $F(t_i)$ of solid phase can be represented as follows:

$$F(t_i) = M_i/M_o \quad (11)$$

where M_o, M_i are the flow rate of solid tracer particles with time at inlet and outlet respectively. The typical result is shown in Fig.5. The age distribution function E(ti) of liquid phase can be represented as below:

$$E(t_i) = vC(t_j)/\sum_{j=0}^{i} vC(t_j) \quad (12)$$

therefore

$$F(t_i) = \sum_{j=0}^{i} E(t_j)\Delta t_j \quad (13)$$

The model parameter k which represents the flow rate between two compartments of the model can be estimated by the optimization based on the experimental data to fit the RTD model. The objective function S during optimization takes as

$$S = \sum_{i=0}^{m} (F(k,t_i) - F(t_i))^2$$

where F(k,ti) is calculated value and F(ti) is the experimental data. The main factors influencing the value of model parameter k are: stirring speed; concentration of solid particles and geometrical parameters of the tank. Therefore the model parameter k can be represented as the following empirical equation

$$k = KN^a H_L^b C_s^c \quad (14)$$

RTD of solid particles is influenced by the above parameters as well as location of exit of stream. The exchange flow rate coefficient k of solid particles between two regions is as follows: when exit located at upper section

$$k = 4.72\, N^{0.92} H_L^{-0.063} C_s^{0.69} \quad (15)$$

when exit located at lower section:

$$k = 7.02\, N^{0.97} H_L^{-0.071} C_s^{0.81} \quad (16)$$

As exchange rate coefficient k of solid particles, that of liquid phase between two regions are as below:
when exit is in upper section

$$k = 12.6 N^{0.028} H_L^{-0.071} C_s^{0.13} \quad (17)$$

when exit is in lower section

$$k = 88.8 N^{0.14} H_L^{-0.012} C_s^{0.013} \quad (18)$$

The comparison of k between experimental data and calculated result by the model is shown in Fig.6,7.

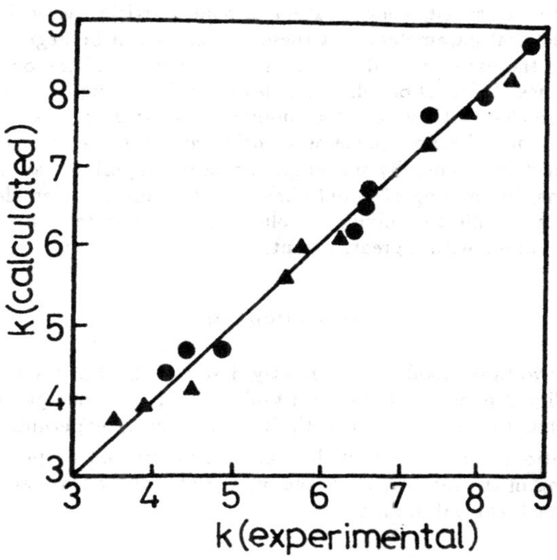

Fig.6 Comparison of k value for solid particle
inlet: ▲ upper, ● upper
outlet: ▲ lower, ● upper

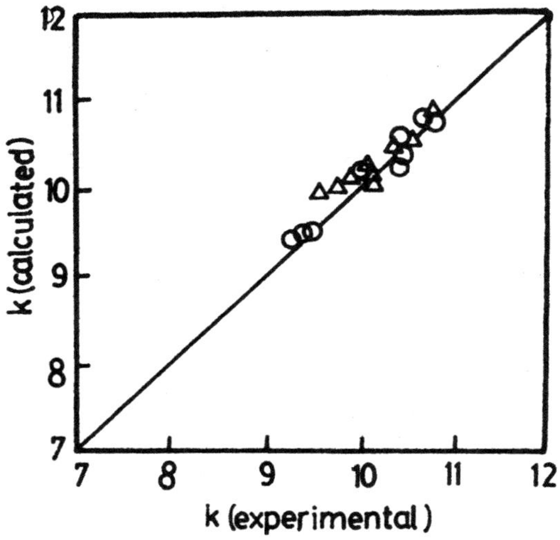

Fig.7 Comparison of k value for liquid phase
inlet: ○ upper, △ upper
outlet: ○ upper, △ lower

According the above results, exchange flow rate coefficient k of solid particles is mainly influenced by stirring speed. The value of k increases with increasing of stirring speed. It is also influenced by the concentration of solid particles and geometrical parameters. During experimental conidition, the solid particles will be more homogeneously suspended in the whole space of the tank, if there is a higher concentration of the solid particle in the reactor.

Exchange flow rate coefficient k of liquid phase is also influenced by stirring speed, concentration of solid particles as well as geometrical parameters, but these influences can be neglected under the experimental condition. The location of the outlet influence the RTD of solid particles as well as of liquid phase. The models with different combinations of two regions used to describe the flow patterns of different locations of exit is satisfactory. Generally speaking, during the experimental conditions, the mixing of liquid phase can be considered as ideal mixing. While the mixing of solid particles deviate from the ideal mixing with a greate extent.

CONCLUSIONS

The two tanks model can correctly describle RTD of the solid and liquid phases in the stirred tank, there is only one parameter has to be estimated in the model. During experimental conditions, the solid particles can not be treated as perfect mixing in the reactor, while the mixing of liquid phase can be assumed as ideal mixing.

Locations of outlet of the stream will influence RTD of both two phases. When a reactor is operatecd as a CSTR, the location of outlet must be carfully arranged in order to obtain the flow pattern as expected.

NOMENCLATURE

a, b, c	constant in eq. (14)	
C	concentration of liquid tracer	g/l
C_0	concentration of tracers in the feed flow	g/l, g/g
C_1	concentration of tracers in upper region of the model	g/l, g/g
C_2	concentration of tracers in lower region of the model	g/l, g/g
d	diameter of agitator	m
E(t)	age distribution of liquid tracer	
F(t)	flow function of liquid or solid tracer	
F(k,t)	calculated value of F(t) by the model	
H_L	height of slurry	m
K	constant in eq. (14)	
k	flow rate coefficient between two regions of the model	
M_i	solid tracer folw rate in the sample at exit	kg/s
M_o	solid tracer feed flow rate	kg/s
N	stirring speed of the agitator	1/s
R	fraction of upper region of the reactor	
t	time	s
V	volume of reactor	m^3
v	flow rate	m^3/s
τ	nominal residence time	τ=V/v, s
θ	dimensionless time	
ν	kinematic viscosity of liquid	$Pas/(kg/m^3)$

LITERATURE CITED

(1). **Zwietering,** Th. N., Chem. Eng. Sci. Vol.8, 244 (1958).
(2). **Porcelli,** J. V. and Marr, G. R., Ind. Eng. Chem. Fund., Vol.1, 172 (1962)
(3). Narayanan, S. et al, Chem. Eng. Sci., Vol.24, 233 (1969)
(4). **Rieger,**F. et al, Proc. Europ. Conf. on Mixing, 1982, 4th, 263
(5). Bohnet, M. et al, Germ. Chem. Eng., Vol.3, 57 (1980)
(6). Tojo, K. et al, Ind. Eng. Chem. Fund, Vol.21, 214 (1982).
(7). Penaz, F. et al, Coll. Czech. Chem. Commun., Vol.43, 848 (1978)
(8). Bowling, K. H. et al, Aust. J. Appl, Sci., Vol.14, 57 (1968)
(9). Krishnaiah, K. et al, Chem. Eng. Sci., Vol.37, 1371 (1982)
(10). Mao, H. H. and Yang, S. Z., Proc. Europ. Conf. on Mixing, 1988, 6th, 281

Answers to the referees' questions

1. The size of your mixing device is very small. Slight construction defects and small changes have considerable effect on the experimental results! The mean residence time of the solution in the stirred tank is very short. Are you quite sure that the sea sand and limestone concentrations were constants during your trials?

   The size of mixing device is small of course, but its residence time is 1.5-3 minutes, it is large enough with respect to the tracer impulse time which is less than 3 seconds. The diameter of the tank is 120 mm, the height is 180 mm, the volume of the reactor is 2.6 liter, which can be assumed there is an error not more than 5% if there is deviation in construction by 6 mm which is not common happened.

   The concentration of solid particle was detected along the verticle axis during the experiment and it had been proved that the concentration of solid particles was constant and there was a concentration distribution along the axial direction.

2. What are the values of H_L, D, C (see offical nomenclature of the Conference) and the exact location (off bottom distance, position with respect to baffles) and diameter of the entrance and exit orifices?
   Where do you carry out the pulse injection of liquid tracer?

   The geometrical parameters of the experimental tank and the position of tracer inlet or pulse injection of liquid tracer are shown in Fig. 3.

3. What is the contents of vessels 1, 2 and 3 on your figure 4?

   Vessel 1, 2 are particle feeders, vessel 3 is premixing tank with a volume equals one half of the experimental tank, which is described yet in the text.

4. What are the experimental conditions corresponding to the points of figures 5, 6 and 7?

   The experimental conditions of the points in Fig. 5, 6 and 7 have been listed out.

5. In relations (15)-(18), you propose dimensionless correlation for the parameter k of your hydrodynamic model though you have only investigated the effects of N, H_L (n and H for you) and Cs on liquid and solid RTD!
   For the solid phase, k is practically proportional to N. For the liquid phase k is almost independent of N and HL because the phase is nearly perfectly mixed. Those results can be consistent. On the contrary, I can't understand why k depends on Cs, in the liquid phase. In relation (18), is the exponent of Cs correct?
   What are the standard deviations of k, a, b and c in your correlations?

   The relations (15)-(18), have been modified with ordinary correlation with respect of N, H C. The exponent of C in relation (18) is correct. Why the k in liquid phase is also influenced by C, which can be understood as that, there is a space will be occupied by the solid particle and the homogenity of the liquid phase will be disturbed by the solid content.

LOCAL WALL HEAT TRANSFER COEFFICIENTS USING SURFACE CALORIMETERS

J.B. FASANO, Chemineer Inc., Dayton, Ohio
R.S. BRODKEY, S.J. HAAM, The Ohio State University, Columbus, USA

INTRODUCTION

Surface calorimeters are thin surface mounted sensors (0.076 to 0.330 mm thick) designed to measure the local heat flux. It does this by differentiating temperature between opposite sides of certain rigid materials thereby allowing a direct measurement of the heat loss or gain through the material surface. By way of a simplified explanation, the heat flux sensor is constructed with two temperature measuring elements physically separated by a thermal insulating material. When the heat transfers through the surface, the thermal energy at the first thermocouple junction generates a small voltage. As the heat continues to transfer through the material it reaches the second thermocouple junction, where it generates a differential voltage with the first thermocouple. Since the temperature differential is proportional to the voltage differential, the heat transfer rate can be directly read out as a function of voltage. Calibration is a constant EMF output for a constant heat flux rate.

These surface calorimeters were used on the process side (agitator side) of agitated heat transfer systems to measure local heat flux. The surface calorimeters used in the studies of this paper also contained a surface thermocouple for surface temperature measurements, and several thermocouples were used in the fluid to obtain the mean bulk temperature at any time t. With known temperatures both at the surface and in the bulk, a temperature differential across the liquid film could be easily obtained. With this temperature differential and the heat flux, Q/A, computation of the local heat transfer coefficient can be easily determined from:

$$h = \frac{\left(\frac{Q}{A}\right)}{(\theta_b - \theta_l)}$$

This technique has been used previously by Bourne and associates[1,2,3,4]

The objective of this work was to develop a system and a technique for the fast determination of process side heat transfer coefficients under various conditions. One of the problems with industrial scale heat transfer systems is that rarely are any two the same. This differentiation is caused by such factors as: mixed impeller systems and impeller spacing; unique heat transfer surfaces such as single or multiple helical coils, various arrangements of vertical tubes, platecoils bundled or uniformly spaced; etc. From an industrial development standpoint, there is a need for a rapidly flexible heat transfer system by which Nusselt type correlations can be quickly obtained for such unique systems.

EXPERIMENTAL EQUIPMENT

The experimental apparatus consists of (3) basic sections:

1. The calorimeters, signal treatment and the computer data acquisition system.
2. The baffled vessel with specified heat transfer zones at the calorimeters.
3. The agitator and mounting stand, including motor, variable frequency motor controller, gear reducer, shaft and impeller(s).

A schematic of the calorimeters, signal treatment and the computer data acquisition system are shown in Figure 1. Rdf Micro-Foil heat flow sensors were used with a typical response time of 0.02 to 0.5 sec. Copper Constantin thermocouples were used to measure the temperature of the wall-film temperature as well as the bulk temperature of the fluid. A MetraByte DAS-8 data acquisition board and a MetraByte EXP-16 multiplexer were installed in a PC. The multiplexer provides signal amplification, filtering and conditioning. Two BASIC programs were used to determine the heat transfer coefficient, the heat flux, the wall-film temperature and the bulk liquid temperature in the tank, Reynolds number, Nusselt number and all physical properties which were programmed in as a function of temperature. One program was used to collect the raw data from the experiments. The second program converted the raw data to appropriate variables and dimensionless numbers. Each reported heat transfer coefficient is an average of approximately three thousand data points

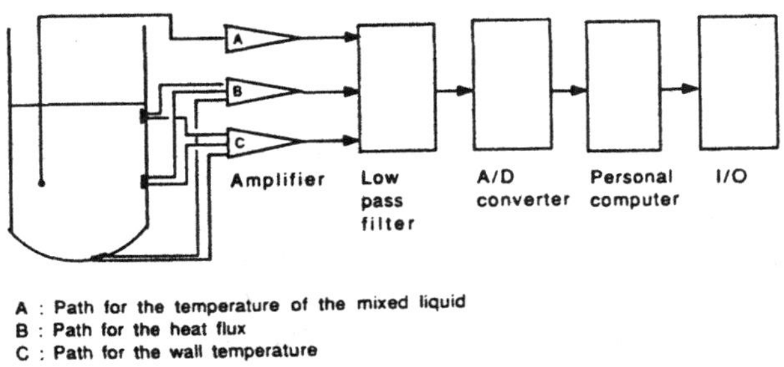

Figure 1

The vessel is 15-1/4" (0.3874 m) diameter with a dished bottom and has a straight side of 35.9" (0.9119 m). The nominal wall thickness was 3/8" (0.00952 m) except in the heat transfer zones where the wall thickness was machined down to 1/16" (0.00159 m). The bottom head calorimeter was located 5-3/8" (0.1365 m) out from the centerline of the vessel. Calorimeters 1, 2, 3, and 4 were located at distances above the lower tangent of 3.7" (0.0940 m), 11.5" (0.292 m), 19.3" (0.490 m) and 27.1" (0.688 m) respectively directly opposite the center of a 2" (0.0508 m) high by 1" (0.0254 m) heat transfer zone. In this manner, the small heat transfer zones allowed the vessel contents to be cooled slowly permitting high fluxes throughout the run. In fact several runs could be made from one heat-up or cool down. The active area of the calorimeter is on the order of 1 square centimeter and provided sufficient heat transfer area to avoid edge effects. Four standard baffles were used each 1.27" (0.0322 m) wide at 90°. The baffles were attached to a rotatable ring in the top of the vessel. This arrangement permitted the baffle system to be rotated with respect to the heat transfer coefficients with respect to radial angular orientation. This vessel system is shown schematically in Figure 2. The specific angular orientations investigated for the baffles is shown in Figure 3, where the location of the calorimeter is defined at the zero degree position.

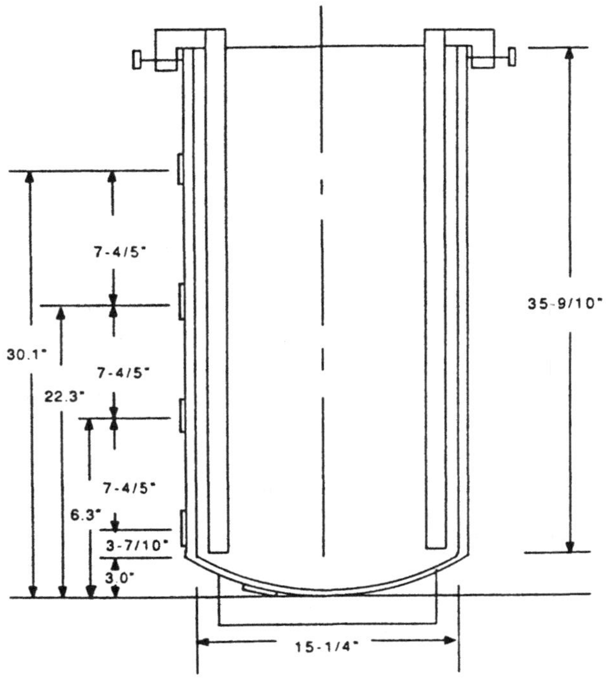

Figure 2

Incremental Baffle Rotation Settings

Figure 3

The agitator consisted of a two horsepower 1800 rpm motor driving through a changeable ratio single reduction gearbox. The highest reduction ratio was 4:1 and the lowest was 1:1 which in combination with a variable frequency motor controller provided for a speed range of 40 to 1800 rpm. Three different impellers were utilized for these studies: the six bladed disc turbine (D-6 Impeller), the four bladed 45° pitched turbine (P-4), and the Chemineer high efficiency impeller (HE-3). These impellers are described schematically in Figure 4. The blade width/diameter ratio and the blade length/diameter ratio of the D-6 impeller were 0.2 and 0.25 respectively, and the disc/diameter ratio was 0.67. The blade width/diameter ratio for the P-4 was also 0.2

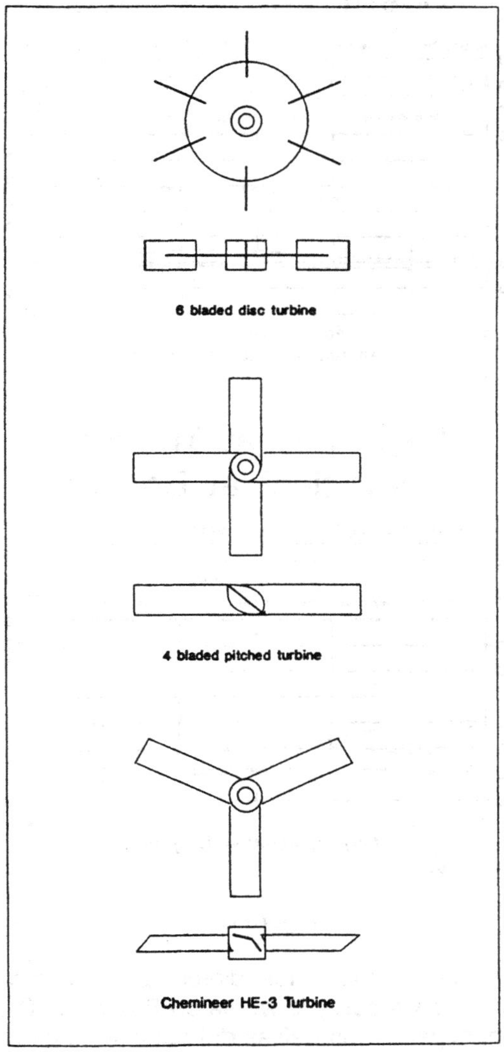

Figure 4

VARIABLES STUDIED

All combinations of the following variables were studied:

1. Impeller type: D-6, P-4, HE-3
2. Reynolds number: 50,000; 160,000
3. Impeller/Tank Diameter: 0.33, 0.50
4. Off-Bottom/Tank Diameter: 0.25, 0.33
5. Baffle Orientation: 7 places over 90°

All of these studies were done at a liquid level to tank diameter ratio, Z/T, of 1.05

EXPERIMENTAL RESULTS

Space does not permit the results of all of these runs to be presented graphically. Complete results however are presented by Haam.[5] Certain interesting aspects moreover, can be examined. Figure 5 examines the effect of vertical location on the heat transfer coefficient. The lower graph represents the data collected at the bottom probe. Probe 1 data, or the first probe above the lower tangent line, is represented in the middle graph and probe 2 data, or the second probe above the lower tangent line, is represented as the top graph. Figure 5 represents data for the D-6, P-4 and HE-3 impellers at three different probes at the bottom and along the tank wall. All of the data are represented at 50,000 Reynolds number. What was interesting throughout all of these studies was that the coefficient near the baffle was as high as, often higher, and only very seldom lower than any of the angular coefficients. When examining these graphs, it will help to picture the fluid moving from left to right, with the two baffles located at 0° and 90°. The front side of the baffle (baffle at 90°) at times produced higher coefficients than the backside of the baffle. Previous concerns for adequate heat transfer in the area of the baffle, at least in turbulent flow for Newtonian fluids, can be ignored. It is important to note that there is also relatively little variation in the heat transfer coefficient between the baffles. The D-6 impeller appears to have a considerable advantage in heat transfer. However, it must be remembered that the power numbers of these impellers are considerably different. The D-6 impeller is investing approximately 17 times the power of the HE-3 impeller and about 4 times the power of the P-4 impeller. Similar results were observed at different combinations of D/T and C/T. As would be expected, the D-6 impeller, having a more concentrated flow profile, has its greatest effect on heat transfer in the vicinity of the impeller. Therefore the greater variation in heat transfer coefficient from bottom to top is to be expected. Likewise the more even flow profile of the HE-3 impeller would be expected to produce a more even distribution of heat transfer coefficients. Examination of Figure 5, as expected, reveals this to be the case.

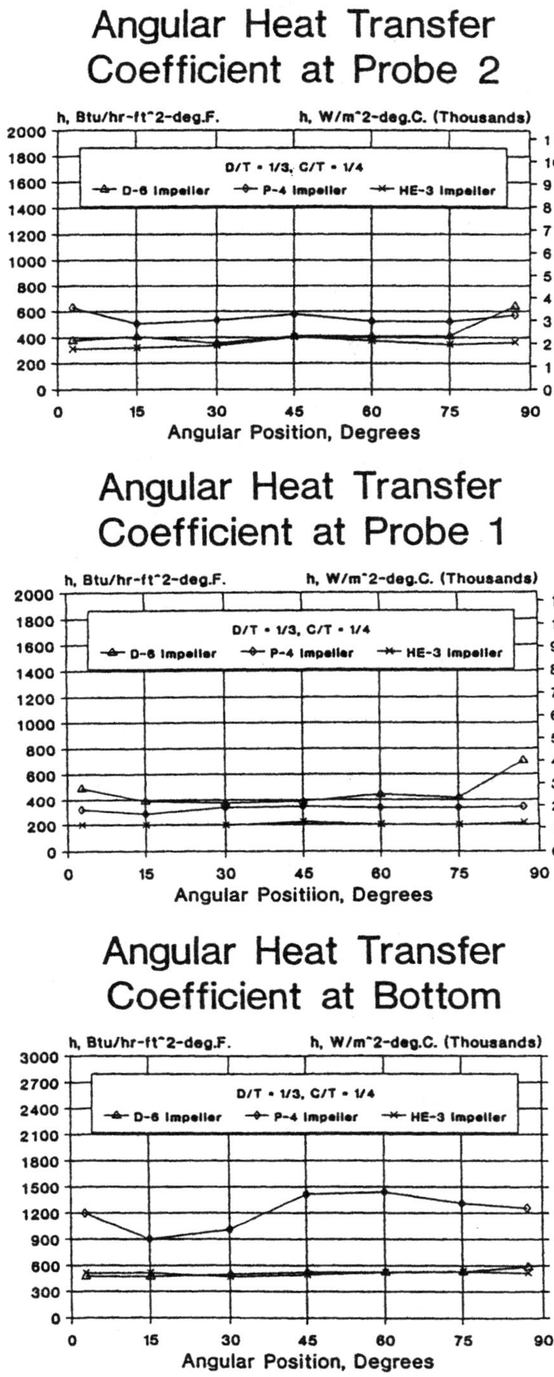

Figure 5

The effects of impeller to tank diameter ratio, D/T, can be observed in Figures 6 and 7. The smaller D/T, 0.33, it must be remembered, inputs about 50% more power. More variability was observed at the higher D/T, especially on the bottom probe. In general, all three impellers exhibited higher heat transfer coefficients at the smaller D/T which corresponds to the 50% increase in power.

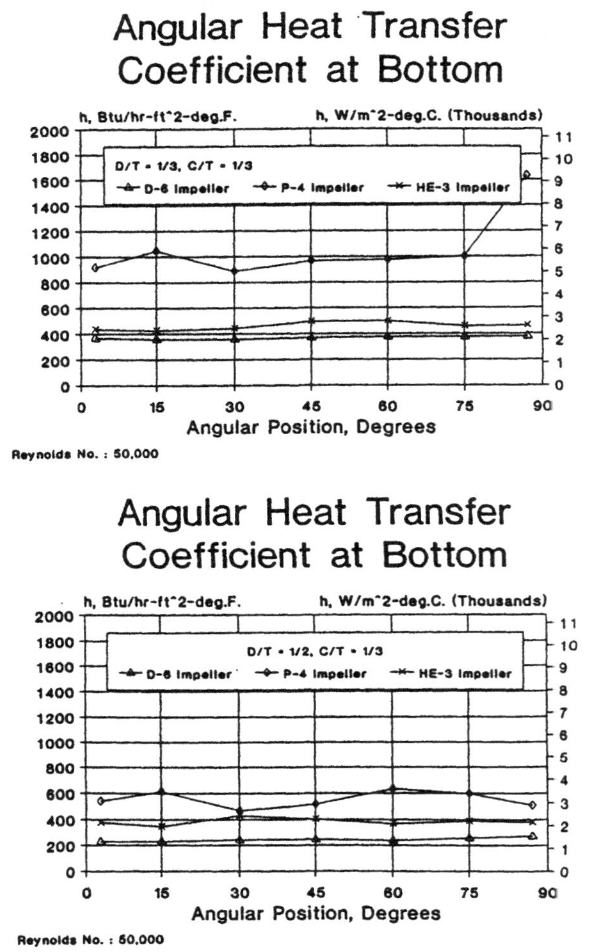

Figure 6

Although Probe 1 is almost directly opposite the impeller, the variability is worse on Probe B. This was not expected as baffles did not extend into the bottom head of the tank. Probe 1 responses can be found in Figure 7. It should also be observed that the P-4 and HE-3 impellers do not depend as significantly as the D-6 impeller on D/T.

In full scale equipment, not only is the invested power significant as it relates directly to operating cost, but also is torque. The torque at the smaller D/T is only 2/3 of the torque at the larger D/T. And as torque is directly related to machine size, it represents an indication of the capital cost. This size difference at constant Reynolds number will usually represent the next smaller agitator size. Thus at smaller D/T's one can expect to have higher horsepowers and smaller machine sizes while the reverse is true for large D/T's.

Angular Heat Transfer Coefficient at Probe 1

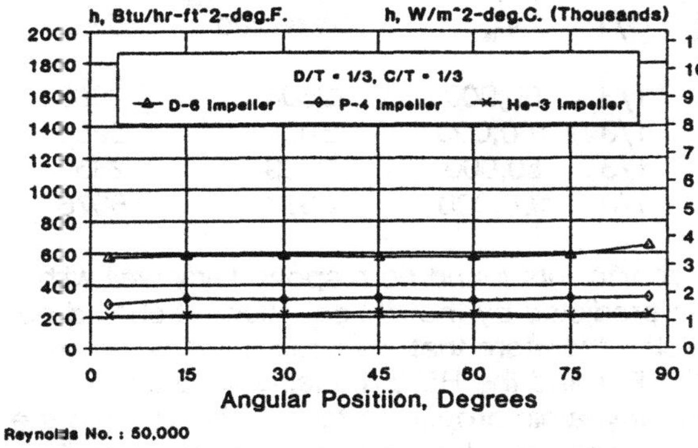

Angular Heat Transfer Coefficient at Probe 1

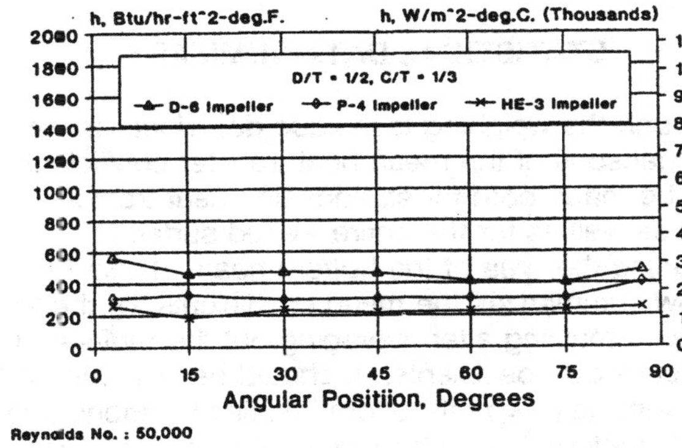

Figure 7

Changes in off-bottom clearance to tank diameter ratio, C/T, displayed predictable characteristics. There are basically two factors which can affect the local heat transfer coefficient:

1. Proximity of impeller to calorimeter
2. Local discharge energy intensity.

These factors are observed in Figure 8. The D-6 impeller discharges radially, and the discharge contains substantial microflow in the form of vortex pairs emanating from each blade.[6,7] Therefore any probe which is moved closer to its discharge will produce greater local changes in heat transfer coefficient. The P-4 impeller also produces a substantial tip vortex but with less energy intensity than the D-6 impeller.[7] And as would be expected, moving a calorimeter closer to the discharge does not produce as large an effect as for the D-6 impeller. The HE-3 impeller produces only a very small tip vortex and an essentially constant velocity profile.[7] And also as would be expected, has the least effect when moved closer to a calorimeter.

Angular Heat Transfer Coefficient at Probe 1

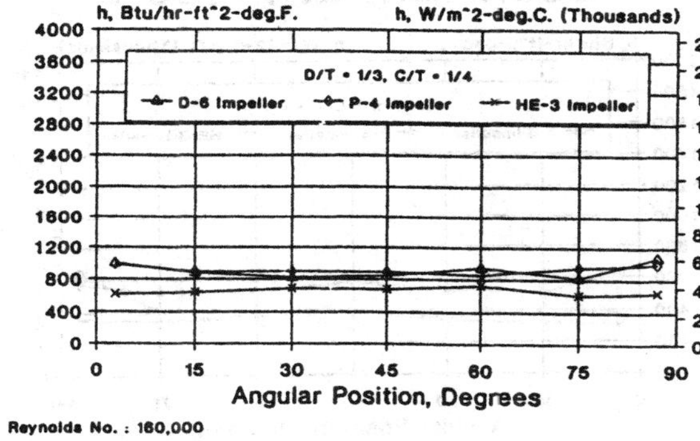

Angular Heat Transfer Coefficient at Probe 1

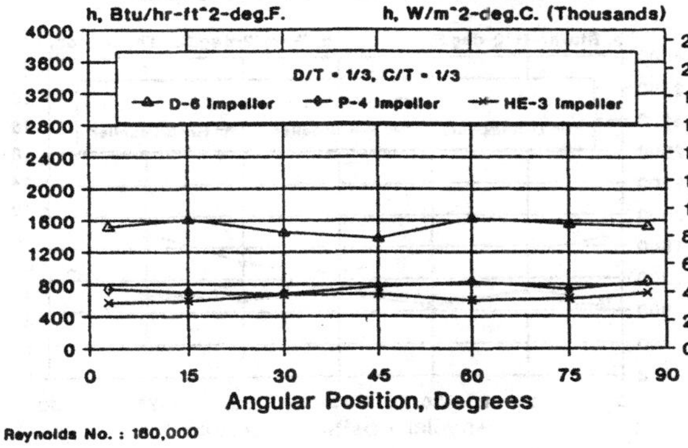

Figure 8

Heat transfer coefficients were rather strong functions of the Reynolds number. The Reynolds number effect can be observed in Figure 9. The functionality is, as expected, less than proportional. The Reynolds number ratio shown is 3.2 and the coefficients have increased somewhat less than that amount. Such data would suggest an exponent on the Reynolds term used in a Nusselt heat transfer equation less than 1. This would tend to agree with numerous prior studies especially with respect to the D-6 impeller. A 2/3 exponent as suggested by Uhl[8] would produce an improvement factor of 2.2.

Even though each data point shown on the graphs in Figure 9 is an average of several thousand, there is more variability at higher Reynolds numbers, despite the fact that the higher Reynolds number represents a higher heat flux rate where the signal to noise ratio is more favorable.

Angular Heat Transfer Coefficient at Probe 2

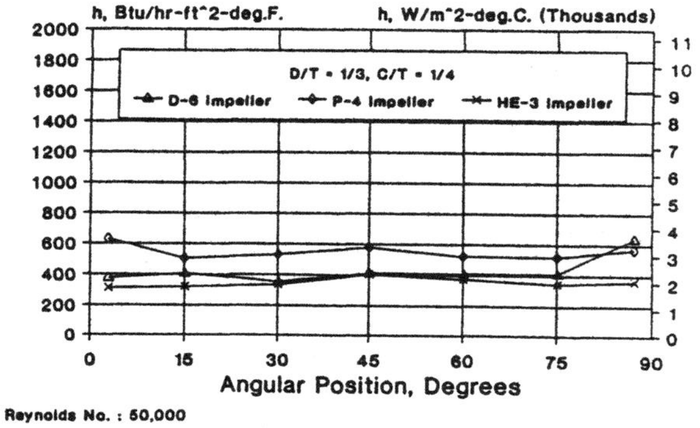

Angular Heat Transfer Coefficient at Probe 2

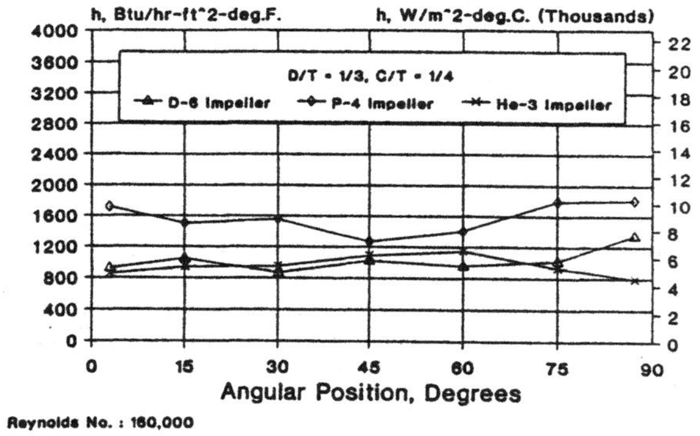

Figure 9

COMPARISON TO LITERATURE DATA

The data collected was weighted by the area each data point represented to arrive at mean jacket-side heat transfer coefficient. This data should compare favorably with reported data for identical or similar systems. Table 1 below compares the mean heat transfer coefficient obtained with those which would be predicted by the equation given by Uhl for a D-6 impeller.[8] These comparisons are made on the basis of vessel straight side heat transfer area as that was the basis upon which Uhl conducted his experiments.

Table 1
Heat Transfer Coefficient Comparison for the D-6 Impeller, W/m^2-°C

D/T	C/T	N_{Re}	h_{mean} Observed	Uhl
1/3	1/3	50,000	2504	2680
1/3	1/3	160,000	6002	5956
1/3	1/3	50,000	2209	2657
1/3	1/3	160,000	6030	6076

The coefficients found correspond fairly well with those predicted by the Uhl equation. It should be expected therefore that the coefficients observed for the P-4 and the HE-3 are fairly accurate. Thermodynamic properties for substitution into the Uhl equation were based on the average for all the runs used constituting the observed mean shown in Table 2.

STATISTICAL DATA SUMMARY

Through the weighing technique described above, comparisons of the mean heat transfer coefficient can be made both for straight side heat transfer only as well as for the entire wetted surface including the area of the bottom head. Table 2 below summarizes the mean coefficients for the 24 cases remaining after averaging out the radial and vertical local coefficients. It should be remembered that with only two off-bottom impeller locations, the local coefficients can be highly influenced by the flow patterns of the impellers and may not represent as good a mean value as might be obtained for more in depth studies.

Table 2

Turbine Type	D/T	C/T	N_{Re}	h_{mean} straight side only, W/m^2-°C	h_{mean} with dished head, W/m^2-°C
D-6	1/3	1/3	50,000	3160	2749
D-6	1/3	1/3	160,000	8752	7513
D-6	1/3	1/4	50,000	2267	2524
D-6	1/3	1/4	160,000	5122	5934
D-6	1/2	1/3	50,000	3068	2419
D-6	1/2	1/3	160,000	9201	6504
D-6	1/2	1/4	50,000	2473	2075
D-6	1/2	1/4	160,000	7500	5863
P-4	1/3	1/3	50,000	2165	3393
P-4	1/3	1/3	160,000	5262	7617
P-4	1/3	1/4	50,000	2366	4267
P-4	1/3	1/4	160,000	8003	16075
P-4	1/2	1/3	50,000	2575	2764
P-4	1/2	1/3	160,000	5453	6871
P-4	1/2	1/4	50,000	2568	3944
P-4	1/2	1/4	160,000	6478	8969
HE-3	1/3	1/3	50,000	1258	1635
HE-3	1/3	1/3	160,000	3260	5202
HE-3	1/3	1/4	50,000	1337	2091
HE-3	1/3	1/4	160,000	3686	10823
HE-3	1/2	1/3	50,000	1135	1441
HE-3	1/2	1/3	160,000	2945	4478
HE-3	1/2	1/4	50,000	1254	1673
HE-3	1/2	1/4	160,000	3271	6607

NOMENCLATURE

A = heat transfer area, m^2 or ft^2
C = impeller to tank bottom, m or ft
D = impeller diameter, m or ft
h = process side heat transfer coefficient W/m^2-°C, or BTU/ft^2-°F
N = rotational speed of impeller, sec^{-1} or min^{-1}
N_{Re} = impeller Reynolds number, $\rho ND^2/\mu$
Q = heat transfer rate, W or BTU/hr
T = tank diameter, m or ft
Z = liquid depth in tank, m or ft
θ_b = temperature in the bulk fluid, °C or °F
θ_j = process film-wall temperature, °C or °F
ρ = fluid density, kg/m^3 or lb$_m$/ft^3
μ = fluid viscosity, Pa-sec or lb$_m$/ft-sec

REFERENCES

1. Bourne, J. R.; Buerli, M.; Regenass, W., "Application of Heat Flow Calorimetry to Measuring Heat Transfer to Agitated Suspensions", **Fluid Mixing**, The Institution of Chemical Engineers Symposium Series No. 64, 1981.
2. Bourne, J. R.; Dossenbach, O.; and Post, T., "Local and Average Mass and Heat Transfer Generated by Pfaudler-type Impellers", **Fluid Mixing II**, 233-242, 1984
3. Bourne, J. R.; Buerli, M.; Regenass, W., "Heat Transfer and Power Measurements in Stirred Tanks Using Heat Flow Calorimetry", **Chemical Engineering Science**, Vol. 36, pp. 347-354, 1981.
4. Bourne, J. R.; Buerli, M.; Regenass, W., "Power and heat transfer to agitated suspensions: Use of heat flow calorimetry", **Chemical Engineering Science**, Vol. 36, pp.782-784, 1981.
5. Haam, S. J., "Full Field Local Wall Heat Transfer Coefficients in a Mixing Vessel as Determined From Measurements with a Heat Flux Sensor", Masters Thesis, The Ohio State University, Columbus, Ohio, 1990.
6. Stoots, C. M.; Calabrese, R. V., "The Trailing Vortex System Behind a Rushton Turbine Blade", **Engineering Foundation Conference - Mixing XII**, Pitosi, MO, USA, August 6-11, 1989.
7. Fasano, J. B., "Flow Visualization Techniques on Rotating Impellers", **Engineering Foundation Conference - Mixing XII**, Pitosi, MO, USA, August 6-11, 1989.
8. Uhl, V. W., "Heat Transfer to Viscous Materials in Jacketed Agitated Kettles", **Chemical Engineering Progress Symposium Series**, No. 17, Vol. 51, pp.93-107, 1952.

Errata

1. Figure 2, height to probe 2, change from 6.3" to 14-1/2"
2. Last two C/T values in Table 1, change from 1/3 to 1/4
3. Replace first sentence under **Comparison to Literature Data** with, "The average coefficient for the 30, 45 and 60 degree positions at each vertical location was vertically area weighted to arrive at an integrated mean process side heat transfer coefficient."

Referees' Comments and Questions and Authors' Reply

1. *Question or Comment*: Generally speaking the figures do not support the text and are poorly described. The literature comparison with straight-side heat transfer coefficient prediction is only good if the probe positions are representative.

   Reply: This one can't really be answered. "Generally speaking . . ." doesn't mean much; specifics are needed to be able to reply. The comment on literature comparison is quite true and this was the specific goal of the research. The fact that the results do check (Table I, 4 points only) means that we did do a good job of selection of the positions and demonstrated experimentally that they were correct.

2. *Question or Comment*: In a given vessel, the characteristic variables which assess the value of the process heat transfer coefficient (wall or coil) are the mean fluid circulation velocity V_c and the distribution curve of V_c near the heat transfer surface (M. Bruxelmane, **Mixing in biological treatment systems**, Chap. 2 in D. L. Wise, **Biotechnology applied to environmental problems**, C.R.C. Press, Inc. Boca Raton, 1986). Thus (if you don't want to carry out a large number of measurements), local wall measurements of heat transfer coefficient can give interesting information only if the liquid flow pattern near the heat transfer surface is not much modified. So the effect on h of parameters such as the blade width, blade length, blade number and rotational speed of impeller can easily determined from local measurements. On the contrary, the use of surface calorimeters to investigate the effects on h of parameters such as those enumerated on your p.1, second column, is very tedious.

   Reply: Point 2 is a misunderstanding of the objectives of the work. The objectives were to show that a simple scheme and hopefully only one local angular position need be measured to be a reasonable representation of the entire vessel at one given vertical level. This was shown to be very feasible, however, the variation with vertical position is more of a problem. Here the results show that care will have to be made to make sure the sensor is in a relatively correct position or that, alternatively, several sensors can be used to obtain an integrated weighted average as we did.

3. *Question or Comment*: What is the clearance of baffle from wall?

   Reply: The term "standard baffling" in the U.S. refers to four baffles each at 90°, each having a width equal to one twelfth the tank diameter and each baffle positioned one seventy second the tank diameter off the wall.

4. *Question or Comment*: Why do you use a non-standard disc turbine (disc/diameter = 0.67 instead of 0.75)?

   Reply: When the disc impeller was first introduced it did have a disc to impeller diameter ratio of 0.75. However, very shortly after its introduction, the preferred geometry was changed to a ratio of 0.67. All of the major US equipment manufacturers have been producing the 0.67 geometry as standard since the mid 1950's. In addition, studies on power and pumping show no statistically significant differences for disc to impeller diameter ratios between 0.5 and 0.75.

5. *Question or Comment*: You compare the impellers tested at constant agitation Reynolds number! It would have been better to compare the impeller at constant process-side heat transfer coefficient!

   Reply: Ultimately, our goal is to arrive at Nusselt expressions of the general form outlined in the referees' question 6. Additional studies are currently underway to generate these models. Once the models are generated, comparisons can be made from a number of different perspectives.

6. *Question or Comment*: For the impellers tested, can you give in the correlation

$$Nu = A\, Re^{x}\, Pr^{y}\, Vi^{z}$$

   or at least, the experimental (mean) values of the parameters x, y and z? Does x depend on the measure point?

   Reply: With the limited data presented in Table 1, accurate determination of the coefficient and exponents simply can't be done. There is no way of getting y and z. Only water was used and the temperature range was too small for accurate determination. Also note that the exponent, x, would be based on only two Reynolds numbers. The coefficient 'A' does not seem to depend on position. However, this cannot be assured as it would only involve two point curves at two positions.

7. *Question or Comment*: Your figures 6 and 7 show that at constant Re, $h \propto (D/T)^{a}$, a being negative. However, all published results give positive values of a (same reference as above). What are the values of a you have observed?

   Reply: The same limited information exists here also. Note again, that one would only have two points to work with to get slopes. We do not believe this is a good idea. We don't believe there is any data for heat transfer to the bottom dish. The D/T ratio has no visible effect for an HE-3 type impeller. On the P-4 the exponent 'a' is about -1.4 and for the D-6 it is about -1.1. However, slopes based on only two points are not valid. We might add that the decrease seems logical if one balances the expected flow as being towards the sensor with the HE-3 and over the sensor from the outside inward for the other two. Position 1 data is more important from this standpoint. There is no change for either the HE-3 or the P-4 as would be expected for sensors on the wall when an axial flow type impeller is used. The flow is upward more or less along the wall and over the sensor. For the D-6, the flow is direct and the coefficient is -0.68. It is not clear why this is so. It may be that the data is in error. New data can resolve such issues, if they are important. However, we are not sure that was our goal in the first place. We wanted to know if the sensors could be used. Finally, the data the referees referred to is probably overall data and not local and the effects might not be locally the same.

Mechanics

FLUID MECHANICS AND ITS APPLICATIONS

Series Editor: R. Moreau

Aims and Scope of the Series

The purpose of this series is to focus on subjects in which fluid mechanics plays a fundamental role. As well as the more traditional applications of aeronautics, hydraulics, heat and mass transfer etc., books will be published dealing with topics which are currently in a state of rapid development, such as turbulence, suspensions and multiphase fluids, super and hypersonic flows and numerical modelling techniques. It is a widely held view that it is the interdisciplinary subjects that will receive intense scientific attention, bringing them to the forefront of technological advancement. Fluids have the ability to transport matter and its properties as well as transmit force, therefore fluid mechanics is a subject that is particularly open to cross fertilisation with other sciences and disciplines of engineering. The subject of fluid mechanics will be highly relevant in domains such as chemical, metallurgical, biological and ecological engineering. This series is particularly open to such new multidisciplinary domains.

1. M. Lesieur: *Turbulence in Fluids.* 2nd rev. ed., 1990 ISBN 0-7923-0645-7
2. O. Métais and M. Lesieur (eds.): *Turbulence and Coherent Structures.* 1991 ISBN 0-7923-0646-5
3. R. Moreau: *Magnetohydrodynamics.* 1990 ISBN 0-7923-0937-5
4. E. Coustols (ed.): *Turbulence Control by Passive Means.* 1990 ISBN 0-7923-1020-9
5. A. A. Borissov (ed.): *Dynamic Structure of Detonation in Gaseous and Dispersed Media.* 1991 ISBN 0-7923-1340-2
6. K.-S. Choi (ed.): *Recent Developments in Turbulence Management.* 1991 ISBN 0-7923-1477-8
7. E.P. Evans and B. Coulbeck (eds.): *Pipeline Systems.* 1992 ISBN 0-7923-1668-1
8. B. Nau (ed.): *Fluid Sealing.* 1992 ISBN 0-7923-1669-X
9. T.K.S. Murthy (ed.): *Computational Methods in Hypersonic Aerodynamics.* 1992 ISBN 0-7923-1673-8
10. R. King (ed.): *Fluid Mechanics of Mixing.* Modelling, Operations and Experimental Techniques. 1992 ISBN 0-7923-1720-3

Kluwer Academic Publishers – Dordrecht / Boston / London

Mechanics

SOLID MECHANICS AND ITS APPLICATIONS

Series Editor: G.M.L. Gladwell

Aims and Scope of the Series

The fundamental questions arising in mechanics are: *Why?*, *How?*, and *How much?* The aim of this series is to provide lucid accounts written by authoritative researchers giving vision and insight in answering these questions on the subject of mechanics as it relates to solids. The scope of the series covers the entire spectrum of solid mechanics. Thus it includes the foundation of mechanics; variational formulations; computational mechanics; statics, kinematics and dynamics of rigid and elastic bodies; vibrations of solids and structures; dynamical systems and chaos; the theories of elasticity, plasticity and viscoelasticity; composite materials; rods, beams, shells and membranes; structural control and stability; soils, rocks and geomechanics; fracture; tribology; experimental mechanics; biomechanics and machine design.

1. R.T. Haftka, Z. Gürdal and M.P. Kamat: *Elements of Structural Optimization*. 2nd rev.ed., 1990 ISBN 0-7923-0608-2
2. J.J. Kalker: *Three-Dimensional Elastic Bodies in Rolling Contact*. 1990 ISBN 0-7923-0712-7
3. P. Karasudhi: *Foundations of Solid Mechanics*. 1991 ISBN 0-7923-0772-0
4. N. Kikuchi: *Computational Methods in Contact Mechanics*. (forthcoming) ISBN 0-7923-0773-9
5. Y.K. Cheung and A.Y.T. Leung: *Finite Element Methods in Dynamics*. (forthcoming) ISBN 0-7923-1313-5
6. J.F. Doyle: *Static and Dynamic Analysis of Structures*. With an Emphasis on Mechanics and Computer Matrix Methods. 1991 ISBN 0-7923-1124-8; Pb 0-7923-1208-2
7. O.O. Ochoa and J.N. Reddy: *Finite Element Modelling of Composite Structures*. (forthcoming) ISBN 0-7923-1125-6
8. M.H. Aliabadi and D.P. Rooke: *Numerical Fracture Mechanics*. ISBN 0-7923-1175-2
9. J. Angeles and C.S. López-Cajún: *Optimization of Cam Mechanisms*. 1991 ISBN 0-7923-1355-0
10. D.E. Grierson, A. Franchi and P. Riva: *Progress in Structural Engineering*. 1991 ISBN 0-7923-1396-8
11. R.T. Haftka and Z. Gürdal: *Elements of Structural Optimization*. 3rd rev. and exp. ed. 1992 ISBN 0-7923-1504-9; Pb 0-7923-1505-7
12. J.R. Barber: *Elasticity*. 1992 ISBN 0-7923-1609-6

Kluwer Academic Publishers – Dordrecht / Boston / London

Mechanics

From 1990, books on the subject of *mechanics* will be published under two series:
FLUID **MECHANICS AND ITS APPLICATIONS**
Series Editor: R.J. Moreau
SOLID **MECHANICS AND ITS APPLICATIONS**
Series Editor: G.M.L. Gladwell

Prior to 1990, the books listed below were published in the respective series indicated below.

MECHANICS: DYNAMICAL SYSTEMS
Editors: L. Meirovitch and G.Æ. Oravas

1. E.H. Dowell: *Aeroelasticity of Plates and Shells.* 1975 ISBN 90-286-0404-9
2. D.G.B. Edelen: *Lagrangian Mechanics of Nonconservative Nonholonomic Systems.* 1977 ISBN 90-286-0077-9
3. J.L. Junkins: *An Introduction to Optimal Estimation of Dynamical Systems.* 1978 ISBN 90-286-0067-1
4. E.H. Dowell (ed.), H.C. Curtiss Jr., R.H. Scanlan and F. Sisto: *A Modern Course in Aeroelasticity.* *Revised and enlarged edition see under Volume 11*
5. L. Meirovitch: *Computational Methods in Structural Dynamics.* 1980 ISBN 90-286-0580-0
6. B. Skalmierski and A. Tylikowski: *Stochastic Processes in Dynamics.* Revised and enlarged translation. 1982 ISBN 90-247-2686-7
7. P.C. Müller and W.O. Schiehlen: *Linear Vibrations.* A Theoretical Treatment of Multi-degree-of-freedom Vibrating Systems. 1985 ISBN 90-247-2983-1
8. Gh. Buzdugan, E. Mihăilescu and M. Radeş: *Vibration Measurement.* 1986 ISBN 90-247-3111-9
9. G.M.L. Gladwell: *Inverse Problems in Vibration.* 1987 ISBN 90-247-3408-8
10. G.I. Schuëller and M. Shinozuka: *Stochastic Methods in Structural Dynamics.* 1987 ISBN 90-247-3611-0
11. E.H. Dowell (ed.), H.C. Curtiss Jr., R.H. Scanlan and F. Sisto: *A Modern Course in Aeroelasticity.* Second revised and enlarged edition (of Volume 4). 1989 ISBN Hb 0-7923-0062-9; Pb 0-7923-0185-4
12. W. Szemplińska-Stupnicka: *The Behavior of Nonlinear Vibrating Systems.* Volume I: Fundamental Concepts and Methods: Applications to Single-Degree-of-Freedom Systems. 1990 ISBN 0-7923-0368-7
13. W. Szemplińska-Stupnicka: *The Behavior of Nonlinear Vibrating Systems.* Volume II: Advanced Concepts and Applications to Multi-Degree-of-Freedom Systems. 1990 ISBN 0-7923-0369-5
Set ISBN (Vols. 12–13) 0-7923-0370-9

MECHANICS OF STRUCTURAL SYSTEMS
Editors: J.S. Przemieniecki and G.Æ. Oravas

1. L. Frýba: *Vibration of Solids and Structures under Moving Loads.* 1970 ISBN 90-01-32420-2
2. K. Marguerre and K. Wölfel: *Mechanics of Vibration.* 1979 ISBN 90-286-0086-8

Mechanics

3. E.B. Magrab: *Vibrations of Elastic Structural Members*. 1979 ISBN 90-286-0207-0
4. R.T. Haftka and M.P. Kamat: *Elements of Structural Optimization*. 1985
   Revised and enlarged edition see under Solid Mechanics and Its Applications, Volume 1
5. J.R. Vinson and R.L. Sierakowski: *The Behavior of Structures Composed of Composite Materials*. 1986 ISBN Hb 90-247-3125-9; Pb 90-247-3578-5
6. B.E. Gatewood: *Virtual Principles in Aircraft Structures*. Volume 1: Analysis. 1989
   ISBN 90-247-3754-0
7. B.E. Gatewood: *Virtual Principles in Aircraft Structures*. Volume 2: Design, Plates, Finite Elements. 1989 ISBN 90-247-3755-9
   Set (Gatewood 1 + 2) ISBN 90-247-3753-2

MECHANICS OF ELASTIC AND INELASTIC SOLIDS
Editors: S. Nemat-Nasser and G.Æ. Oravas

---

1. G.M.L. Gladwell: *Contact Problems in the Classical Theory of Elasticity*. 1980
   ISBN Hb 90-286-0440-5; Pb 90-286-0760-9
2. G. Wempner: *Mechanics of Solids with Applications to Thin Bodies*. 1981
   ISBN 90-286-0880-X
3. T. Mura: *Micromechanics of Defects in Solids*. 2nd revised edition, 1987
   ISBN 90-247-3343-X
4. R.G. Payton: *Elastic Wave Propagation in Transversely Isotropic Media*. 1983
   ISBN 90-247-2843-6
5. S. Nemat-Nasser, H. Abé and S. Hirakawa (eds.): *Hydraulic Fracturing and Geothermal Energy*. 1983 ISBN 90-247-2855-X
6. S. Nemat-Nasser, R.J. Asaro and G.A. Hegemier (eds.): *Theoretical Foundation for Large-scale Computations of Nonlinear Material Behavior*. 1984 ISBN 90-247-3092-9
7. N. Cristescu: *Rock Rheology*. 1988 ISBN 90-247-3660-9
8. G.I.N. Rozvany: *Structural Design via Optimality Criteria*. The Prager Approach to Structural Optimization. 1989 ISBN 90-247-3613-7

MECHANICS OF SURFACE STRUCTURES
Editors: W.A. Nash and G.Æ. Oravas

---

1. P. Seide: *Small Elastic Deformations of Thin Shells*. 1975 ISBN 90-286-0064-7
2. V. Panc: *Theories of Elastic Plates*. 1975 ISBN 90-286-0104-X
3. J.L. Nowinski: *Theory of Thermoelasticity with Applications*. 1978
   ISBN 90-286-0457-X
4. S. Łukasiewicz: *Local Loads in Plates and Shells*. 1979 ISBN 90-286-0047-7
5. C. Fiřt: *Statics, Formfinding and Dynamics of Air-supported Membrane Structures*. 1983 ISBN 90-247-2672-7
6. Y. Kai-yuan (ed.): *Progress in Applied Mechanics*. The Chien Wei-zang Anniversary Volume. 1987 ISBN 90-247-3249-2
7. R. Negruţiu: *Elastic Analysis of Slab Structures*. 1987 ISBN 90-247-3367-7
8. J.R. Vinson: *The Behavior of Thin Walled Structures*. Beams, Plates, and Shells. 1988
   ISBN Hb 90-247-3663-3; Pb 90-247-3664-1

Mechanics

MECHANICS OF FLUIDS AND TRANSPORT PROCESSES
Editors: R.J. Moreau and G.Æ. Oravas

1. J. Happel and H. Brenner: *Low Reynolds Number Hydrodynamics*. With Special Applications to Particular Media. 1983 ISBN Hb 90-01-37115-9; Pb 90-247-2877-0
2. S. Zahorski: *Mechanics of Viscoelastic Fluids*. 1982 ISBN 90-247-2687-5
3. J.A. Sparenberg: *Elements of Hydrodynamics Propulsion*. 1984 ISBN 90-247-2871-1
4. B.K. Shivamoggi: *Theoretical Fluid Dynamics*. 1984 ISBN 90-247-2999-8
5. R. Timman, A.J. Hermans and G.C. Hsiao: *Water Waves and Ship Hydrodynamics*. An Introduction. 1985 ISBN 90-247-3218-2
6. M. Lesieur: *Turbulence in Fluids*. Stochastic and Numerical Modelling. 1987 ISBN 90-247-3470-3
7. L.A. Lliboutry: *Very Slow Flows of Solids*. Basics of Modeling in Geodynamics and Glaciology. 1987 ISBN 90-247-3482-7
8. B.K. Shivamoggi: *Introduction to Nonlinear Fluid-Plasma Waves*. 1988 ISBN 90-247-3662-5
9. V. Bojarevičs, Ya. Freibergs, E.I. Shilova and E.V. Shcherbinin: *Electrically Induced Vortical Flows*. 1989 ISBN 90-247-3712-5
10. J. Lielpeteris and R. Moreau (eds.): *Liquid Metal Magnetohydrodynamics*. 1989 ISBN 0-7923-0344-X

MECHANICS OF ELASTIC STABILITY
Editors: H. Leipholz and G.Æ. Oravas

1. H. Leipholz: *Theory of Elasticity*. 1974 ISBN 90-286-0193-7
2. L. Librescu: *Elastostatics and Kinetics of Aniosotropic and Heterogeneous Shell-type Structures*. 1975 ISBN 90-286-0035-3
3. C.L. Dym: *Stability Theory and Its Applications to Structural Mechanics*. 1974 ISBN 90-286-0094-9
4. K. Huseyin: *Nonlinear Theory of Elastic Stability*. 1975 ISBN 90-286-0344-1
5. H. Leipholz: *Direct Variational Methods and Eigenvalue Problems in Engineering*. 1977 ISBN 90-286-0106-6
6. K. Huseyin: *Vibrations and Stability of Multiple Parameter Systems*. 1978 ISBN 90-286-0136-8
7. H. Leipholz: *Stability of Elastic Systems*. 1980 ISBN 90-286-0050-7
8. V.V. Bolotin: *Random Vibrations of Elastic Systems*. 1984 ISBN 90-247-2981-5
9. D. Bushnell: *Computerized Buckling Analysis of Shells*. 1985 ISBN 90-247-3099-6
10. L.M. Kachanov: *Introduction to Continuum Damage Mechanics*. 1986 ISBN 90-247-3319-7
11. H.H.E. Leipholz and M. Abdel-Rohman: *Control of Structures*. 1986 ISBN 90-247-3321-9
12. H.E. Lindberg and A.L. Florence: *Dynamic Pulse Buckling*. Theory and Experiment. 1987 ISBN 90-247-3566-1
13. A. Gajewski and M. Zyczkowski: *Optimal Structural Design under Stability Constraints*. 1988 ISBN 90-247-3612-9

Mechanics

MECHANICS: ANALYSIS

Editors: V.J. Mizel and G.Æ. Oravas

1. M.A. Krasnoselskii, P.P. Zabreiko, E.I. Pustylnik and P.E. Sbolevskii: *Integral Operators in Spaces of Summable Functions*. 1976 ISBN 90-286-0294-1
2. V.V. Ivanov: *The Theory of Approximate Methods and Their Application to the Numerical Solution of Singular Integral Equations*. 1976 ISBN 90-286-0036-1
3. A. Kufner, O. John and S. Pučík: *Function Spaces*. 1977 ISBN 90-286-0015-9
4. S.G. Mikhlin: *Approximation on a Rectangular Grid*. With Application to Finite Element Methods and Other Problems. 1979 ISBN 90-286-0008-6
5. D.G.B. Edelen: *Isovector Methods for Equations of Balance*. With Programs for Computer Assistance in Operator Calculations and an Exposition of Practical Topics of the Exterior Calculus. 1980 ISBN 90-286-0420-0
6. R.S. Anderssen, F.R. de Hoog and M.A. Lukas (eds.): *The Application and Numerical Solution of Integral Equations*. 1980 ISBN 90-286-0450-2
7. R.Z. Ḩas'minskiĭ: *Stochastic Stability of Differential Equations*. 1980 ISBN 90-286-0100-7
8. A.I. Vol'pert and S.I. Hudjaev: *Analysis in Classes of Discontinuous Functions and Equations of Mathematical Physics*. 1985 ISBN 90-247-3109-7
9. A. Georgescu: *Hydrodynamic Stability Theory*. 1985 ISBN 90-247-3120-8
10. W. Noll: *Finite-dimensional Spaces*. Algebra, Geometry and Analysis. Volume I. 1987 ISBN Hb 90-247-3581-5; Pb 90-247-3582-3

MECHANICS: COMPUTATIONAL MECHANICS

Editors: M. Stern and G.Æ. Oravas

1. T.A. Cruse: *Boundary Element Analysis in Computational Fracture Mechanics*. 1988 ISBN 90-247-3614-5

MECHANICS: GENESIS AND METHOD

Editor: G.Æ. Oravas

1. P.-M.-M. Duhem: *The Evolution of Mechanics*. 1980 ISBN 90-286-0688-2

MECHANICS OF CONTINUA

Editors: W.O. Williams and G.Æ. Oravas

1. C.-C. Wang and C. Truesdell: *Introduction to Rational Elasticity*. 1973 ISBN 90-01-93710-1
2. P.J. Chen: *Selected Topics in Wave Propagation*. 1976 ISBN 90-286-0515-0
3. P. Villaggio: *Qualitative Methods in Elasticity*. 1977 ISBN 90-286-0007-8

Mechanics

MECHANICS OF FRACTURE
Editors: G.C. Sih

1. G.C. Sih (ed.): *Methods of Analysis and Solutions of Crack Problems*. 1973 ISBN 90-01-79860-8
2. M.K. Kassir and G.C. Sih (eds.): *Three-dimensional Crack Problems*. A New Solution of Crack Solutions in Three-dimensional Elasticity. 1975 ISBN 90-286-0414-6
3. G.C. Sih (ed.): *Plates and Shells with Cracks*. 1977 ISBN 90-286-0146-5
4. G.C. Sih (ed.): *Elastodynamic Crack Problems*. 1977 ISBN 90-286-0156-2
5. G.C. Sih (ed.): *Stress Analysis of Notch Problems*. Stress Solutions to a Variety of Notch Geometries used in Engineering Design. 1978 ISBN 90-286-0166-X
6. G.C. Sih and E.P. Chen (eds.): *Cracks in Composite Materials*. A Compilation of Stress Solutions for Composite System with Cracks. 1981 ISBN 90-247-2559-3
7. G.C. Sih (ed.): *Experimental Evaluation of Stress Concentration and Intensity Factors*. Useful Methods and Solutions to Experimentalists in Fracture Mechanics. 1981 ISBN 90-247-2558-5

MECHANICS OF PLASTIC SOLIDS
Editors: J. Schroeder and G.Æ. Oravas

1. A. Sawczuk (ed.): *Foundations of Plasticity*. 1973 ISBN 90-01-77570-5
2. A. Sawczuk (ed.): *Problems of Plasticity*. 1974 ISBN 90-286-0233-X
3. W. Szczepiński: *Introduction to the Mechanics of Plastic Forming of Metals*. 1979 ISBN 90-286-0126-0
4. D.A. Gokhfeld and O.F. Cherniavsky: *Limit Analysis of Structures at Thermal Cycling*. 1980 ISBN 90-286-0455-3
5. N. Cristescu and I. Suliciu: *Viscoplasticity*. 1982 ISBN 90-247-2777-4

Kluwer Academic Publishers - Dordrecht / Boston / London

9780792317203